Environmental Engineering: Design, Sustainability and Management

Environmental Engineering: Design, Sustainability and Management

Editor: Katie Nixon

Callisto Reference
www.callistoreference.com

Callisto Reference,
118-35 Queens Blvd., Suite 400,
Forest Hills, NY 11375, USA

Visit us on the World Wide Web at:
www.callistoreference.com

ISBN: 978-1-64116-603-4 (Hardback)

Cataloging-in-Publication Data

Environmental engineering : design, sustainability and management / edited by Katie Nixon.
p. cm.
Includes bibliographical references and index.
ISBN 978-1-64116-603-4
1. Environmental engineering. 2. Environmental engineering--Design and construction. 3. Sustainable engineering.
4. Environmental management. I. Nixon, Katie.
TD145 .E58 2022
628--dc23

Table of Contents

Preface VII

Chapter 1 **Impact of Municipal Solid Waste Landfill on Environment** 1
Magdalena Daria Vaverková, Dana Adamcová, Jan Zloch, Maja Radziemska, Agnieszka Boas Berg, Stanislava Voběrková and Alžbeta Maxianová

Chapter 2 **Air Pollution Reduction and Environment Protection using Methane Fuel for Turbocharged CI Engines** 15
Osama Hasib Ghazal

Chapter 3 **Priority Areas for Initiating Land Consolidations Related to Erosion and Water Retention in the Landscape, Czech Republic** 22
Petr Karásek, Jana Konečná, Michal Pochop, Josef Kučera and Jana Podhrázská

Chapter 4 **Investigation of Waste Properties of Subway Construction as a Potential Component of Soil Layer** 35
Andrey Vladimirovich Ivanov, Yuri Dmitrievich Smirnov and Georgiy Igorevich Petrov

Chapter 5 **Legal Regulations and Methods Neutralising Expired Food Products** 46
Joanna Kazimierowicz

Chapter 6 **Surface and Subsurface Water Runoff and Selected Matter Components from the Forested Loess Slope** 54
Andrzej Mazur

Chapter 7 **Monitoring of Road Noise in the Urban Environment** 62
Agnieszka Malec and Gabriel Borowski

Chapter 8 **Poznań Air Pollution Analysis for 2015–2017** 70
Anna Smurzyńska, Wojciech Czekała, Piotr Hektus, Stanisław Marks, Jakub Mazurkiewicz, Michał Brzoski, Dawid Chełkowski and Kamil Kozłowski

Chapter 9 **Reduction of Odour Nuisance of Industrial Plants - Biofiltering in Composting Plants** 78
Monika Gałwa-Widera, Anna Kwarciak-Kozłowska and January Bień

Chapter 10 **Stabilization/Solidification of Waste Containing Heavy Metals and Hydrocarbons using OPC and Land Trass Cement** 87
Masrullita, R.Y. Perry Burhan and dan Yulinah Trihadiningrum

Chapter 11 **Status Assessment of Heavy Metals in Water of the Lepenci River Basin** 96
Pajtim Bytyçi, Osman Fetoshi, Bujar H. Durmishi, Ferdije Zhushi Etemi, Hazir Çadraku, Murtezan Ismaili and Albona Shala Abazi

Chapter 12 **The Content of Heavy Metals in Bottom Sediments of Selected City Rivers of the Podlasie Province** ... 110
Mirosław Skorbiłowicz, Elżbieta Skorbiłowicz and Małgorzata Górska

Chapter 13 **The Effect of Organic Substrates Application on the Amylolytic Activity of Urban Soils** ... 120
Urszula Wydro, Elżbieta Wołejko, Agata Jabłońska-Trypuć, Andrzej Butarewicz and Tadeusz Łoboda

Chapter 14 **The Spatial Distribution of Heavy Metal Lead and Cadmium Pollution and Coliform Abundance of Waters and Surface Sediment** ... 129
Agus Tjahjono and Djoko Suwarno

Chapter 15 **Water Quality Assessment of the Morava e Binçës River based on the Physicochemical Parameters and Water Quality Index** ... 141
Milaim Musliu, Astrit Bilalli, Bujar Durmishi, Murtezan Ismaili and Halil Ibrahimi

Chapter 16 **Heavy Metal Concentrations along the Banks of the Sitnica River and in Four Types of Herbaceous Plants** ... 150
Albona Shala Abazi, Fatbardh Sallaku, Pajtim Bytyçi, Mimoza Hyseni Spahiu and Fadil Millaku

Chapter 17 **Assessment of Diesel Particulate Matter Exposure of Underground Miners** ... 159
Arif Susanto, Purwanto Purwanto, Henna R Sunoko and Onny Setiani

Chapter 18 **Effectiveness of Carbon Monoxide Concentration Reduction on Active Carbon Contact System in Burning Polystyrene Foam** ... 168
Elanda Fikri and Aprilia Veronica

Chapter 19 **Epidemiology of Selected Diseases Related to Air Pollution** ... 174
Agnieszka Petryk

Chapter 20 **Examination of Dust Concentration in the Interior Air of a Teaching Room** ... 182
Halina Marczak

Chapter 21 **Effectiveness of Measures Taken to Protect Waters by Example of Soda Production Plant** ... 190
Adam Gołub and Janina Piekutin

Chapter 22 **Assessment of Ground Ozone Level under the Physiological Strain Conditions** ... 196
Robert Krzysztof Sobolewski and Robert Kalbarczyk

Chapter 23 **Bacterial ACC Deaminase Activity in Promoting Plant Growth on Areas Contaminated with Heavy Metals** ... 206
Anna Grobelak, Paulina Kokot, Jakub Świątek, Marta Jaskulak and Agnieszka Rorat

Chapter 24 **Application of Steel-Smelting Slags as Material for Reclamation of Degraded Lands** ... 214
Vera Matveeva, Tatiana Lytaeva and Aleksandr Danilov

Permissions

List of Contributors

Index

Preface

The world is advancing at a fast pace like never before. Therefore, the need is to keep up with the latest developments. This book was an idea that came to fruition when the specialists in the area realized the need to coordinate together and document essential themes in the subject. That's when I was requested to be the editor. Editing this book has been an honour as it brings together diverse authors researching on different streams of the field. The book collates essential materials contributed by veterans in the area which can be utilized by students and researchers alike.

Environmental engineering is a sub-discipline of civil engineering and chemical engineering. It is the domain of engineering that makes use of diverse branches of science like chemistry, biology, ecology, geology, hydraulics, hydrology, microbiology and mathematics. The aim of environmental engineering is to create solutions that will protect the health of living organisms, improve the quality of environment and protect the ecosystem. It also focuses on waste water management, air and water pollution control, water disposal, etc. This discipline also studies the effects of technologies on the environment. Some of the effects include acid rain, global warming, water pollution, air pollution, automobile exhaust, etc. This book contains some path-breaking studies in the field of environmental engineering. Some of the diverse topics covered herein address the varied branches that fall under this category. It is a vital tool for all researching and studying this field.

Each chapter is a sole-standing publication that reflects each author´s interpretation. Thus, the book displays a multi-facetted picture of our current understanding of application, resources and aspects of the field. I would like to thank the contributors of this book and my family for their endless support.

Editor

Preface

1

Impact of Municipal Solid Waste Landfill on Environment

Magdalena Daria Vaverková[1*], Dana Adamcová[1], Jan Zloch[1], Maja Radziemska[2], Agnieszka Boas Berg[3], Stanislava Voběrková[4], Alžbeta Maxianová[1]

[1] Department of Applied and Landscape Ecology, Faculty of AgriSciences, Mendel University in Brno, Zemědělská 1, 613 00 Brno, Czech Republic

[2] Department of Environmental Improvement, Faculty of Civil and Environmental Engineering, Warsaw University of Life Sciences, Nowoursynowska 159, 02-776 Warsaw, Poland

[3] Shada BV, Kanaal Noord 350, NL-7323 Am Apeldoorn, Holland

[4] Department of Chemistry and Biochemistry, Faculty of Agronomy, Mendel University in Brno, Zemědělská 1, 613 00, Czech Republic

[*] Corresponding author's e-mail: magda.vaverkova@uake.cz

ABSTRACT

Landfill is the oldest and the most common form of removal and disposal of waste, constituting the final disposal method of municipal solid waste (MSW). It is well known that the impact of MSW landfills can cause pollution of all environmental components. The negative environmental impact of landfills can be reduced by applying protection technologies and appropriate solid waste management technologies. The present study undertook monitoring of the Štěpánovice MSW landfill influence on the environment. The spatial characteristics of the area influenced by the landfill and the possible impact of leachate using phytotoxicity tests were carried out. The landfill itself is engineered, with bottom liner and leachate and landfill gas collection and treatment system. The landfill has not had any direct and significant influence on the surrounding environment and water quality so far.

Keywords: landfill influence, waste disposal site, contamination, phytotoxicity, landfill leachate

INTRODUCTION

Each year, about $10 \cdot 10^3$ Kg of municipal solid waste (MSW) is produced globally (Shu et al., 2018). As the final method disposal of MSW, landfill, is the oldest and the most common form of removal and disposal of waste (Koda et al., 2015). Waste management is a civilizational problem (Wolny-Koładka and Malinowski, 2015) and landfills have been widely applied due to a number of advantages, including simplicity, low investment, large handling capacity and low operating cost (Li et al., 2017). Some studies indicated that almost 95% of MSW was disposed of by landfilling worldwide (Ghosh et al., 2015); in the European Union (EU), most of the member states dispose of more than 50% of their waste in landfills (Cuartas et al., 2017). However, with the progress of urbanization and the residents' demands of higher living environment quality, the pollution of landfill sites has increasingly attracted attention in the world, because landfill contains a great deal of potential toxic compounds, some of which may threaten the safety of the surrounding environment. It is known that the impact of MSW landfills can cause pollution of all environmental components (Makarenko and Budak, 2017). In recent years, some researchers studied the influence of landfill sites on the groundwater. For instance, El-Salam and Abu-Zuid (2015) analysed the environmental impacts associated with MSW landfilling, leachate and groundwater quality near the landfills in Egypt. Nagarajan et al. (2012) studied the possible impact of leachate percolation on the groundwater quality. Another study, provided by Koda et al. (2017) assessed the groundwater quality in a landfill and a waste management site, with special regard to the levels of organic pollution. There have been studies in the literature reporting more generally the influence of landfill sites on nearest surrounding. For example, Gworek et al. (2015) focused on assessing the influence of a 35-year-old MSW landfill on the environmental mercury pollution in the soil profile, groundwater, and the plants. Palmiotto et al. (2014) conducted integrated risk assessment for emissions of hazard compounds and odour nuisance to describe the environmental quality in the landfill proximity.

The negative environmental impact of landfills can be reduced by applying protection technologies in landfill design, such as the use of an appropriate impermeable material for bottom and top capping (Dajić et al., 2016; Cuartas et al., 2017). These negative environmental impacts must be controlled both during the operational phase of a landfill, and post-closure (known as the 'aftercare' period) until they no longer pose an unacceptable risk to the environment. Our research team has been involved in the investigation of the environmental problems of contaminants produced and released from landfill facilities. The present study undertook the monitoring of the Štěpánovice MSW landfill influence on the environment in order to: (i) study the spatial characteristics of the area influenced by the landfill, (ii) better understand the possible impact of leachate using phytotoxicity tests, (iii) and to assess landfill influence on water quality. The research results provide scientific information for protecting and managing the environment in the landfill surrounding and similar areas, and we hope it provides a good example for further environmental research.

MATERIAL AND METHODS

The research focuses on several characteristics describing the MSW Štěpánovice: historical development of the cadastre illustrated using maps, analysis of the environmental conditions and landfill reclamations. The research likewise focuses on the assessment of the phytotoxicity of the leachate and waters from the active body of MSW Štěpánovice landfill. The research involves assessing the correlation between the toxicity of the tested water samples and the precipitation volumes in the given area as well as the resulting environmental conditions.

Study area

The study area is located in the Pilsen Region (Štěpánovice, 49°26'15.934"N, 13°16'55.352"E), western part of the Czech Republic (Figure 1). It is being used for disposal of mixed municipal waste. The landfill is formed by three sub-landfills: landfill A (closed in 2003, area of 8,750 m²); landfill B (working from 2003–2016, area of 26,000 m²); landfill C (still in operation). The total volume of both (A, B) parts of the landfill is 289,000 m³ (Heralová, 2017; Adamcová et al., 2017).

Every day, up to 37.5 · 10³ kg of waste is authorized for landfilling. The disposed waste includes: municipal solid waste, non-hazardous

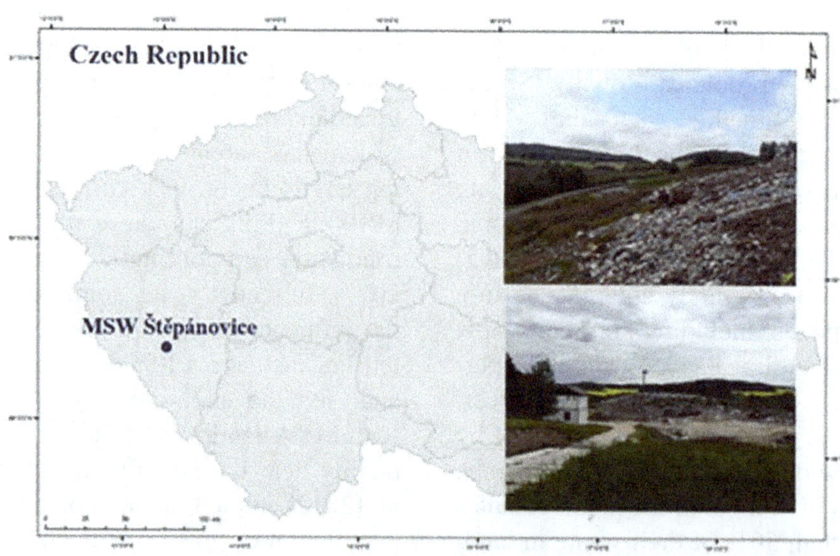

Figure 1. Location of the landfill

wastes and the material for landfill cover. Wastes may include scraps of paper, plastics and metals, packing, spent tires, textile products, building materials, ashes from municipal solid waste incinerators, polluted terrain from environment reclamation, etc. (Heralová, 2017) (the amount of waste landfilled in years 2015 and 2016 is presented in Figure 2). In 2015, a total of 18,291.3 × 10^3 kg of waste was deposited on the landfill and 16,142.9 × 10^3 kg of wastes in 2016. In last years, the production of landfilled waste has been reduced.

The landfill site is located over an impermeable natural clay layer; the bottom and side boundaries may vary according to the period of cultivation; however, they generally include several protective layers, such as a compact clay layer (100 cm), geotextile membranes, gravel (50 cm), geomembranes (1.5 mm) non-woven fabric (1200 g/m^2), pulper products. The landfill covers (top and side) are formed by a waste layer to stabilize the surface, drainage systems, compact clay (20 cm), soil and a vegetative soil layer (up to 100 cm). A grassy mantle and/or forestation with local vegetation will complete the recovery of the environment after closing of each parcel. The systems for leachate treatment, and gas recovery, collection and treatment are in operation. The landfill is situated in the north part of widely opened valley directed towards W-E. The landfill is surrounded to the N and S by a vegetation belt, predominantly comprising *Pinus sylvestris*. The hilly landscape in the western part of the study area is used for agriculture, similarly to the eastern lowland (Adamcová et al., 2017).

The history of solid municipal waste landfill Štěpánovice

The development of the Štěpánovice landfill cadastre is illustrated using historical maps between 1950 and the present. The site has been described in terms of cadastral changes. The landfill is located in the cadastres of Dehtín and Štěpánovice municipalities. Figure 3 shows the borderline between municipal cadastres using a red line. The landfill is located at the Šumava foothills.

Natural conditions

The environmental conditions around the landfill were described in terms of their pedological, geological and geomorphological features. The characteristics of the environmental conditions also comprise the description of protected landscape areas and territorial systems of ecological stability. The territorial systems of ecological stability are defined by the Act No. 114/1992 Coll., on Nature and Landscape Protection as a mutually integrated complex of natural and altered, although nearly natural, ecosystems which maintain natural stability. The aim of the territorial systems of ecological stability is to set up a network of environmentally relatively stable locations, preservation or recovery of the landscape genetic resources, as well as preservation or bolstering of the diversity of indigenous species and their communities (biodiversity). Setting up territorial systems of ecological stability is, by law, a public interest shared by the owners of the land, the municipality and the state.

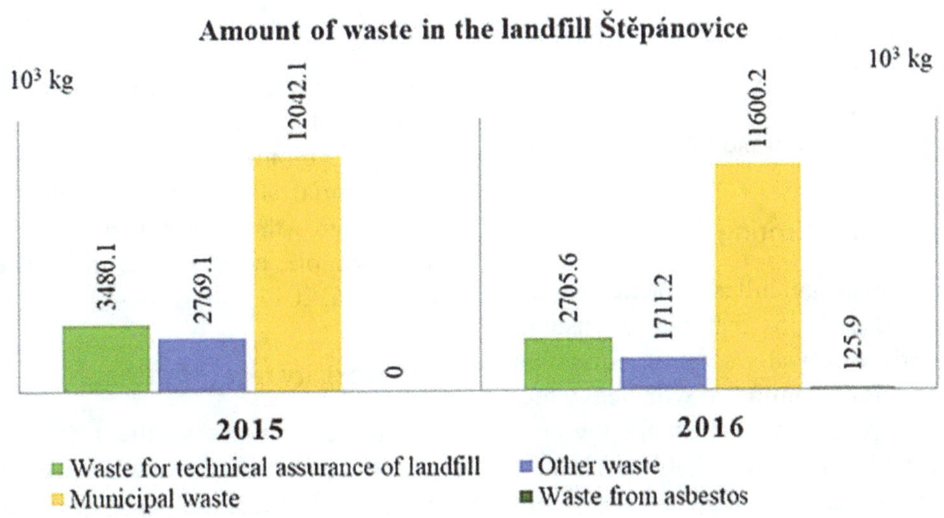

Figure 2. Amount of waste in the years 2015–2016

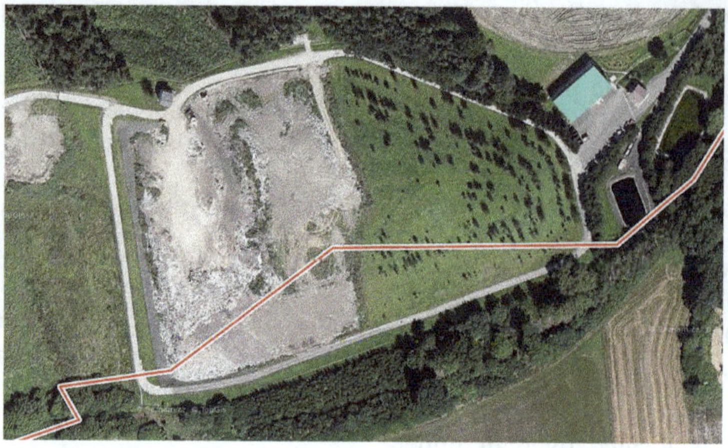

Figure 3. The map of land borders (Heralová, 2017)

Reclamation

Closing and reclamation of a landfill is defined by the CSN 83 8035 standard as the aggregate of measures and works carried out at the landfill having filled its volume and finished dumping waste in compliance with the conditions set by the standard. The binding measures include: regulating the shape of the landfill body, closure and reclamation of the surface with subsequent maintenance of the closed landfill including implementing prescribed monitoring in association with the set approved integrated pollution prevention and control (IPPC). Landfill closure potentially prevents the pollution or negative influence on individual environmental components in close vicinity to the landfill. Reclamation results in the best possible securing and setting up of suitable technical conditions for successive utilisation of the given site and apt incorporation of the reclamation site into the surrounding countryside.

The technical and biological reclamation at the landfill was carried out in 2009. The research focused only on the biological reclamation between 2009 and 2017. The course of biological reclamation was captured in photographs over the specified years (Vaverková and Adamcová, 2018).

Leachate toxicity monitoring

Leachate from the landfill is a potential source of environmental pollution that can for instance have a toxic effect on water sources, animals or plants. Leachate is contaminated with heavy metals which can represent a problem for the environment and the health of its inhabitants. Heavy metals are not biodegradable and have a toxic impact on living organisms. As a result, there is a potential risk of heavy metal pollution and their incorporation into the food chain. It is therefore essential to monitor the quality of leachate. Pursuant to legislation, the leachate, the underground and surface water by the landfill undergo regular monitoring twice a year. As part of the research, the samples of leachate from the leachate sink were taken. The collection of samples was made once a month between April and October 2016 (7 months). The leachate sample was collected once, directly from the active body of the landfill. Precipitation in the given period (mm) was monitored simultaneously. Figure 4 shows the leachate collection sites. Upon collection of leachate, the values below were measured at the site: pH, current temperature (°C) and conductivity (mS/m). Sterile 2 litre sample containers were used to collect the leachate samples and transport them for analysis in a laboratory under the stipulated conditions of 4°C temperature and in the dark. Every leachate sample was subsequently subjected to a toxicity test. A semi-chronic test using *Sinapis alba* L seeds was used. The main reason for carrying out the test was ascertaining the growth inhibition or leachate stimulation on plants and their potential to irrigate crops. The principle of the test is monitoring *Sinapis alba* L. reactions in the initial stages of development (sprouting and the growth of roots) in the tested solution of the sample, compared to the nutrient solution (Heralová, 2017).

Phytotoxicity test of leachate

In order to assess the toxicity of landfill leachate, test organism, i.e. white mustard (*Sinapis alba* L.) was used. The phytotoxicity of the leachate was determined by calculating

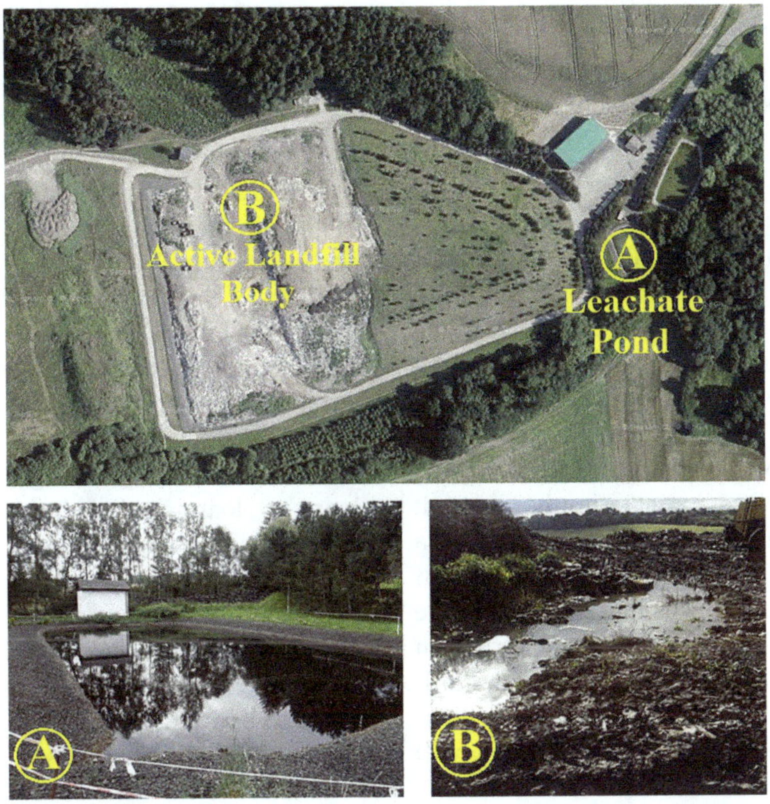

Figure 4. Sampling points; A – Leachate pond; B – Active landfill body

the germination index. The germination index is widely used for the assessment of phytotoxicity (Del Moro et al., 2014; Vaverková et al., 2017). On the basis of the studies conducted by previous authors, each leachate sample was diluted to give final leachate concentrations of 25%, 50%, 75%, 90% and 100%. Each concentration of the dilution series was tested with two replicate samples. The test organisms were exposed to the leachate solutions for a total of 72 h (Vaverková et al., 2017). The seeds of *Sinapis alba* L. were germinated in Petri dishes with a 14 cm diameter on filter paper on the bottom. The setup of the experiment is presented in Figure 5. The hydroponic solution (distilled water with the following chemical ingredients (mg/L): $Ca(NO_3)_2$ 0.8, KH_2PO_4 0.2, KNO_3 0.2, $MgSO_4 \cdot 7\ H_2O$ 0.2, KCl 0.2, $FeSO_4$ 0.01, pH=5.2) with tested liquid was added into each dish, and 15 healthy looking seeds of similar size were evenly spread onto the surface of the filter paper. The Petri dishes were covered by a glass cap to prevent loss due to evaporation and were stored in darkness, inside an Ecocell incubator (t=24°C). After 72 h, the root length was measured (Vaverková et al., 2017).

Calculations and data analysis

The bioassays were performed in two replicates. The percent of root growth inhibition (RI) were calculated with the formula (Eq. 1) (Oleszczuk, 2010; Jośko and Oleszczuk, 2013; Vaverková et al., 2017):

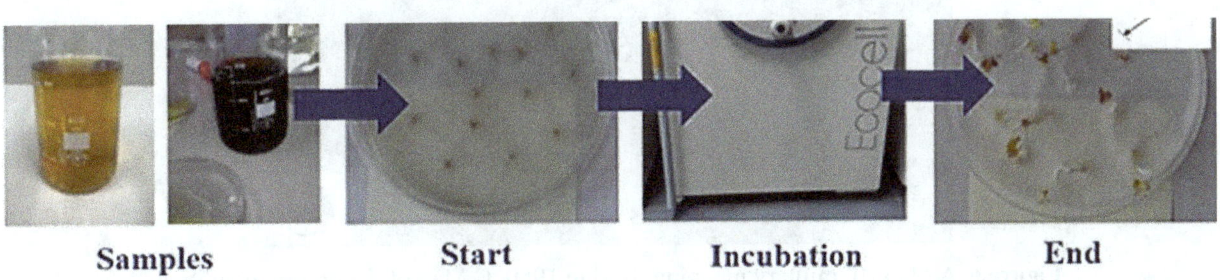

Figure 5. The course of the experiment

$$RI = (A-B)/A \times 100 \quad (1)$$

where: *A* means root length in the control;
B means root length in the tested leachate sample (Vaverková et al., 2017).

RESULTS AND DISCUSSION

The development of the location in question is evident from the analysis of the historical map of the military mapping III (A), then the 1950 map (B) and the present map (C). The maps are depicted in Figure 6. Since the first mapping, the forest vegetation has spread, countryside mosaic has transformed, and a landfill was built (Heralová, 2017). The site of the Štěpánovice landfill was marked in red. The first stage of MSW Štěpánovice construction started in 1995. In 1998 and 2010, the landfill was expanded with 2nd and 3rd stages. At present, the fourth stage is under way and it will enable running the landfill until 2030.

In terms of the pedological conditions, the prevailing type of soil is mesobasic cambisol, or respectively dystric cambisol. The landfill lies in sand-loam to loam-sand sediment and on a rock called chert. There are sands, loam and gravel located at the landfill (Heralová, 2017). Figure 7 shows the location of the landfill in pedological (A) and geological (B) maps. The maps illustrate diverse types of soil (map A) and rock (map B).

KAd(RNm) dystric cambisol (RN modal); KAga' mesobasic ogleyic cambisol; KAa' mesobasic cambisol; RNt lithic ranker; HNg' mildly oglyic brown soil; LUg ogleyic luvisol; PGm modal pseudogley; HNlg' luvic mildly ogleyic brown soil; GLm modal gley; 1 chert; 2 sand-loam to loam-sand sediment; 3 alluvial sediments; 4 silicity

In terms of geomorphic ranking, the landfill site belongs to the Czech Highland province, Broun system, Švihov Highland unit, Radyň Highland microregion, Kamýk Highland region. Štěpánovice and its vicinity are located in the mild warm climate MT 10 (very dry winters with short-term snow cover and a mildly warm course, dry and warm long-lasting summers). The average annual temperature is 8.0°C with

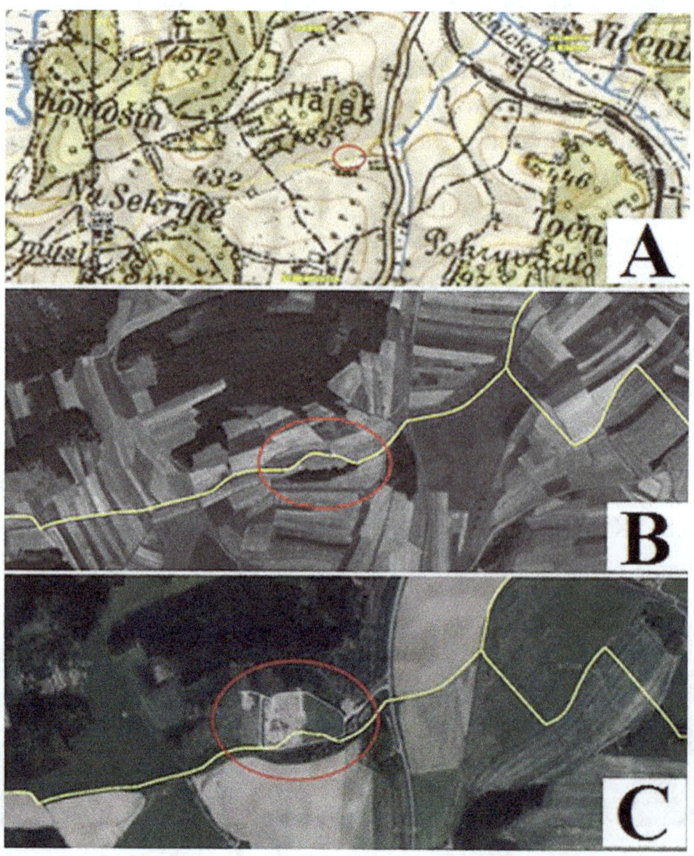

Figure 6. A Map III. military mapping, B Map 1950, C Map of the present condition (Heralová, 2017, modified Zloch, 2018)

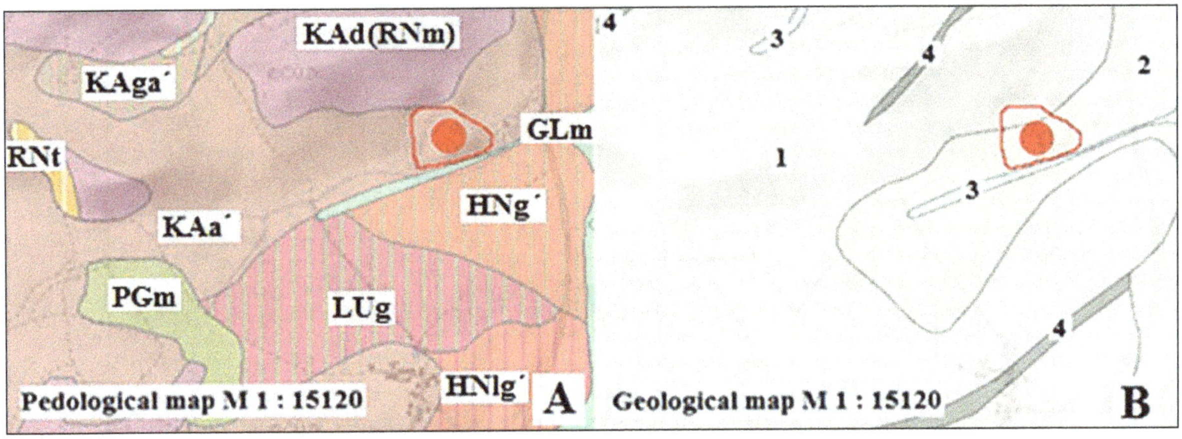

Figure 7. Geological and pedological map of the interest site

precipitation of 582 mm per year. The Úhlava river, Točnický brook, Štěpánovice pond and Nový pond are located in the vicinity of the landfill. The surrounding wetlands are classified in the territorial system of ecological stability as of supra-regional importance. The site falls under the biospherical unit of Central Bohemian deciduous forests. Several small areas under special protection surround the landfill: Stará Úhlava, Bělč, Bělyšov, Americká zahrada and Chudenická bažantnice. The landfill is located between the three protected landscape areas of Šumava, Brdy and Český les (Figure 8) (Heralová, 2017).

Green corridors and habitats are significant components of the territorial system of ecological stability surrounding the landfill. The location of the landfill, in respect to the components of territorial system of ecological stability, is demonstrated in Figure 9, the components of territorial system of ecological stability are distinguished in colour. The green corridor and habitats of regional and supra-regional significance are situated near the landfill (Heralová, 2017).

The technical and biological (forestry) reclamation of the landfill's first stage was completed in 2009. The forestry reclamation incorporated the landfill into the territorial systems of ecological stability. The biological reclamation was implemented by hydro-seeding, application of compost, and planting shrubs and trees. The biological reclamation layer was monitored up until 2012. Trees were planted, weeds eradicated and fertilisers reapplied repeatedly. The reclamation was successful as the mortality of planted trees was low. Figure 10 shows the pictures demarking the reclaimed areas of 1st and 2nd stage landfills between 2009 and 2017 (Vaverková and Adamcová, 2018).

Characterization of the leachate

Landfill leachate is one of the major sources of pollutions discharged into the environment. It is composed from a complex mixture of chemicals and handling typically involves treatment, either on-site or at a wastewater treatment (Vaverková et al., 2017). The physical and chemi-

Figure 8. Map of protected landscape areas (Heralová, 2017)

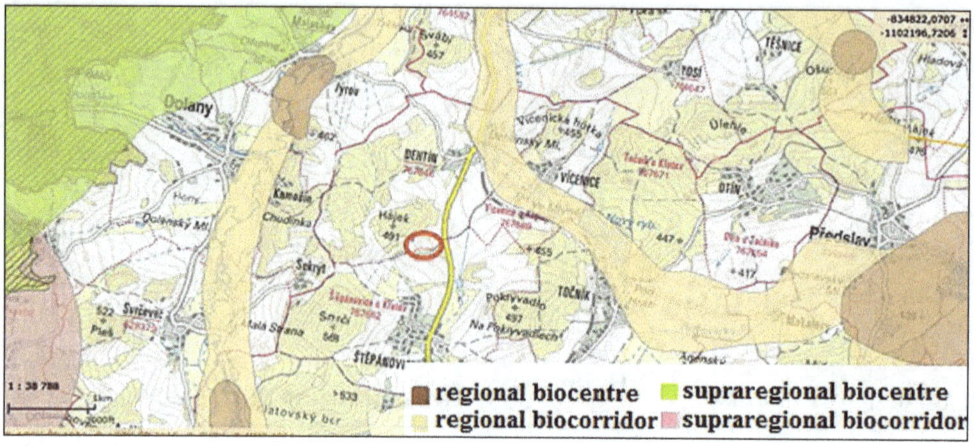

Figure 9. Map of the Territorial System of Environmental Stability (Heralová, 2017)

Figure 10. Reclaimed part of the landfill MSW Štěpánovice (Vaverková and Adamcová, 2018)

cal analysis of raw leachate sample is shown in Table 1. The leachate samples were analysed for pH, electrical conductivity (EC), chemical oxygen demand (COD), other parameters and a series of metals (Cd, Hg, Pb, Zn, Cr). Mean values for landfill leachate (month April – October 2016) are listed in Table 1.

Results of growth inhibition of leachate samples

The toxicity of the collected leachate samples was ascertained using the semi-chronic method opting for the seeds of *Sinapis alba* L. plant. The leachate concentrations were tested in all sam-

Table 1. Physical and chemical characteristics of raw leachate

Parameters	Unit	Values
pH	-	8.92
Temperature	°C	4
BOD	mg/L	35
COD_{Cr}	mg/L	489
NH_4^+	mg/L	3.11
NO_3^-	mg/L	198
Cl^-	mg/L	*
Cd	mg/L	*
Hg	mg/L	0.0001
Pb	mg/L	<0.05
Zn	mg/L	*
Cr^{6+}	mg/L	*
N_{total}	mg/L	*
PO_4^{3-}	mg/L	*
SO_4^{2-}	mg/L	*
EC	mS/m	522

* immeasurable value

ples. Once the incubation term was over, the root length was measured for all seeds in the tested samples. The results were recorded and further processed. The obtained data was used to calculate the growth inhibition. The growth inhibition data for individual tested concentrations are processed into a graphic form (Figure 11–16). Likewise, the precipitation volume is provided for the periods under scrutiny.

The growth inhibition for a 25% leachate concentration (Figure 11) moved in positive values in the period under scrutiny, i.e. 25% leachate concentration shows an inhibitory effect on *Sinapis alba* L. seeds. The growth inhibition figures for 25% concentration oscillated between 13.51% and 53.27%. The average figure for growth inhibition under 25% was 36.51%. The lowest values of growth inhibition were measured in the samples of leachate collected in June (13.51%) and July (26.46%), at the same time, the highest precipitation volumes over the whole monitored period were measured in these months. The volume of precipitation affected the leachate concentration and subsequently also the growth inhibition intensity for 25% leachate concentration.

The growth inhibition for 50% leachate concentration (Figure 12) moved in positive values in the period under scrutiny, i.e. 50% leachate concentration shows an inhibitory effect on *Sinapis alba* L. seeds. The growth inhibition values oscillated between 10.19% and 72.94%. The average value for growth inhibition for 50% concentration was 41.59 %. The lowest growth inhibition figures were measured for the leachate samples collected in June (20.05%) and July (10.19%), at the same time the highest precipitation volumes over the whole monitored period were measured in these months. The volume of precipitation affected the leachate concentration and subsequently also the growth inhibition intensity for 50% leachate concentration. The 50% concentration showed a greater growth inhibition on average than the 25% leachate concentration, by 5.08%.

The growth inhibition for 75% leachate concentration (Figure 13) moved in positive values in the period under scrutiny, i.e. 75% leachate concentration shows an inhibitory effect on *Sinapis*

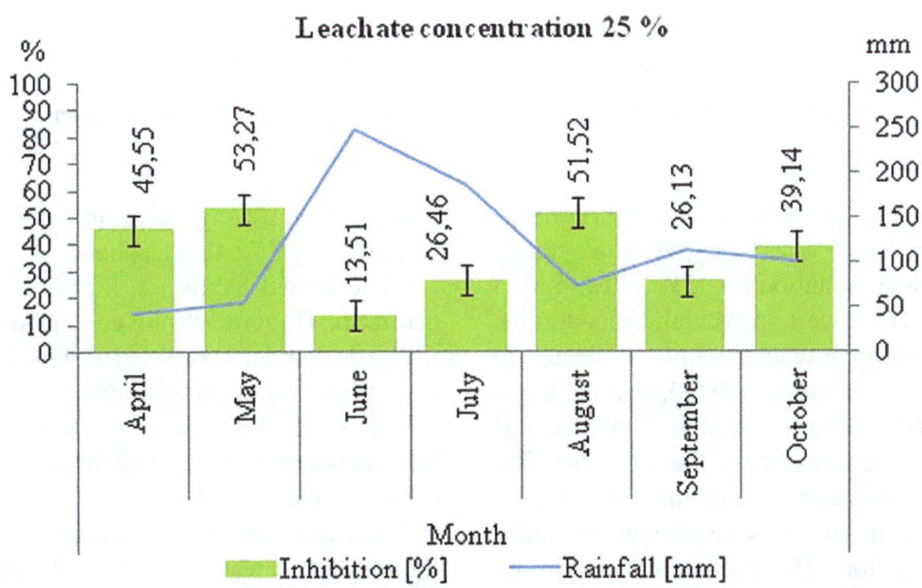

Figure 11. Inhibition and rainfall relationship at 25% leachate concentration

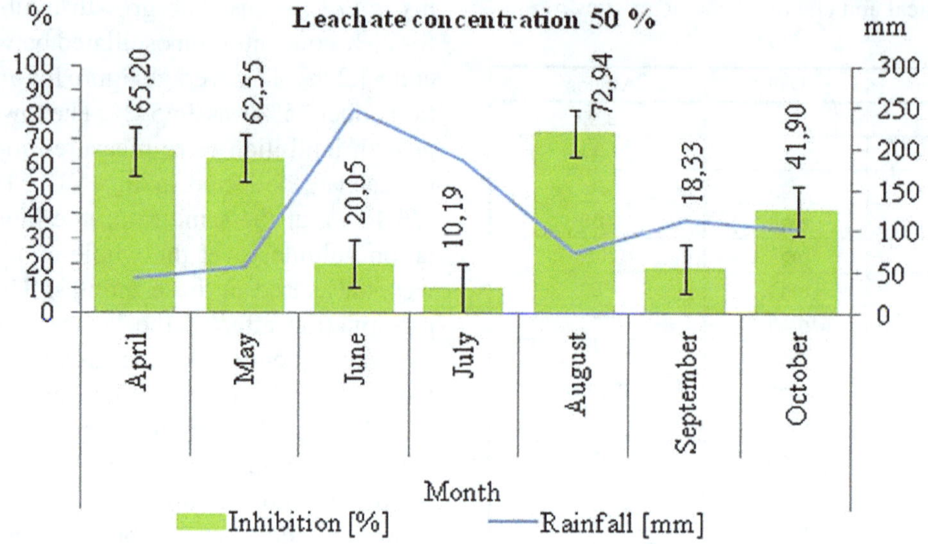

Figure 12. Inhibition and rainfall relationship at 50% leachate concentration

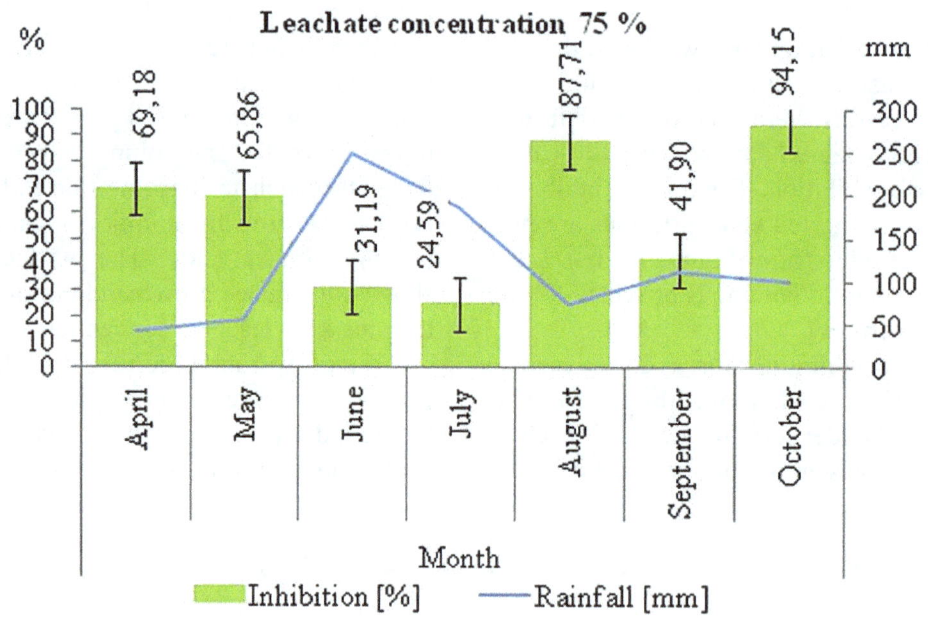

Figure 13. Inhibition and rainfall relationship at 75% leachate concentration

alba L. seeds. The growth inhibition figures oscillated between 24.59 % and 94.15%. The average value for growth inhibition for 75% concentration was 59.23%. The lowest growth inhibition figures were measured for leachate samples collected in June (31.19%) and July (24.59%), and the highest precipitation volumes over the whole monitored period were measured in these months. The volume of precipitation affected the leachate concentration and subsequently also the growth inhibition intensity for 75% leachate concentration. The 75% concentration showed a greater growth inhibition on average than the 25% (by 22.72%) and 50% (by 17.64%) leachate.

The growth inhibition for 90% leachate concentration (Figure 14) moved in positive values in the period under scrutiny, i.e. 90% leachate concentration shows an inhibition effect on *Sinapis alba* L. seeds. The growth inhibition values oscillated between 38.64 % and 98.44%. The average value for the growth inhibition for 90% concentration was 67.36 %. The lowest growth inhibition figures were measured for the leachate samples collected in June (47.41%) and July (38.64%),

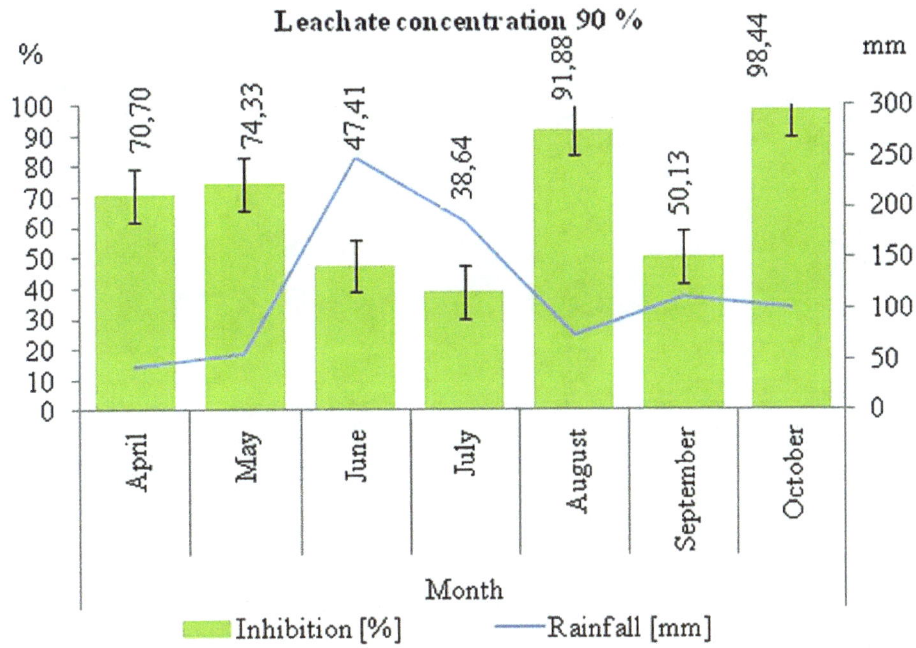

Figure 14. Inhibition and rainfall relationship at 90 % leachate concentration

while at the same time, the highest precipitation volumes over the whole monitored period were measured in these months. The volume of precipitation affected the leachate concentration and subsequently also the growth inhibition intensity for 90% leachate concentration, just like with 25, 50 and 75% concentrations. On average, the 90% concentration shows a greater growth inhibition than 25% (by 30.85%), 50% (by 25.77%) as well as 75% (by 8.13%) of leachate concentrations.

The growth inhibition for 100% leachate concentration (Figure 15) moved in positive values in the period under scrutiny, i.e. 100% leachate concentration shows an inhibitory effect on *Sinapis alba* L. seeds. The growth inhibition values oscillated between 43.91 % and 99.48%. The average value for the growth inhibition for 100% concentration was 73.26 %. The lowest growth inhibition figures were measured for the leachate samples collected in June (51.69%) and July (43.91%), while at the same time the highest precipitation volumes over the whole monitored period were measured in these months. The volume of precipitation affected the leachate concentration and subsequently also the growth inhibition intensity for 100% leachate concentration, just like with all the previously tested concentrations. The 100% leachate concentration showed the greatest overall growth inhibition (against 25% concentration by 36.75%, 50% by 31.67%, 75% by 14.03% and by 5.9% in the case of 90% concentration).

Figure 16 shows the values of the growth inhibition for the leachate sample that was collected directly from the active body of the landfill as a one off, this water was tested at 25%, 50%, 75%, 90% and 100% concentration. The growth inhibition for all monitored concentrations showed positive figures, all tested concentrations had an inhibiting effect on the tested seeds of *Sinapis alba* L. The growth inhibition oscillated between 45.01% and 67.76%. The growth inhibition increased along with the concentration. All concentrations showed significant growth inhibition.

The toxicity of a leachate is given not only by the character of the stored waste but also the amount of precipitation that reaches the active body of the landfill.

With increasing use of MSW landfills as the most widely utilized method of solid waste disposal, the potential contamination from municipal landfills has become a serious problem. Therefore, it is important to study the toxic effects of operating and/or closed landfills on plants, mammals and humans. As it was mentioned above, landfills can be a source of environmental pollution mainly through landfill leachate, which is wastewater highly saturated with various compounds leaching out of decomposing MSW (Ohe et al., 2004; Paskuliakova et al., 2018). Its composition is very complex because of the wide range of toxicants it contains (Cecilia and Junestedt, 2008, Paskuliakova et al., 2018). The biological testing has been used as an indicatory means of evaluating the

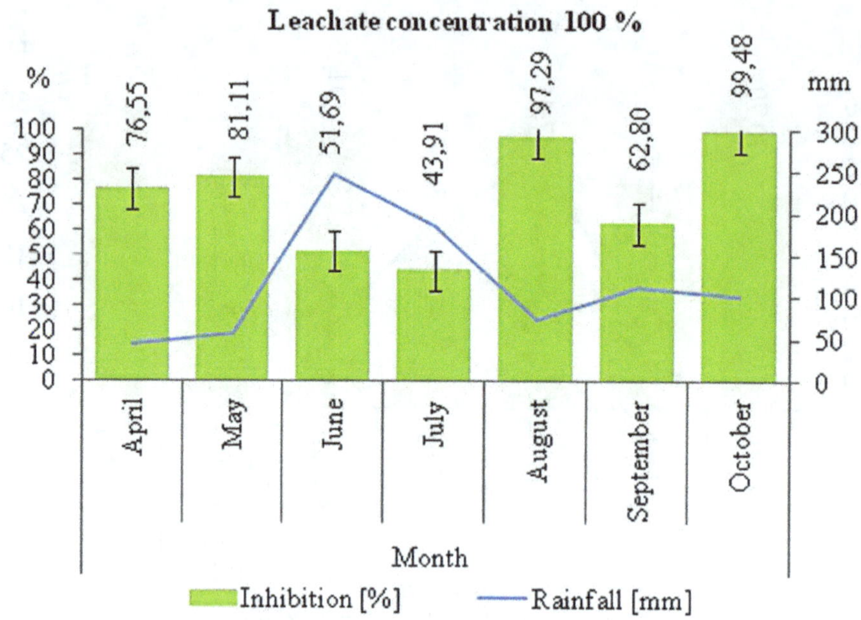

Figure 15. Inhibition and rainfall relationship at 100% leachate concentration

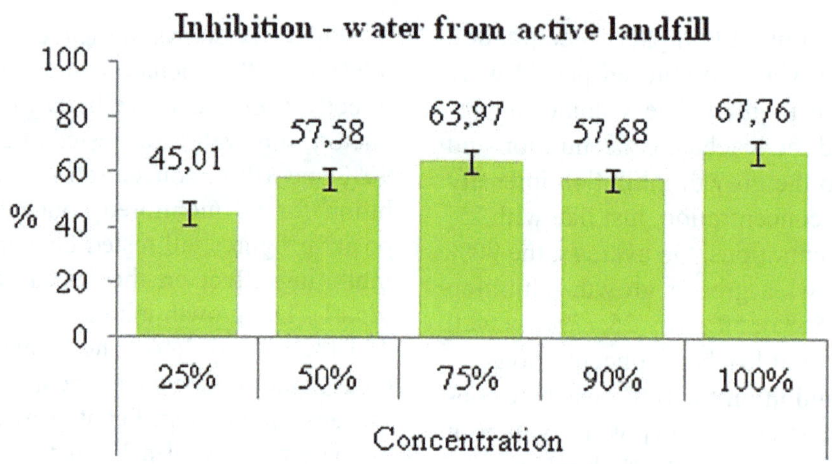

Figure 16. Inhibition of water from active landfill

toxicological impact of landfill leachate (Paskuliakova et al., 2018). The toxicity of raw landfill leachates varies greatly depending on the site and its environmental circumstances. Mostly, they are very toxic (Płaza et al., 2011; Paskuliakova et al., 2018; Zloch et al., 2018). Bortolotto et al. (2009) evaluated the toxic of landfill leachates from Southern Brazil using *A. cepa* as an indicator species. The inhibition of root growth and alkaline comet assay were used to assess the toxicity and genotoxicity potentials, respectively. The untreated leachate resulted in a significant inhibition of root growth. In the study carried out by Li et al. (2008) the physiological and genetic toxicity of the leachate, generated from Xingou Municipal Landfill in China, were investigated with *Triticum aestivum* (wheat) bioassay. The results from this study indicate that the lower leachate concentrations stimulated the germination and growth; however, the components of leachate from the landfill may be genotoxic in plant cells, and exposure to leachate in the aquatic environment may pose a potential genotoxic risk to organisms. In another study (Sang et al., 2006), examine the genotoxicity of leachate with the *Hordeum vulgare*. The results confirmed that components of leachate might be genotoxic in plant cells. The leachate toxicity also varies seasonally. As highlighted by Sang and Li (2004), seasonal difference in genotoxicity induced by leachate was observed and landfill leachates collected in different seasons vary in respect to their toxicity.

CONCLUSIONS

Many studies have shown that the leachate from municipal solid waste landfills can be a source of contaminations. In this study it was observed that the contaminants in the leachate were diluted by rainfall, resulting in lower toxicity of leachate samples. The leachate at higher levels of concentrations heavily inhibited the plant growth, except for the leachate samples from June, July and September. In these three months, the rainfall was the largest so the toxicity of leachate samples was lower. Generally, the water at the landfill site is not toxic. The landfill itself is engineered, with bottom liner and leachate and landfill gas collection and treatment system. Therefore, leachate cannot find a path into the subsurface environment and migrate. The landfill has no direct and significant influence on the surrounding environment and water quality because: (i) the closed part of landfill is covered with an impermeable compact clay layer, geotextile membranes, nonwoven fabric and pulper products in order to prevent the rainwater from infiltrating the waste site and reaching the landfill base, (ii) the leachate from the landfill base is collected so that the leachate is entering the leachate pond, (iii) the evapotranspiration rate was increased by the biological recultivation (vegetative soil layer) – planting of a vegetation cover over the landfill in order to reduce the leachate production.

Acknowledgements

This study was supported by the IGA – Internal Grant Agency Faculty of AgriSciences MENDELU No. TP 003/2018.

REFERENCES

1. Adamcová D., Radziemska M., Ridošková A., Bartoň S., Pelcová P., Elbl J., Kynický J., Brtnický M., Vaverková M.D., 2017. Environmental assessment of the effects of a municipal landfill on the content and distribution of heavy metals in Tanacetum vulgare L. Chemosphere 185, 1011 – 1018.
2. Bortolotto T., Bertoldo J.B., da Silveira F.Z., Defaveri T.M., Silvano J., Pich C.T. 2009. Evaluation of the toxic and genotoxic potential of landfill leachates using bioassays. Environmental Toxicology and Pharmacology, 28(2), 288–293.
3. Cecilia B.O., Junestedt C. 2008. Chemical characterization of landfill leachates – 400 parameters and compounds. Waste Management, 28(10), 1876–1891.
4. Cuartas M., López A., Pérez F., Lobo A. 2017. Analysis of landfill design variables based on scientific computing. Waste Management, 71, 287–300.
5. Dajić A., Mihajlović M., Jovanović M., Karanac M., Stevanović D., Jovanović J. 2016. Landfill design: need for improvement of water and soil protection requirements in EU Landfill Directive. Clean Technologies and Environmental Policy, 18(3), 753–764.
6. Del Moro G., Barca E., Cassano D., Di Iaconi C., Mascolo G., Brunetti G. 2014. Landfill wall revegetation combined with leachate recirculation: a convenient procedure for management of closed landfills. Environmental Science and Pollution Research, 21 (15), 9366–9375.
7. El-Salam M. M. A., Abu-Zuid G. I. 2015. Impact of landfill leachate on the groundwater quality: A case study in Egypt. Journal of Advanced Research, 6(4), 579–586.
8. Ghosh P., Gupta A., Thakur I.S. 2015. Combined chemical and toxicological evaluation of leachate from municipal solid waste landfill sites of Delhi, India. Environmental Science and Pollution Research, 22(12), 9148–9158.
9. Gworek B., Dmuchowski W., Gozdowski D., Koda E., Osiecka R., Borzyszkowski J. 2015. Influence of a Municipal Waste Landfill on the Spatial Distribution of Mercury in the Environment PLoS One, 10(7).
10. Heralová K. 2017. Phytotocity leachate from landfills Štěpánovice. Mgr. Thesis, Mendel University, Brno.
11. Jośko I., Oleszczuk P. 2013. The influence of ZnO and TiO2 nanoparticles on the toxicity of sewage sludges. Environmental Science: Processes & Impacts, 15(1), 296–306.
12. Koda E., Miszkowska A., Sieczka A. 2017. Levels of Organic Pollution Indicators in Groundwater at the Old Landfill and Waste Management Site. Applied Sciences, 7(6), 638.
13. Koda E., Osinski P., Sieczka A., Wychowaniak D. 2015. Areal Distribution of Ammonium Contamination of Soil-Water Environment in the Vicinity of Old Municipal Landfill Site with Vertical Barrier. Water 7(6), 2656–2672.
14. Li G., Yun Y., Li H., Sang N. 2008. Effect of landfill leachate on cell cycle, micronucleus, and sister chromatid exchange in Triticum aestivum. Journal of Hazardous Materials, 155(1–2), 10–16.
15. Li J., Wang Ch., Du L., Lv Z., Li X., Hu X., Niu Z., Zhang Y. 2017. Did municipal solid waste landfill have obvious influence on polychlorinated dibenzo-p-dioxins and polychlorinated dibenzofurans (PCDD/Fs) in ambient air: A case study in East China. Waste Management, 62, 169–176.

16. Makarenko N., Budak O. 2017. Waste management in Ukraine: Municipal solid waste landfills and their impact on rural areas. Annals of Agrarian Science, 15(1), 80–87.

17. Nagarajan R., Thirumalaisamy S., Lakshumanan E. 2012. Impact of leachate on groundwater pollution due to non-engineered municipal solid waste landfill sites of erode city, Tamil Nadu, India. Iranian J Journal of Environmental Health Science & Engineering, 9(1), 35.

18. Ohe T., Watanabe T., Wakabayashic K. 2004. Mutagens in surface waters: a review. Mutation Research, 567(2–3), 109–149.

19. Oleszczuk P. 2010. Toxicity of Light Soil Fertilized by Sewage Sludge or Compost in Relation to PAHs Content. Water, Air and Soil Pollution, 210(1–4), 347–356.

20. Palmiotto M., Fattore E., Paiano V., Celeste G., Colombo A., Davoli E. 2014. Influence of a municipal solid waste landfill in the surrounding environment: Toxicological risk and odor nuisance effects. Environment International, 68, 16–24.

21. Paskuliakova A., Mc Gowan T., Tonry S., Touzet N. 2018. Phycoremediation of landfill leachate with the chlorophyte Chlamydomonas sp. SW15aRL and evaluation of toxicity pre and post treatment. Ecotoxicology and Environmental Safety. 147, 622–630.

22. Płaza A., Nałe G., Ulfig K. 2011. Estimation of the environmental risk posed by landfills using chemical, microbiological and ecotoxicological testing of leachates. Chemosphere, 82(7), 1017–1023.

23. Sang N., Li G., Xin X. 2006. Municipal landfill leachate induces cytogenetic damage in root tips of Hordeum vulgare. Ecotoxicology and Environmental Safety, 63(3), 469–473.

24. Sang N., Li G. 2004. Genotoxicity of municipal landfill leachate on root tips of Vicia faba. Mutation Research/Genetic Toxicology and Environmental Mutagenesis, 560(2), 159–165.

25. Shu S., Zhu W., Wang S., Wang Ch., Ng C.W.W., Chen Y., Chiu A.C.F. 2018. Leachate breakthrough mechanism and key pollutant indicator of municipal solid waste landfill barrier systems: Centrifuge and numerical modelling approach. Science of The Total Environment, 612, 1123–1131.

26. Vaverková M.D., Zloch J., Adamcová D., Radziemska M., Vyhnánek T., Trojan V., Winkler J., Đorđević B., Elbl J., Brtnický M. 2017. Landfill Leachate Effects on Germination and Seedling Growth of Hemp Cultivars (Cannabis sativa L.). Waste and Biomass Valorization, https://doi.org/10.1007/s12649–017–0058-z

27. Vaverková M.D., Adamcová D. 2018. Case study of landfill reclamation at Czech landfill site. Environmental Engineering and Management Journal, In print.

28. Wolny-Koładka K., Malinowski M. 2015. Assessment of the microbiological contamination of air in a municipal solid waste treatment company. Ecological Chemistry and Engineering A, 22(2), 175–183.

29. Zloch J., Vaverková M.D., Adamcová D., Radziemska M., Vyhnánek T., Trojan V., Đorđević B., Brtnický M. 2018. Seasonal alterations of landfill leachate composition and toxic potency of leachate for White mustard (Sinapis alba L.). Acta Universitatis Agriculturae et Silviculturae Mendelianae Brunensis, 66(1), 235–242.

2

Air Pollution Reduction and Environment Protection using Methane Fuel for Turbocharged CI Engines

Osama Hasib Ghazal[1]

[1] Mechanical Engineering Department, Applied Science Private University, Al-Arab Street 21, Amman 11931 Shafa Badran, Jordan, e-mail: o_ghazal@asu.edu.jo

ABSTRACT

The internal combustion engine is considered as one of the main sources for air pollution due to hydrocarbon fuel combustion. The increased land transport usage requires improvement of the engine efficiency and combustion process technology to reduce the engine emissions. A turbocharged engine and the gaseous fuel replacement are the green tools proposed by researchers to enhance fuel saving and emissions reduction. In this paper, both methods were investigated. The methane is a preferred gaseous fuel due to its lower carbon to hydrogen ratio, resulting in lesser HC and CO emissions. In this paper, a turbocharged compression ignition engine with methane/diesel dual fuel is simulated using professional GT-power code to investigate the effect of methane percentage in mixture on the engine performance and emissions. A turbocharged 6 cylinders compression ignition engine has been built and investigated. During the simulation, the methane/diesel ratios were varied from pure diesel with zero percent methane to 90% methane concentration by mass with 10% increment every run. The results show that the engine brake power and specific fuel consumption increased while the thermal efficiency decreased for lower CH_4 concentration. For higher CH_4 percentage, the brake power and thermal efficiency increased while specific fuel consumption decreased. Moreover, NO emission has 35% reduction compared to neat diesel fuel when 50% of methane was added to the mixture. Conversely, the CO and HC concentration increased when the methane ratio is less than 50% compared to neat diesel combustion. In general, the engine efficiency improved when methane was added to diesel fuel in compression ignition engine with turbocharger boost, resulting in lesser emissions and cleaner environment.

Keywords: air pollution, clean combustion, methane emissions, heavy-duty diesel engines

INTRODUCTION

One of the main challenges which faced by the automobile manufacturers is to reduce the engine emissions without serious modifications of engine systems in order to decrease the production cost and time. Many solutions were proposed by the researchers in the last decades to achieve green internal combustion engine. The use of gaseous fuel in compression ignition (CI) engine has received a great attention by researchers [Rakopoulos et al., 2001; Rakopoulos et al., 2006; Stone et al., 1991]. Several papers considered the performance of diesel engine operating on dual fuel mode [Ishida et al., 2000; Singh et al., 2004; Abd Alla et al., 2002; Lee et al., 2003]. Methane, as a second fuel in conventional compression ignition engine dual fuel (methane/diesel), is a promising gaseous fuel due to its clean combustion and low price compared with diesel fuel. In addition, many works reported that using methane/diesel mixture considerably improved the NOx-Soot trade off compared to neat diesel engine [Agarwal et al., 1998; Nwafor et al., 2000; Lee et al., 2003]. Moreover, methane has a high auto-ignition temperature, suitable for use in diesel engine with high compression ratio without the issue of knocking. The main drawbacks of using methane/diesel in dual fuel mode are the higher amounts of CO and HC emissions with lower engine efficiency compared to conventional diesel engine particularly at partial engine load, as reported by researchers [Krishnan et al., 2002; Ling et al., 2005; Shenghua et al., 2003, Nwafor et al., 2000]. Carlucci et al. [2004] reported that using a CH_4/diesel mixture in a compression engine at low load increases the HC and CO emission due to the lean methane/air mixture results in deficiency of combustion process. Korakianitis et al. [2011] observed that due do flame quenching caused by cylinder wall temperature, more methane is emitted to the atmosphere. Ansari et al. [2016] reported that the NG/diesel mixture extends the combustion duration and a decrease in the cylinder temperature results in lower NOx emission. The difference reactivity between NG and diesel decreases the flame temperature and, consequently, the NO formation [Nieman et al., 2012]. They confirmed that the multi-injection technique of DI diesel fuel increase the engine efficiency and decrease the emissions for different engine load. Zoldaket al. [2014] studied theoretically the effect of NG/diesel dual mode on the engine performance and emissions. The concluded that the cylinder pressure and pressure rate are increased compared to neat diesel fuel. Doosje et al. [2014] controlled the soot and NO_x emissions to match the Euro standards for NG/diesel engine operating at different load. Dahodwala et al. [2014] investigated the RCCI technique on CNG/Diesel engine. They concluded that the lower engine emissions can be achieved by higher CNG replacement at low load. Peykani et al. [2015] studied the effect of injection mode on the engine efficiency. They observed that the first and second start of injection timing and the amount of injected fuel have a major influence on the engine emissions. Another green technique to reduce the exhaust emissions and save fuel in a diesel engine is using turbocharge system to recover the waste heat from the exhaust gases [Rongchao et al. 2014]. The turbocharged compression ignition engine operating with methane fuel has a higher thermal efficiency and lesser emissions due to lean combustion characteristic [Sunyoup et al., 2014].

In this study, the effect of methane addition to turbocharged diesel engine on the performance and emission has been investigated using professional code GT-power. A six cylinder compression engine has been built and simulated. The methane/diesel ratio varied from (10–90%) by mass. Additionally, neat diesel was also combusted to compare with the CH_4/diesel mixture. The model has been validated using the available data from the literature and a good agreement was obtained. The engine output such as: brake power, thermal efficiency, fuel consumption, cylinder pressure and temperature, NO, HC and CO emissions were calculated and fully discussed.

SIMULATION MODEL

A six cylinder turbocharger with intercooler system for compression ignition engine has been built and simulated using professional code GT-power. The CH_4/diesel ratio varied from (0–90%) by mass. The methane was injected to the intake port with various mass flow rates to mix with air and create homogeneous charge. Liquid diesel fuel was injected directly to the combustion chamber with rail pressure = 2000 bar and the diesel fuel temperature equaled 330 K. The properties of methane and diesel are included in the software and are presented in Table 1. During the simulation, the total fuel mass (methane and diesel) was kept constant = 140 mg and the methane energy share was calculated using Eq. (1).

$$CH_4\ energy\ share = \frac{\dot{m}_{CH_4} \cdot LHV_{CH_4}}{\dot{m}_{CH4} \cdot LHV_{CH4} + \dot{m}_{diesel} \cdot LHV_{diesel}} \tag{1}$$

The turbocharger operating parameters such as speed, pressure ratio and mass flow rate are determined by a standard performance map specified by the user. The flow model required the solution of Naviera-Stokes equations. The conservation of momentum, energy and continuity are solved in one dimensional model with average values across the flow. Compressor and turbine are modeled using the map-based approach created by the user. The map describes the efficiency of the turbocharger system for each operating point. It should determine the operating conditions such as speed, mass flow rate and thermodynamics parameters for both turbine and compressor. The power consumed/produced by the turbomachinery can be calculated due to the enthalpy change. The program uses different heat transfer model. In this simulation, the WoschniGT model has been used to predict the heat transfer inside the cylinder. The program support dual fuel combustion. The combustion model used to predict the burn rate of direct injection diesel fuel is called DIPulse. The CH_4/air mixture burn rates were predict using EngCylCombSITurb and standard turbulent flame speed model. The system was equipped with throttle controller to control the throttle angle and the methane/air mixture flow rate and consequently the engine load. All the simulations were run with full engine load and without exhaust gases recirculation. The engine speed was kept constant for all the runs. The engine specifications and engine operation conditions are presented in Table 2 and 3. The model has been validated using available data from the literature. The results from this model compared with references [Saito et al., 2001; Zhou et al., 2014; Yu et al., 1986; Gatts et al., 2012]. The theoretical and experimental data showed a good agreement.

RESULTS AND DICSUSSION

The author investigated analytically the performance and emission characteristics of a turbocharged (DI) diesel engine operating on methane/diesel in dual fuel mode. The methane content varied from (10–90%) by mass. The neat diesel fuel was also calculated to compare with the dual mode to find out the effect of methane addition to diesel on the engine performance and emissions. Another technique used in the simulation to reduce emissions was by means of a turbocharger system. The engine ran at full load with constant lambda.

Table 1. Fuel properties

Properties	Units	Methane	Diesel
Carbon atoms per molecule	–	1	13.5
Hydrogen atoms per molecule	–	4	23.6
Oxygen atoms per molecule	–	0	0
Lower heating value	MJ/kg	50	43
Critical temperature]	K	190.4	569.4
Critical pressure	bar	46	24.6
Absolute entropy at 298k	J/kg-K	11618	3445.47
Laminar flame speed	m/s	0.38	0.3
Auto ignition temperature	k	813	530

Table 2. Engine geometry

Parameter	Unit	Value
Bore	mm	120
Stroke	mm	170
Conn. rod length	mm	290
Piston pin offset	mm	0
Displacement/cyl	L	1.9
Total displacement	L	11.4
Number of cylinders		6
Compression ratio		13.5
Bore/stroke		0.71
IVC, IVO	°CA	-122, 310
EVO, EVC	°CA	105, 390

Table 3. Engine operation conditions

Parameter	Unit	Value
Initial pressure	bar	1
Initial temperature	K	298
Speed	rpm	1800
Combustion start	°CA	-1.5
Injection start	°CA	-6
Vol. eff. ref. pressure	bar	1.7
Vol. eff. ref. temperature	K	315
Mean piston velocity	m/s	11

Combustion characteristics

Figure 1 shows the cylinder pressure as a function of methane percentage under full engine load. As shown in the figure, the cylinder pressure increased along with the CH_4 mass in the mixture. Moreover, the addition of 70% of CH_4 by mass increased the cylinder pressure up to 30% compared to neat diesel fuel. In turn, the addition of more than 80% methane results in a decrease of the cylinder pressure compared to diesel fuel. In addition, the ignition delay increased along with the methane fraction in the mixture due to the decrease in the oxygen concentration which affected the diesel combustion process [Saito et al., 2001]. The considerable increase of peak cylinder pressure caused by uncontrolled combustion of CH_4/diesel mixture compared with diesel fuel combustion was observed. This is due to high cylinder temperature resulting in multipoint methane combustion and consequently uncontrolled heat release rate. Moreover, the ignition delay increased when the methane ratio increased in the mixture due to the extended of premixed combustion period [Zhou et al., 2014]. Figure 2 shows the maximum in-cylinder temperature versus gas substitu-

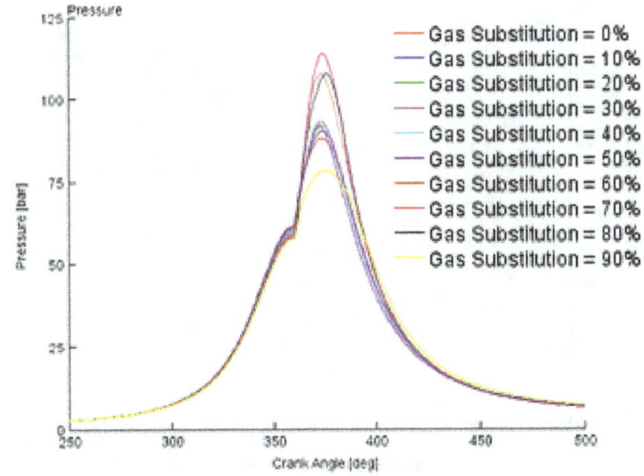

Figure 1. Maximum cylinder pressure versus methane concentration

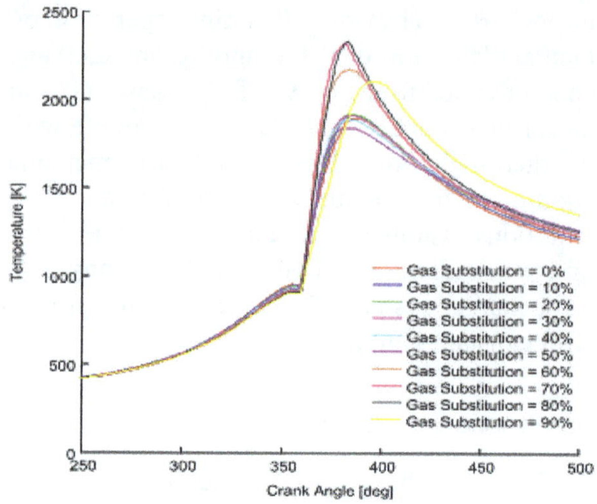

Figure 2. In-cylinder temperature versus methane concentration

tion. The results show that the increase of CH_4 by 80% causes a 22% increase in cylinder temperature compared to conventional diesel fuel.

The effect of methane addition on the heat release rate is presented in Figure 3. It is shown that the addition of methane between (60–70%) increases the heat release rate drastically compared to neat diesel fuel. Moreover, the heat release is retarded when CH_4 is added to diesel. The results agree with [Yu et al., 1986]. They performed that the increase of hydrogen mass in hydrocarbon fuels results in increasing the laminar flame speed and consequently, the cylinder pressure and heat release rate. For methane/diesel combustion, compared to neat diesel fuel, the heat release rate increased between (15–30%) depending on the methane concentration in the fuel mixture. In addition, the accumulation of heat by methane and diesel resulted in a remarkable increase in the heat release rate. Figure 4 shows the burn rate for different methane/diesel mixtures. The burn rate increased along with the methane ratio in the mixture due to high flame speed of methane.

Engine performance

The major engine outputs are presented in Figure 5. The variation of engine brake power versus CH_4 concentration is presented in Figure 5a. As shown, the increase of methane concentration raises the brake power approximately by 20%, compared to diesel baseline. Moreover, the addition of (40–50%) of CH_4 decreases the power output. The effect of CH_4 concentration on the thermal efficiency is presented in Figure 5b. The brake thermal efficiency can be calculated using Eq. (2).

$$b.thermal\ efficiency = \frac{brake\ power}{\dot{m}_{CH4} \cdot LHV_{CH4} + \dot{m}_{diesel} \cdot LHV_{diesel}} \quad (2)$$

The engine efficiency decreases when the CH_4 ratio is lesser than 50% compared to neat diesel, due to incomplete combustion [Gattes et al., 2012] as well as the high amount of input energy and low output power [Ghazal, 2013]. On the other hand, the efficiency increased when the methane ratio was higher than 60% due to raised indicated mean effective pressure (IMEP). The effect of methane addition on the fuel consumption and indicated mean effective pressure are illustrated in Figure 5c and Fig-

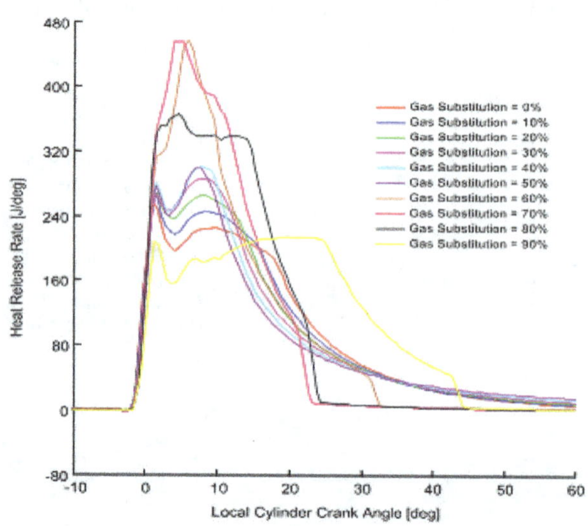

Figure 3. Heat release rate versus methane concentration mixture

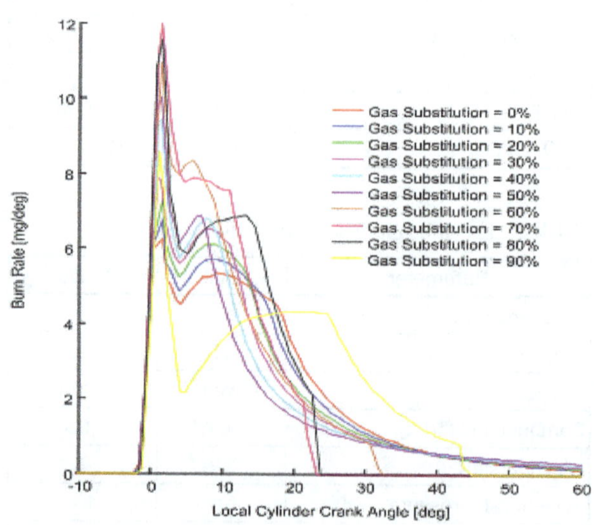

Figure 4. The burn rate of the CH_4/diesel versus gas substitution

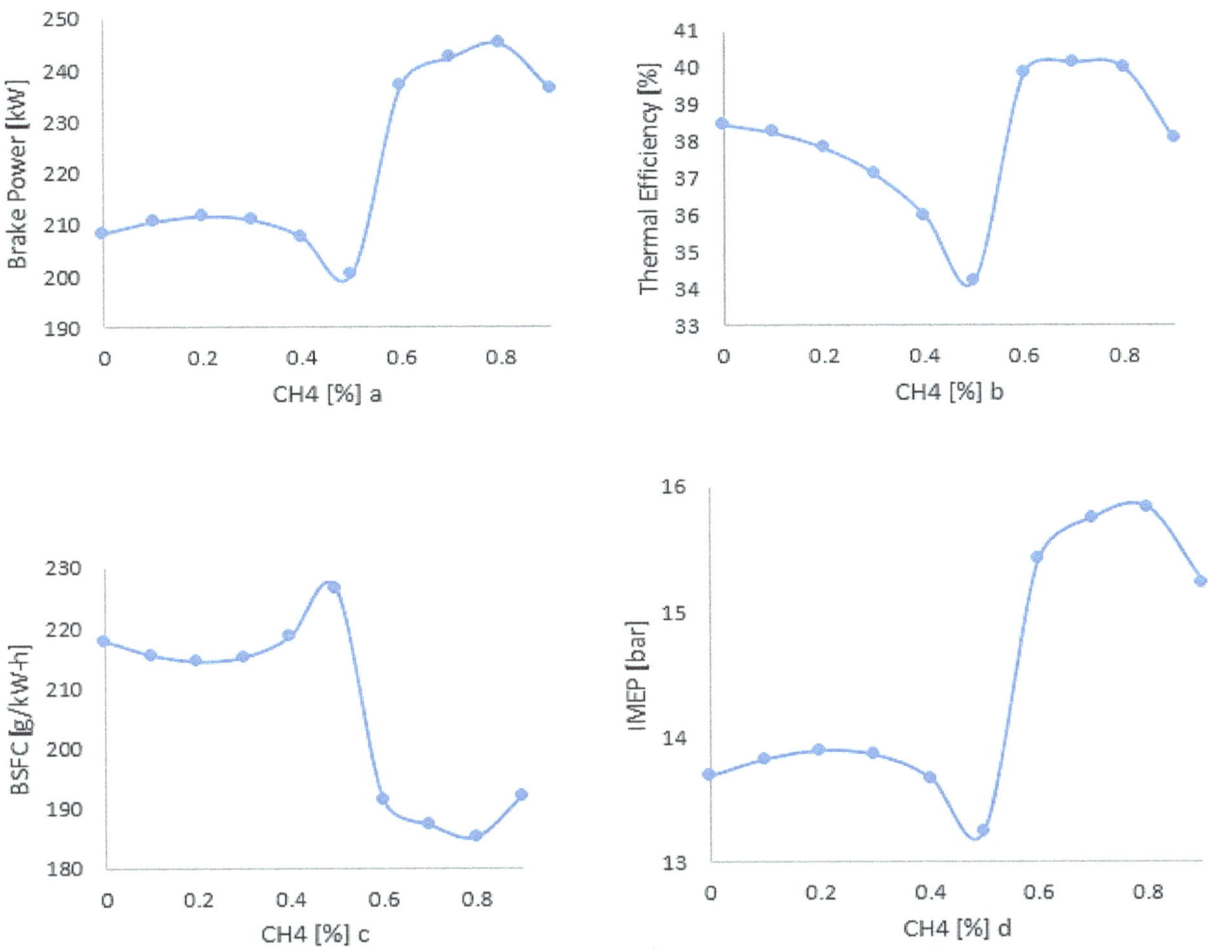

Figure 5. The engine b. power, efficiency, fuel consumption and effective pressure versus CH_4 fraction

ure 5d. The fuel efficiency slightly decreased when CH_4 added was less than 50% in the mixture at full engine load. On the other hand, the increase of methane percentage in the mixture of more than 60% decreases the fuel consumption up to 15% compared to neat diesel combustion. Additionally, the high methane concentration results in an increase in the mean effective pressure. The same results have been obtained by [Ausquiza et al., 2011]. They reported that when the NG ratio increased, the fuel consumption increased as well, both at low load and at high load.

Engine emissions

The exhaust emissions characteristics are presented in Figure 6. The NO emission decrease drastically when gaseous fuel is added to the mixture. The 35% NO reduction can be achieved when 50% methane is added compared to neat diesel fuel as shown in Figure 6a and 6b. The NO concentration is presented in Figure 6a as a function of CH_4 mass fraction. The results show a reduction in the NO emissions when methane fraction is increased in the mixture. However, the NO formation increased when methane was added to diesel due to high combustion temperature, but the NO oxidation rate also increased resulting in lower NO exhaust emissions. Conversely, the CO concentration increased slightly compared to neat diesel combustion when the engine was operating at full load and the methane ratio increased up to 50%. On the other hand, the CO concentration is lesser than base diesel line when the CH_4 ratio increased up to 90% as presented in Figure 6b. This is due to incomplete combustion of methane. The same results were demonstrated by [Gatts et al., 2012]. The HC production characteristic is illustrated in the Figure 6c. The increase of CH_4 concentration up to 50% results in an increased HC amount in the exhaust gases.

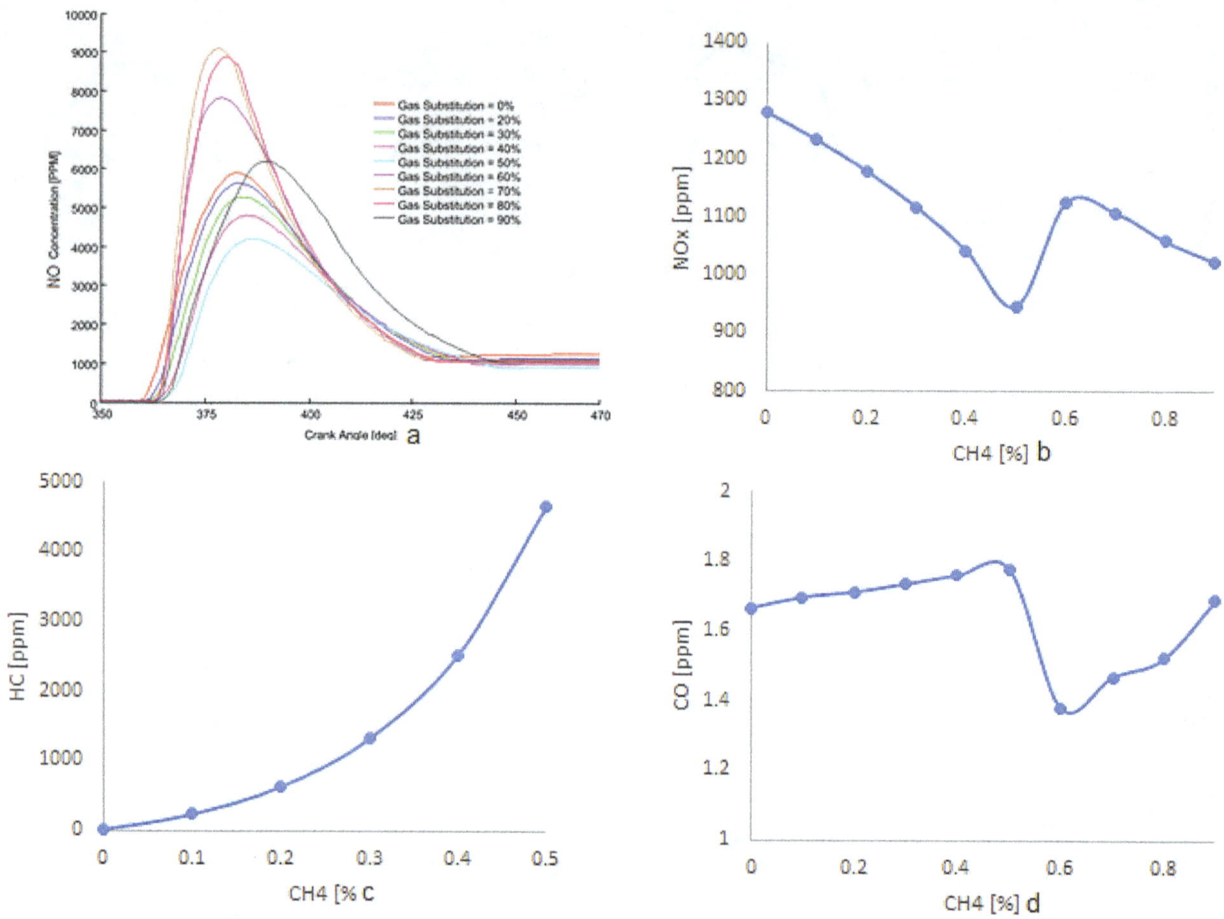

Figure 6. The effect of CH_4 concentration on No, CO and HC emissions

CONCLUSIONS

In this study the effect of methane gaseous fuel on the performance and emissions of turbocharged compression ignition engine was investigated. The engine ran with various CH_4/diesel ratios under full load. The results show that the brake power and thermal efficiency increased while the fuel consumption decreased for high CH_4 concentration in the mixture. Moreover, NO emission reduced by 35% compared to neat diesel fuel when 50% of methane was added to the mixture. Conversely, the CO and HC concentration increased when the methane ratio was less than 50% compared to neat diesel combustion. To conclude, the turbocharger and alternative fuel are promising green techniques to reduce the engine emissions and air pollution.

REFERENCES

1. Abd Alla GH, Soliman HA, Badr MF, Abd Rabbo MF. 2002. Effect of injection timing on the performance of a dual fuel engine. Energ Convers Manage, 43, 269–277.
2. Ansari E, Poorghasemi K, Irdmousa BK, Shahbakhti M, Naber J. 2016. Efficiency and emissions mapping of a light duty diesel-natural gas engine operating in conventional diesel and RCCI modes. In: SAE international powertrain, fuel and lubricants conference, 2016-01-2309.
3. Carlucci AP, Ficarella A, Laforgia D. 2004. Experimental comparison of different strategies for natural gas addition in a common rail diesel engine. In: Proceedings of FISITA 2004, Barcelona, Spain, May 23–27, Paper F2004V136.
4. Dahodwala M, Joshi S, Koehler E, Franke M. 2014. Investigation of diesel and CNG combustion in a dual fuel regime and as an enabler to achieve RCCI combustion. SAE technical paper 2014-01-1308, http://dx.doi.org/10.4271/2014-01-1308.
5. Doosje E, Willems F, Baert R. 2014. Experimental demonstration of RCCI in heavy-duty engines using diesel and natural gas. SAE technical paper 2014-01-1318. http://dx.doi.org/10.4271/2014-01-1318.
6. Eg'ausquiza, J., Braga, S., and Braga, C. 2011. Experimental Investigation of a Diesel Engine Operating on Natural Gas/Diesel Dual-Fuel Mode.

SAE Technical Paper 2011-36-0351. https://doi.org/10.4271/2011-36-0351.

7. Gatts T, Liu SY, Liew C, Ralston B, Bell C, Li HL. 2012. An experimental investigation of incomplete combustion of gaseous fuels of a heavy-duty diesel engine supplemented with hydrogen and natural gas. Int J Hydrogen Energy, 37, 7848-59.
8. Ghazal OH. 2013. Performance and combustion characteristic of CI engine fueled with hydrogen enriched diesel. Int J Hydrogen Energy, 38, 15469-15476.
9. Ishida M, Amimoto N, Tagai T, Sakaguchi D. 2001. Effect of EGR and preheating on natural gas combustion assisted with gas-–oil in a diesel engine. Proceeding of the 5th International Symposium of COMODIA, Nagoya, 382–9.
10. J.H. Zhou, C.S. Cheung, C.W. Leung. 2014. Combustion, performance and emissions of a diesel engine with H_2, CH_4 and H_2-CH_4 addition. International journal of hydrogen energy, 39, 4611-4621.
11. Karim GA. 2003. Combustion in gas fueled compression ignition engines of the dual fuel type. J Eng Gas Turb Power, 125, 827–36.
12. Korakianitis T, Namasivayam AM, Crookes RJ. 2011. Natural-gas fueled spark-ignition (SI) and compression-ignition (CI) engine performance and emissions. Prog Eng Combust Sci, 37, 89-112.
13. Krishnan SR, Biruduganti M, Mo Y, Bell SR, Midkiff KC. 2002. Performance and heat release analysis of a pilot-ignited natural gas engine. Int J Engine Research, 3, 171–84.
14. Krishnan S.R., Srinivasan K.K., Singh S., Bell S.R., Midkiff K.C., Gong W. 2002. Strategies for reduced NOx emissions in pilot-ignited natural gas engines. Proceedings of the ASME-WA-ICEF Meeting, New Orleans USA, September 8–11, 39, 361–367.
15. Lee CS, Lee KH, Kim DS. 2003. Experimental and numerical study on the combustion characteristics of partially premixed charge compression ignition engine with dual fuel. Fuel, 82, 553–60.
16. Ling S, Longbao Z, Shenghua L, Hui Z. 2005. Decreasing hydrocarbon and carbon monoxide emissions of a natural-gas engine operating in the quasi-homogeneous charge compression ignition mode at low loads. P I Mech Eng D-J Aut, 219, 1125–31.
17. Nieman D, Dempsey A, Reitz RD. 2012. Heavy-duty RCCI operation using natural gas and diesel. SAE Int J Engines, 5, 270–85. http://dx.doi.org/10.4271/2012-01-0379.
18. Nwafor OMI. 2000. Effect of choice of pilot fuel on the performance of natural gas in diesel engines. Renew Energy, 21, 495–504.
19. Paykani A, Kakaee AH, Rahnama P, Reitz RD. 2012. Effects of diesel injection strategy on natural gas/diesel reactivity controlled compression ignition combustion Energy.
20. Rakopoulos CD, Kyritsis DC. 2001. Comparative second-law analysis of internal combustion engine operation for methane, methanol and dodecane fuels. Energy, 26, 705–22.
21. Rakopoulos CD, Kyritsis DC. 2006. Hydrogen enrichment effects on the second law analysis of natural and landfill gas combustion in engine cylinders. Hydrogen Energy, 31, 1384–93.
22. Rongchao Zhao, Weilin Zhuge, Yangjun Zhang, Yong Yin , Zhen Chen, Zhigang Li. 2014. Parametric study of power turbine for diesel engine waste heat recovery. Applied Thermal Engineering, 67, 308-319.
23. Saito H, Sakurai T, Sakonji T, Hirashima T, Kanno K. 2001. Study on lean burn gas engine using pilot oil as the ignition source. SAE Tech Pap 2001-01-0143.
24. Shenghua L, Longbao Z, Ziyan W, Jiang R. 2003. Combustion characteristics of compressed natural gas/diesel dual-fuel turbocharged compressed ignition engine. P Mech Eng D-J, 217, 833–8.
25. Singh S, Kong S-C, Reitz RD, Krishnan SR, Midkiff KC. 2004. Modeling and experiments of dual-fuel engine combustion and emissions. SAE Paper No.2004-01-0092.
26. Srinivasan KK, Krishnan SR, Midkiff KC. 2006. Improving low load combustion, stability and emissions in pilot-ignited natural gas engines. P I Mech Eng D-J Aut, 220, 229–39.
27. Stone CR, Ladommatos N. 1991. Design and evaluation of a fast-burn spark-ignition combustion system for gaseous fuels at high compression ratios. J Inst Energy, 64, 202–11.
28. Stone CR, Gould J, Ladommatos N. 1993, Analysis of bio-gas combustion in spark-ignition engines, by means of experimental data and a computer simulation. J Inst Energy, 66,180–7.
29. Sunyoup Lee, Changgi Kim, Young Choi, Gihun Lim, Cheolwoong Park. 2014, Emissions and fuel consumption characteristics of an HCNG-fueled heavy-duty engine at idle. International journal of hydrogen energy, 39, 8078-8086
30. Xu M, Cheng W, Li Z, Zhang H, An T, Meng Z. 2016. Pre-injection strategy for pilot diesel compression ignition natural gas engine. Appl Energy, 179, 1185–93.
31. Yu G, Law CK, Wu CK. 1986, Laminar flame speeds of hydrocarbon of air mixtures with hydrogen addition. Combust Flame, 63, 339-47.
32. Zoldak P., Sobiesiak A., Bergin M., Wickman D. 2014. Computational study of reactivity controlled compression ignition (RCCI) combustion in a heavy-duty diesel engine using natural gas. SAE technical paper 2014-01-1321. http://dx.doi.org/10.4271/2014-01-1321.

3

Priority Areas for Initiating Land Consolidations Related to Erosion and Water Retention in the Landscape, Czech Republic

Petr Karásek[1], Jana Konečná[1], Michal Pochop[1], Josef Kučera[1], Jana Podhrázská[1,2]

[1] Research Institute for Soil and Water Conservation, Department of Land Consolidations and Land Use Planning, Lidická 25/27, 602 00 Brno, Czech Republic

[2] Faculty of Agronomy, Mendel University in Brno, Zemědělská 1, 613 00 Brno, Czech Republic

[*] Corresponding author's e-mail: karasek.petr@vumop.cz

ABSTRACT

Land consolidations represent a unique tool for managing the problems of the rural space. They help to settle the land property rights and meet the current requirements associated with the protection and formation of a stable, permanently sustainable rural landscape. In the Czech Republic, complex land consolidations have been completed in 1,965 cadastral areas out of the total 13,100, and in 2,134 more they are in the preparation phase (up to 1st September 2015). It means that about 60% of cadastral areas in CR still await land consolidation. Our study aimed to localize the priority localities (cadastral areas and regions) for preferential implementation of land consolidation. The authors selected three thematic factors to determine the risk of their occurrence in particular localities and, consequently, the need to initiate the land consolidation. The analysis is based on the assumption that the cadastral areas where land consolidations have been completed or are under way have already dealt with the above-mentioned risk factors or will do so soon. Therefore, such areas have not been included into our study. The key selected thematic factors relevant for the preferential implementation of land consolidation include: the risk of water and wind erosion in arable land, water management, and water retention in the landscape. For each of these three factors, a map of preferential localities (based on the degree of risk/suitability) was processed for initiation of land consolidation. At the same time, a simple multi-criteria analysis of the extent of the Czech Republic's erosion (water and wind) has been prepared. The result of this analysis is the ranking of regions according to the degree of risk of erosion. The most vulnerable regions of the Czech Republic include Jihomoravský (South Moravian Region), Středočeský (Central Bohemia) and Pardubice.

Keywords: land consolidation; rural areas; water erosion; wind erosion; water retention; GIS

INTRODUCTION

Land consolidation is a tool for creation of sustainable rural areas and improving the effectiveness of land cultivation as well as supporting the rural development in different countries, including the Czech Republic (CR) (Sklenička, 2006; Pašakarnis and Maliene 2010). Land consolidation can be also defined as a comprehensive reallocation procedure in the rural area consisting of fragmented agricultural holding, or their parts Vitikainen (2004).

In the countries of the former Eastern Bloc, the land consolidation brings about land reform or de-collectivisation in response to a long period of suppression of the land use and ownership (Podhrázská et al., 2015).

Similarly as in other Central and Eastern European countries, land ownership fragmentation, when the non-contiguous plots of individual owners are scattered around the area of one or more cadastres, is an important issue for farmers in the Czech Republic. After forty years of interrupted farming based on land ownership, a significant number of plots are not accessible by field roads. Additionally, the plot shapes and sizes are not suitable because they reflect the conditions in the first half of the 20th century. These factors, together with other driving forces, primarily the massive exodus of workers from agriculture to industry between 1950 and 1990, and the present-day mismatch between the small holding size and the large scale agricultural machinery, drainage and irrigation systems, are at the root of the sharp distinction between the land ownership and land use (Sklenička et al., 2006).

Land consolidation not only deals with land redistribution, but also arranges the optimal shape and size of plots (Leń, Noga, 2018). At the same time, the environmental conditions, soil protection, water management and ecological stability of the landscape (in particular flood and erosion control measures or systems of ecological stability of landscape) are improved through the application of the land consolidation process.

The land consolidation programme is not a specific Czech project. Land consolidations are applied in various forms all over the world. Sky (2015) reports that the land consolidation processes are similar between countries. Depending on the political, socio-economic and environmental demands of the particular countries or regions, the land consolidation process places stress on land reclamation, nature and landscape conservation (Nikodemus et al. 2005), or possibly it constitutes a remedial platform for the social and economic development of the countryside (Miranda et al. 2006; Sklenička et al. 2006). The basic idea of land consolidations in advanced countries in EU remains the same, i.e. to preserve and improve the rural space in such a way as to make it stable and permanently sustainable for further generations (Podhrázská et al., 2015).

The majority of current land consolidations in CR deal with insensitive human interventions into the landscape in the second half of the 20th century. The main efforts of the Socialist regime were focused on intensifying the agricultural production. The agricultural production was collectivized and large agricultural cooperation was established. The landscape was subjected to extensive devastation of natural eco-stabilizing elements, destruction of spinneys, balks, drainage of river floodplains and wetlands. Large land blocks of hundreds of hectares in size were created, destroying the field roads. The landscape became impassable. These factors led to a massive development of wind and water erosion and to the problems connected with low water retention in the landscape, as well as flash floods from heavy rainfalls. The communist era showed to be problematic and in some aspects, the sustainable development of rural areas was impossible (Thomas, 2006; Hartvigsen, 2014).

The process of land consolidation in the Czech Republic, as it is known today, has been occurring since 1991 and currently is controlled by the State Land Office (SLO). Land consolidations are regulated by Act No. 139/2002 Coll. The law stipulates that land consolidations create the conditions for rational management by landowners, improve the environmental protection and water management. It is a long-lasting process.

The project and design phase of a land consolidation takes 4–5 years on average, according to the size and difficulty of the project implementation (SLO, 2015). The gradual execution of the shared facilities proceeds after agreement with the owners, depending on the financial support from the state budget and possible EU support at different time periods (Podhrázská et al., 2015). In isolated cases, the duration of the land consolidation may even exceed double average length (6 years or more).

At present, designers of land consolidations are bound by several methodological guidelines existing in CR. These methodological guidelines are recommended for use by the State Land Office. They include the methodologies for evaluation of water erosion (Janeček et al., 2012), wind erosion (Podhrázská et al., 2011), design of land consolidations (SLO, 2016), evaluation of the landscape (Tlapáková et al., 2013), assessment of the effectiveness and benefits of implemented land consolidations (Konečná et al., 2014), and other.

The design and application of a land consolidation represents a rather demanding process, involving both the restoration of the cadastral map and reorganization of land parcels, including a proposal of the measures aimed at the recovery of a stable agricultural landscape. According to the statistics of the State Land Office (SLO, 2015), land consolidations have as yet been completed in 1,965 cadastral areas (data of September 1, 2015). This figure represents about 15% cadastral areas in CR. In 16% cadastral areas, the land consolidations are in the initiation phase or under way. That leaves about 60% cadastral areas in CR still awaiting land consolidations.

In 2015, the State Land Office issued the "Concept of Land Consolidations in CR for the period 2016 – 2020" (SLO, 2015). This material assumes the application of 200 land consolidations per year. In the current situation, with land consolidations still lacking in about 9 thousands cadastral areas, this would represent a time horizon of 50 years. From the long-term point of view, this means that in order to meet the needs of nature and landscape protection, we must establish priority localities for preferential initiation of land consolidations. In CR, the major environmental problems are the risk of water and wind erosion, as well as inadequate water retention in the landscape.

Soil erosion is a natural process, which is accelerated by insensitive interventions of man into the natural environment and landscape and by intensive management of the agricultural land. In CR, water erosion potentially threatens almost 50% of agricultural land, 18% of which is at extreme or high risk; about 20% is at risk of wind erosion, almost 5% of which – at extreme or high risk (Ministry of Agriculture of CR, 2015). At present, the maximum soil loss in CR is estimated to be approximately 21 million tons of arable land per year, which may be expressed as an economic loss of CZK 4.3 billion per year. According to the Situation and Prospective Report of the Ministry of Agriculture CR (Ministry of Agriculture of CR, 2015), about 40% of soil in CR has above-average productivity. Due to the land degradation processes, namely erosion, these most productive regions gradually depreciate (Podhrázská et al., 2015).

The issue of soil degradation and the need of soil conservation have been among intensively debated topics for decades. One of the ways to improve the present situation may be the process of land consolidation, because land consolidations include the erosion and flood control measures together with the limitation of soil erosion in particular districts.

Protection of soil and water resources should be one of the state's priorities. The conservation of natural resources is one the main prerequisites for permanently sustainable development and healthy population. The loss of high-quality agricultural land is perceived worldwide as an extensive problem touching not only the developing countries. The main threat to the environment and agriculture is water erosion. This phenomenon is defined as a process of release, transport and deposit of soil particles by flowing water. Soil erosion leads to the loss of soil fertility, reduction of depth of the rooting zone, and loss of nutrients and moisture. The direct consequence is the fall in prices of agricultural land due to the loss of its quality and fertility (Podhrázská et al., 2015). Long-term economic analyses have shown that prevention of the negative effects of soil and water degradation is more advantageous than solving their consequences (Konečná et al., 2014).

The potential negative impacts of the continuing climatic changes on water sources attract attention of both scientists and policy makers worldwide. Their manifestation in the territory of CR is twofold – the growing occurrence and intensity of extreme precipitation brings about the damage to the soil by water erosion. The data from the monitoring of the Czech Hydrometeorological Institute, along with model calcula-

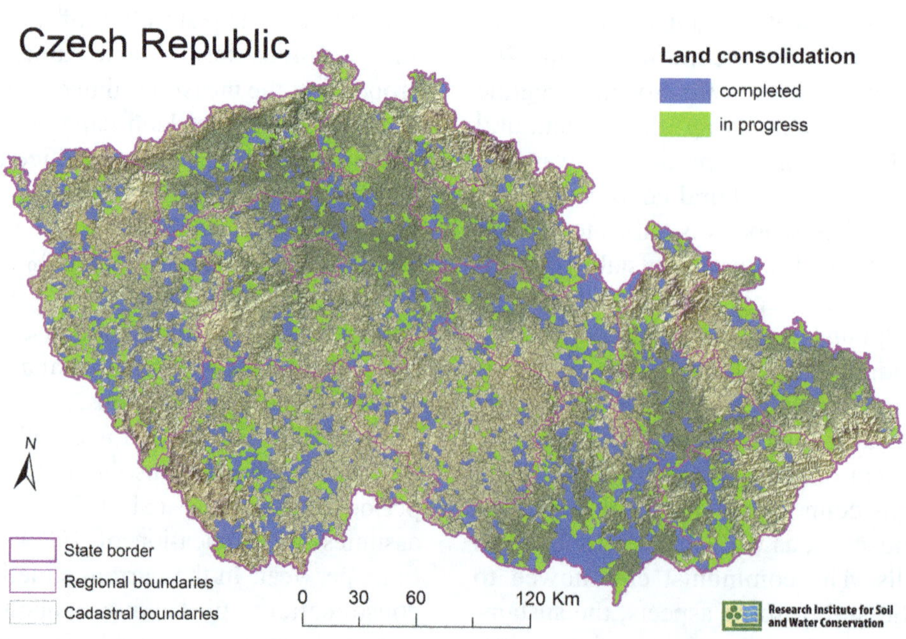

Fig. 1. State of land consolidations in CR (1st September 2015)

tions, show that due to the climatic changes, the territory of CR will probably experience more frequent droughts, both as a consequence of the precipitation deficit and the constant rise in the air temperature and evaporation (Rožnovský et al., 2010). The risk of drought requires the elevation of rainfall water retention in the landscape.

Land consolidations represent a unique tool for designing the landscape in many aspects. A plan of priority localities for initiating land consolidations should come into existence. Particularly in these localities, information should be disseminated about the land consolidations, their course and benefits. On the basis of a questionnaire campaign, a study by Karásek et al. (2014) has shown that a large part of the population still lacks the adequate information on the land consolidations.

METHODOLOGY AND DATA

In our research, we focused on the water and wind erosion and water retention in a landscape. The reason was that the erosion control and water management are key issues which are important under the conditions of the Czech Republic and subsequently they are widely employed in the frame of land consolidations. The methods were applied using the geographical information system (GIS) in the following subsequent steps:

- collection of input geodata (information on land use, pedological characteristics of the land, morphological characteristics, statistical data on demography, data on land consolidation in CR),
- processing of input data into a unified GIS geodatabase,
- creating criterions for evaluation of the obtained values,
- creating information layers (land consolidation, water and wind erosion, soil suitable for construction of retention reservoirs and wetlands),
- processing of simple multi-criteria analysis of the region erosion risk evaluation of the information layers
- processing map outputs and evaluation.

The evaluation of the priority localities for the initiation of land consolidations was done in individual regions of CR (Fig. 2).

Vulnerability of soil by water erosion

The analysis of soil vulnerability by water erosion was based on the method published in the methodological manual "Protection of Agricultural Land against Erosion" (Janeček et al., 2012). The calculation was done using the USLE 2D method (Janeček et al. 2012). The input data were taken from the valid pedological database for CR (Ministry of Agriculture, 2015), the DMT created

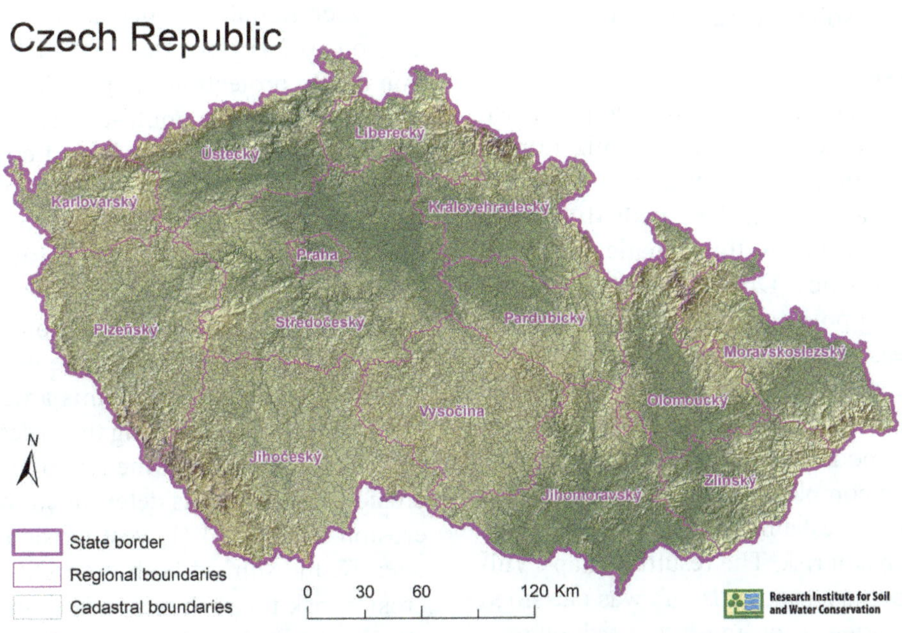

Fig. 2. Regional boundaries in CR

from the main base of geographical data (ZABAGED 2015 – Czech Geographical and Cadastral Office), the database of land blocks and their use (LPIS, Ministry of Agriculture, 2015), and the data on water courses and water surfaces from the valid database for CR (DIBAVOD). Subsequently, the value of average long term soil loss was calculated and the output layer of erosion threat was created. According to the valid guidelines of the State Land Office and the methodology of erosion control (Janeček et al., 2012), the permissible soil loss by water erosion is 4 t/ha/year. Any higher calculated value is considered as an excessive loss. Any locality with erosive loss exceeding this permissible limit requires the application of appropriate erosion control measures. Each cadastral area in CR has a calculated acreage of agriculturally exploited land. We compared this acreage with the acreage of the land susceptible to erosion (land with calculated erosion risk higher than the appropriate permissible limit). Then, for each cadastral area with incomplete or not yet initiated land consolidation we calculated the value (percentage) of agriculturally exploited land in which the erosion risk was assessed as higher than the appropriate permissible erosive wash out. The results were expressed in the map "Vulnerability of soil by water erosion". This map classifies the cadastral areas into five categories, according to the water erosion risk of agriculturally exploited land. These categories are: 0–20; 20–40; 40–60; 60–80; 80–100% of agricultural land with excessive erosive wash.

Vulnerability of soil to wind erosion

The analysis of land potentially vulnerable to wind erosion was based on the method published in the methodological manual "Optimization of Windbreak Functions in Agricultural Landscape" (Podhrázská et al., 2008). The method assesses the localities according to the climatic and pedological characteristics. Dry and warm localities are at the highest potential risk, similarly to light, sandy soil areas. The calculation was done using the GIS method, with resulting weighted means of the potential erosion risk for individual land blocks in the model localities. For each cadastral area with incomplete or not yet initiated land consolidation, we established the average value of the wind erosion risk. The resulting map "Vulnerability of soil by wind erosion" was classified into five categories according to the risk degree, ranging from low risk to high risk.

Suitable locations for construction of retention reservoirs or wetlands

The map was processed based on the evaluation of soil characteristics according to valued pedological-ecological units (BPEJ) obtained during the research projects by Novák et al., 2003. We selected the soil types characterizing extensively waterlogged, very heavy land, difficult to drain, with a high underground water level, in favourable relief organization, in particular: Modal Stagnogleys, Modal Gleys, Aqui-gleysols, Gleyed Fluvisols.

These soil types are not suitable for further agricultural management due to their overall incapacity and ineffectiveness. They can be further used for the construction of retention reservoirs, ponds, and wetlands. The selection of these soil types was done in the cadastral areas where land consolidations have not yet been completed or developed. For each cadastral area, we calculated the total acreage of these localities suitable for increasing the water retention in the landscape. The resulting map "The surface area (ha) of suitable localities for construction of retention reservoirs or wetlands" classifies the cadastral areas into five categories (0; 0.1–10; 10–100; 100–200; >200 ha) according to the acreage of these localities.

Priority localities for initiation of land consolidation due to the risk of water and wind erosion in the Czech Republic

For the purpose of identifying the regions in the Czech Republic which are most threatened by erosion (priority for the start of land consolidation due to protection of agricultural land against erosion), a simple multi-criterion analysis was conducted. The water and wind erosion vulnerability analyses did not include the areas where the land consolidation had already been processed. The multicriterial analysis consisted in assessing the vulnerability of the Czech Republic to water erosion and threatened regions of the Czech Republic by wind erosion. The map outputs and analyses processed within this article were used. It consisted of determining the order of individual regions (depending on the risk of water and wind erosion). The order is determined from 1 (highest erosion risk) to 14 (lowest erosion risk). Subsequently, the sum of both orders is made and the erosion risk for each region is evaluated.

On the basis of this evaluation, a table and a map showing the overall vulnerability of the regions of the Czech Republic by erosion (a region without land treatment) were created.

RESULTS AND DISCUSSION

Priority localities for initiation of land consolidation due to the risk of water erosion

The cadastral areas in CR are threatened by water erosion to various degrees. The results of our evaluation pertaining to the average long term soil loss evocated with water erosion are given in Table 1 and Figure 3. The table contains the numbers of cadastral areas without land consolidation for each CR region in individual categories of risk of water erosion. The risk of water erosion is expressed as the percentage of agriculturally exploited land with erosive loss higher than the permissible limit (4 t/ha/year). In 4,150 cadastral areas (42% of cadastres), the risk was assessed as mild, with the permissible loss exceeded in up to 20% of agricultural land acreage.

These areas do not represent priority localities for initiating land consolidations due to water erosion. The remaining 5,545 cadastral areas (56.4 of cadastres) are at high risk of water erosion according to the analysis (more than 20% agricultural land at risk of excessive erosive loss), while 803 cadastral areas (8.2%) were classified into the category of 60–80 % agricultural land with excessive erosive loss and 157 (1.6%) even into the category of 80–100% agricultural land with excessive erosive loss. These localities were assessed as highly suitable for initiation of land consolidation, with high need of managing the water erosion problems.

The highest number of strongly vulnerable cadastral areas is found in the regions of Jihomoravský, Vysočina, and Středočeský. These most vulnerable regions are shown in red in Figure 4. In the Jihomoravský region, 79% of cadastral areas without completed or developed land consolidation are at high risk of water erosion (more than 20% of agricultural land displays erosion vulnerability higher than the permissible soil loss). In the Vysočina region, this concerns 81% of cadastral areas, and in Středočeský 69%. These regions are intensively exploited agriculturally, with large blocks of arable land. The land parcels often exceed 100 ha and are managed by large agricultural cooperatives. In contrast, the lowest numbers of cadastral areas with excessive erosion vulnerability (more than 20% of agricultural land acreage within the cadastre) are situated in the regions of Karlovarský (20% – green colour, lowest risk), Liberecký (36%), and Ústecký (41%).

The risk of water erosion is a frequent and current topic of debate in CR. A number of studies and research activities are related to the issue of soil degradation by water erosion and its subsequent sedimentation in the river beds and water bodies (Krása et al. 2013, Uhlířová et al, 2009). Novák, Batysta and Havelková (2013) refer to

Table 1. Number of cadastral areas in CR threatened by the water erosion

Region	Number of cadastral areas without land consolidation threatened by water erosion						
	not rated	Expressed as % of agricultural land in the cadastral areas with G > 4 t/ha/year					Total
		0 – 20	20 – 40	40 – 60	60 – 80	80 – 100	
Praha	30	26	26	14	5	7	78
Jihočeský	4	627	353	152	20	0	1152
Jihomoravský	18	101	144	153	107	54	559
Karlovarský	9	358	61	24	3	2	448
Vysočina		192	515	287	32	0	1026
Královehradecký	4	328	191	139	54	5	717
Liberecký	9	256	98	45	7	1	407
Moravskoslezský	10	240	114	107	35	5	501
Olomocký	5	288	83	96	86	17	570
Pardubický	3	261	158	135	46	3	603
Plzeňský	6	383	234	231	100	5	953
Středočeský	4	470	440	425	193	13	1541
Ústecký	31	485	154	131	65	11	846
Zlínský	2	135	33	42	50	34	294
Total	**135**	**4150**	**2604**	**1981**	**803**	**157**	**9695**

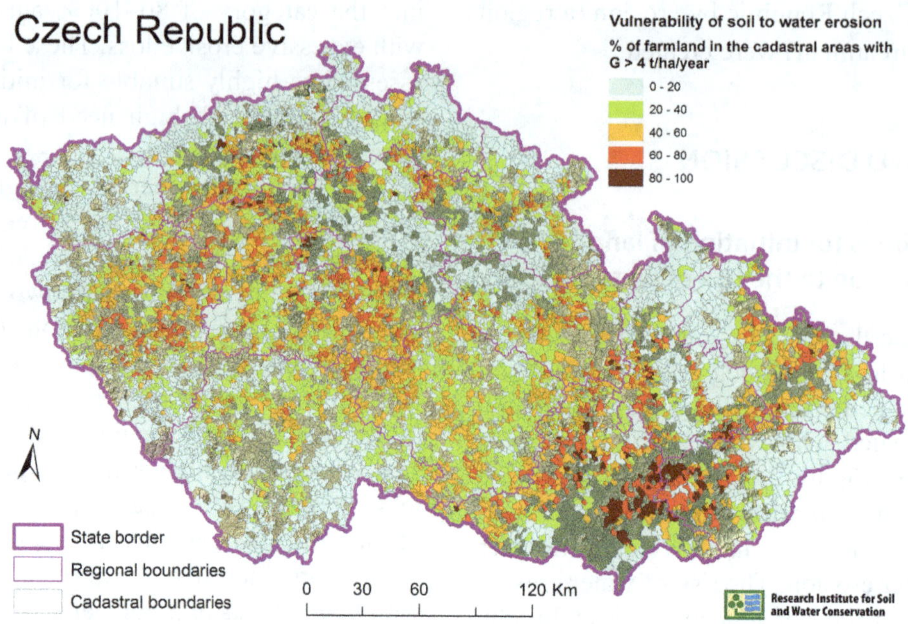

Fig. 3. Vulnerability of soil to water erosion in the cadastral areas without land consolidation

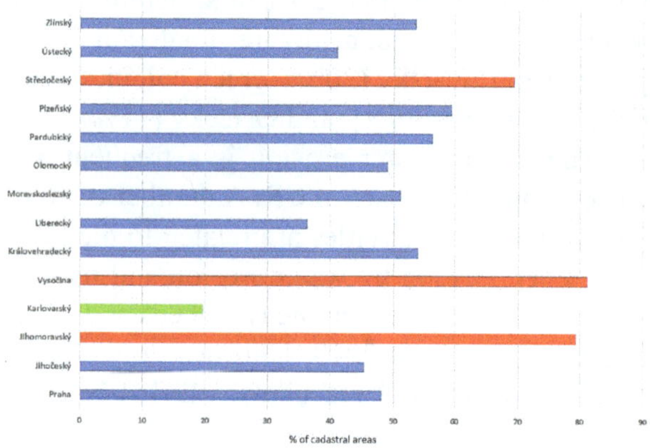

Fig. 4. Percentage of cadastral areas (without land consolidation) threatened by water erosion in individual regions of CR

the extreme threat of water erosion in some regions in CR. They found out that the erosion in Southern Moravia (Jihomoravský region) leads to the degradation of soil profiles and changes of soil types. The negative effects of water erosion were documented by Konečná et al. (2011) in the Vysočina region.

Priority localities for initiation of land consolidation due to the risk of wind erosion

We found out that out of 9,830 cadastral areas without land consolidation, 18% are vulnerable to the effects of wind erosion (ca 1,774). The degree of risk is given in Table 2 and documented in Figure 5.

Table 2 shows the numbers of cadastral areas without land consolidation for each CR region in individual categories of risk of wind erosion. From all considered cadastres, 12.5% (1,231) are classified as being at low or moderate risk. The wind erosion vulnerability in these areas is mild and so they do not represent priority localities for initiation of land consolidations due to the wind erosion risk.

The remaining 5.5% of cadastral areas (total 543) have been assessed as being at medium risk (378), high risk (126) and very high risk (39). In these areas, wind erosion should be managed by appropriate erosion control measures. Therefore, they should be considered as priority localities for initiation of land consolidations.

The vulnerability to wind erosion represents a frequently underestimated risk of soil degradation. In CR, wind erosion threatens 26% of agricultural land. The assessment of vulnerability to wind erosion is a standard element of land consolidation, forming an integral part in the form of proposed measures to be adopted against the wind erosion effects. Appropriate erosion control measures are given in Podhrázská et al., 2008, Janeček, et al., 2012.

The highest number of cadastral areas with a significant risk of wind erosion (medium, high, very high risk categories) is situated in the region of Jihomoravský. In total, they encompass 102 cadastres (18% of all cadastres without land consolidation in the region). In the region of Ústecký, this concerns 17% of cadastral areas, and in

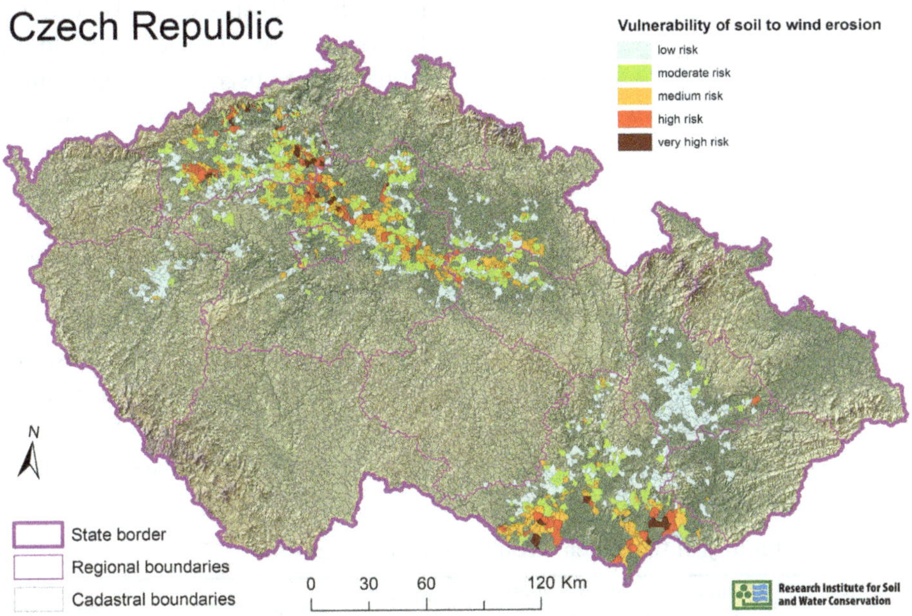

Fig. 5. Vulnerability of soil to wind erosion in cadastral areas without land consolidation

Table 2. Number of cadastral areas in CR threatened by wind erosion

Region	Number of cadastral areas without land consolidation threatened by wind erosion					Total
	low risk	moderate risk	medium risk	high risk	very high risk	
Praha	15	32	7	0	0	54
Jihočeský	0	0	0	0	0	0
Jihomoravský	99	80	74	19	9	281
Karlovarský	1	0	0	0	0	1
Vysočina	4	0	1	0	0	5
Královehradecký	64	25	19	1	0	109
Liberecký	2	1	0	0	1	4
Moravskoslezský	0	0	0	0	0	0
Olomocký	129	15	1	1	0	146
Pardubický	38	62	46	9	0	155
Plzeňský	44	6	2	0	0	52
Středočeský	154	187	151	43	7	542
Ústecký	108	130	75	53	22	388
Zlínský	33	2	2	0	0	37
Total	691	540	378	126	39	1774

Středočeský – 13%. These 3 most vulnerable regions are shown in red in Figure 6.

Wind erosion is one of serious degradation factors, particularly in the localities with most fertile soils (southern Moravia, the Elbe valley). Its harmful effects are reflected in the loss of arable land, deterioration of both physical and chemical soil properties, decrease of hectare yields, and increase of dust pollution in the environment (Podhrázská et al., 2015). The regions of CR not directly threatened by wind erosion are: Jihočeský, Karlovarský, Liberecký, Moravskoslezský, Olomocký, Plzeňský, Zlínský, Vysočina (shown in green colour in Fig. 6).

Priority localities for initiation of land consolidation due to the risk of water and wind erosion

A simple multicriterial analysis was developed to determine the vulnerability of the regions of the Czech Republic to erosion (both water and wind). It consisted of determining the order of individual regions (depending on the risk of wa-

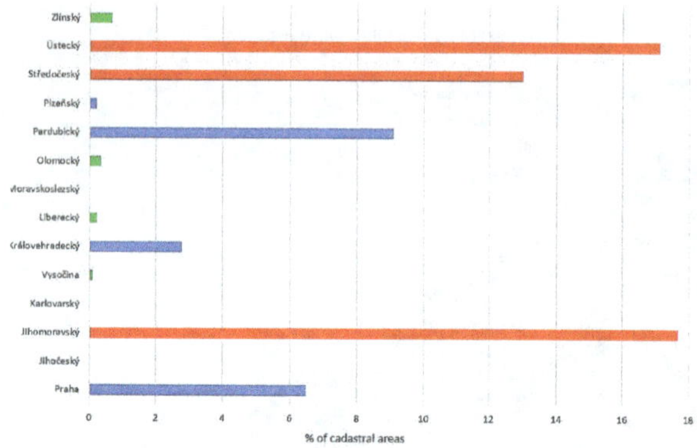

Fig. 6. Percentage of cadastral areas without land consolidation threatened by wind erosion in individual regions of CR

ter and wind erosion). The order is determined from 1 (highest erosion risk) to 14 (lowest erosion risk). Subsequently, the sum of both orders is made and the erosion risk (total order) for each region is evaluated (Table 3, Fig. 7).

The most vulnerable areas of erosion in the Czech Republic is Jihomoravský kraj (South Moravian Region). Then, the region of Středočeský (Central Bohemia) and Pardubice follow. In these areas priority should be given to the erosion control through the state policy instruments. The regions Karlovarský, Jihočeský, Liberecký are at the other end of the ranking of erosion threat, being least vulnerable to erosion.

Table 3. Analysis of vulnerability of the Czech Republic regions by erosion

Region of Czech Republic	The order of the regions of the Czech Republic by erosion threat (1 – highest risk; 14 – lowest risk)		Summary	Overall order of regions
	Water erosion threat	Wind erosion threat	Water erosion threat + Wind erosion threat	1 – the highest risk of erosion; 14 – the lowest risk of erosion
Jihomoravský	2	1	3	1
Středočeský	3	3	6	2
Pardubický	5	4	9	3
Královehradecký	6	6	12	4
Vysočina	1	11	12	4
Plzeňský	4	10	14	6
Ústecký	12	2	14	6
Zlínský	7	7	14	6
Praha	10	5	15	9
Olomocký	9	8	17	10
Moravskoslezský	8	12	20	11
Liberecký	13	9	22	12
Jihočeský	11	12	23	13
Karlovarský	14	12	26	14

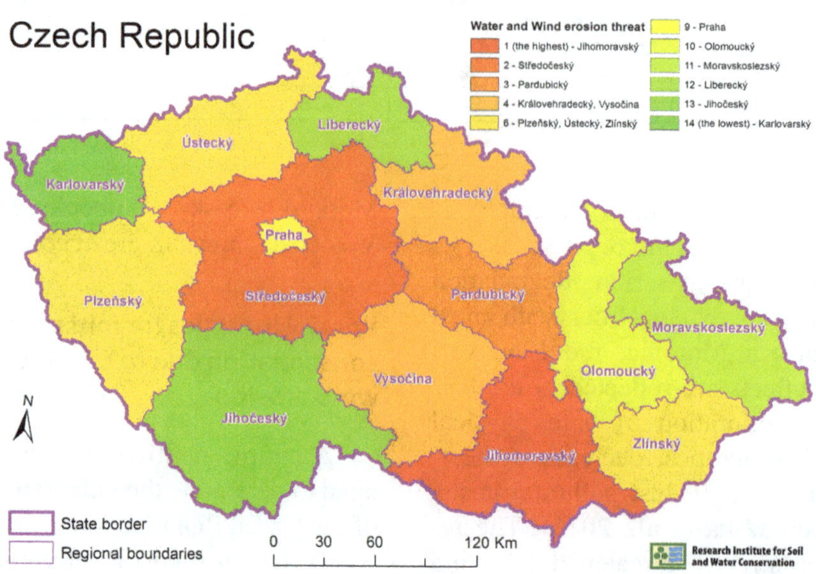

Fig. 7. The threat to the regions of the Czech Republic by erosion (water and wind)

Priority localities for initiation of land consolidations aimed at increasing water retention in the landscape

The analyses presented above assessed the landscape of CR from the aspect of land degradation by water and wind erosion. Additional consequences of the climatic changes that we have to face are drought impacts and growing risk of occurrence of dry periods in the course of the year (Kyselý, Beranová, 2009). During the second half of the 20th century, CR experienced intensive drainage of agricultural land at almost 26% surface (1.1 mil. ha) (Kulhavý et al., 2007). Drainage and fertilization were frequently applied in unsuitable localities, resulting in profound changes in the landscape (Langhammer et al., 2008; Karásek et al., 2015). Both the drainage and land degradation due to water and wind erosion negatively impact the water retention in the landscape. Our analysis was aimed at locating potentially suitable localities (inappropriate for agricultural production) that could be used for increasing the water retention in the landscape by construction of retention reservoirs and wetlands. The analysis has shown that 67% of cadastral areas in CR without land consolidation contain such potentially suitable localities.

Table 4 shows the number of cadastral areas without land consolidation classified into categories according to the acreage of potentially suitable localities for the construction of retention reservoirs and wetlands for all regions of CR. These localities can be found in 6,621 of cadastral areas (67% of all cadastres without land consolidation). In 2,500 cadastral areas (25%), the acreage of these suitable localities ranges between 0.01–10 ha (in each cadastre). In theory, some potentially suitable surfaces exist in these areas, but they are not numerous. In contrast, in 3,906 of cadastral areas (40%), the acreage reaches 10–100 ha. The highest category includes 165 of cadastral areas (2%) with suitable localities of 100–200 ha and 46 of cadastral areas (0.5%) in the category over 200 ha. The growing need of increasing the water retention in the landscape predetermines the required land consolidations to be initiated in these localities.

The highest surface area of potentially suitable localities for the construction of retention reservoirs and wetlands was found in the region of Jihočeský (350 km^2). Historically, this region is a place of numerous fish ponds and artificial reservoirs (Pavelková et al., 2017). In the region of Vysočina, this surface area is 266 km^2, in the Plzeňský region 223 km^2, in Středočeský 176 km^2 (green colour in Fig. 9). In contrast, in the region of Ústecký, these surfaces are markedly smaller (46 km^2), and in the region of Zlínský they encompass 57 km^2. Figure 10 presents the order of regions of the Czech Republic according to the area of potentially suitable sites for the construction of retention reservoirs and wetlands.

Table 4. Number of cadastral areas in CR with suitable localities for construction of retention reservoirs or wetlands

Region	Number of cadastral areas (without land consolidation) with suitable locations for construction of retention reservoirs or wetlands (ha)				Total
	0,01 – 10	10 – 100	100 – 200	> 200 ha	
Praha	20	11	0	0	31
Jihočeský	250	770	45	7	1072
Jihomoravský	139	111	7	10	267
Karlovarský	146	222	14	0	382
Vysočina	266	676	24	2	968
Královehradecký	227	220	5	0	452
Liberecký	112	122	7	2	243
Moravskoslezský	121	145	17	5	288
Olomocký	139	127	16	8	290
Pardubický	199	162	7	0	368
Plzeňský	256	578	13	4	851
Středočeský	381	539	5	0	925
Ústecký	183	148	1	0	332
Zlínský	65	75	4	8	152
Total	2504	3906	165	46	6621

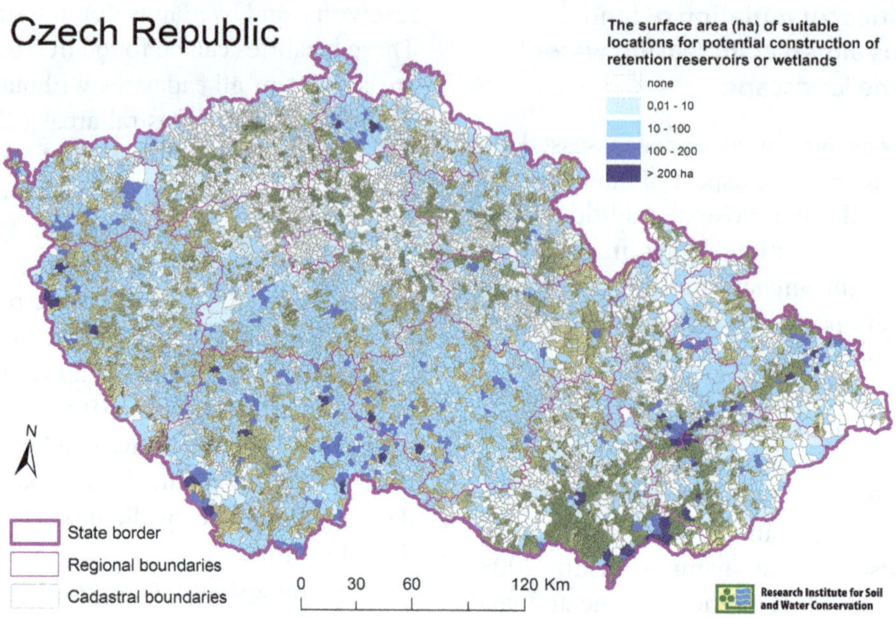

Fig. 8. The surface area of suitable locations for potential construction of retention reservoirs or wetlands in the cadastral areas without land consolidation

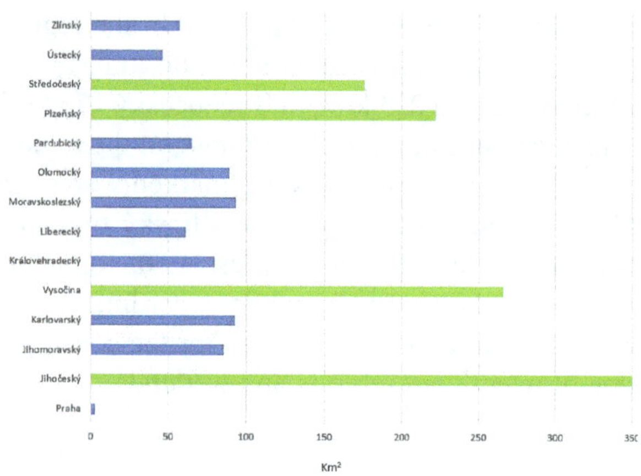

Fig. 9. The surface area of agriculturally exploited areas potentially suitable for construction of retention reservoirs or wetlands in cadastral areas without land consolidation

CONCLUSION

The process of land consolidation represents a unique tool of state policy for the promotion and application of protective erosion control and water management measures in CR. Since land consolidations have not yet been started in 60% of the CR territory, suitable localities (cadastral areas) for their preferential initiation should be defined. Land consolidations deal with a large spectrum of issues, and virtually any land consolidation is unique because it reflects the current and local needs of the area. In this study we selected three thematic factors according to which we analysed the degree of risk/suitability for initiation of land consolidation in the cadastral areas of CR. The first two factors represented the vulnerability of land to water and wind erosion. Every year, these processes result in massive degradation and loss of high-quality arable land. The third selected factor was related to the water retention in the landscape, which should be solved by finding potentially suitable localities for construction of retention reservoirs and wetlands.

The results of our analyses confirmed that CR is strongly threatened by water erosion. This fact was also confirmed at a local level by Podhrázská et al. 2015 who analysed the degradation of soil characteristics in the period 1978–2013 in a land block of 100 ha in southern Moravia and found that erosion resulted in a soil loss of 70,000 t.

Out of 9,830 cadastral areas without land consolidation, 56 % have exceeded the permissible erosive loss of 4 t/ha/year in more than 20% of agricultural land. The regions at the highest risk are Jihomoravský, Středočeský and Vysočina. Wind erosion threatens 18% of the cadastral areas without land consolidation, i.e. 1,174 of them. Manifestations of wind erosion are not as evident those of water erosion. However, the consequences represented by the loss of the upper layer of fertile soil by wind are serious. In the future, attention must be drawn this issue as well, and proper measures must be adopted. As reported by Podhrázská et al. (2008), land consolidations propose networks of wind breakers as appropriate

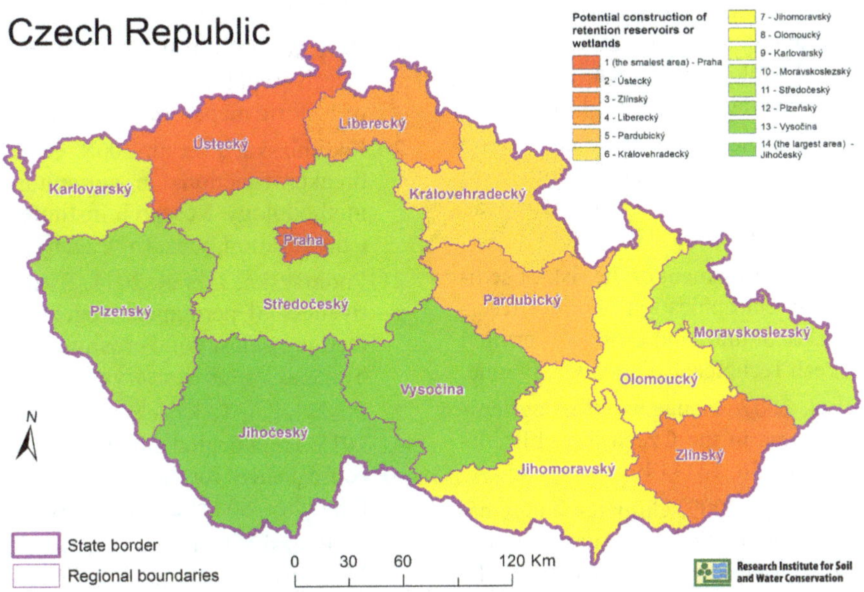

Fig. 10. Potentially suitable sites for the construction of retention reservoirs and wetlands

control of wind erosion. The localities at highest risk of wind erosion in CR are situated in the regions of Jihomoravský, Ústecký and Středočeský.

On the basis of the elaborated multicriteriaal analysis of the Czech Republic's vulnerability to water and wind erosion, the ranking of regions was evaluated according to the risk of erosion risk. In this way, the priority regions were identified as suitable for the intensive solution of the problem of protection against erosion through land consolidations. The regions of Jihomoravský, Středočeský and Pardubický are evidently the regions that are most vulnerable to water and wind erosion in CR. In these regions, the process of land consolidation should be accelerated and the erosion risk should be managed preferentially.

The pressing problems associated with the climatic changes and droughts have emerged in CR in recent years. Both 2015 and 2016 were exceptionally dry. Drought and the need of increasing water retention in the landscape have become the main question of debate at the level of government and state strategy (Ministry of Agriculture of CR, 2016). Identification of the localities which are unsuitable for agricultural production but potentially favourable for the construction of retention reservoirs and wetlands is now a hot topic. Such localities, pedologically unsuitable for agricultural production, are numerous in CR. These potentially suitable surfaces of 10–100 ha can be found in 40% of the cadastral areas without land consolidation. Most of them are situated in the regions of Jihočeský, Vysočina, Plzeňský, and Středočeský.

The thematic issues that could serve as a basis for initiating land consolidation are numerous. It is due to the variability of land consolidations, which comprise the rural space and landscape as a whole, as an integrated functional system. Our study is a partial contribution to solve the most relevant and urgent tasks in the rural landscape of CR.

Acknowledgement

This study was supported by the research project MZE RO0218 and QK 1720303.

REFERENCES

1. Hartvigsen, M. 2014. Land mobility in Central and Eastern Europe land consolidation context. Nordic Journal of Surveying and Real Estate Research, 10(1), 23–46.
2. Janeček, M. et al. 2012. Protection of agricultural land against erosion. Methodology. Prague: Powerprint, s.r.o. (in Czech).
3. Karásek, P., Stejskalová, D., Ulčák, Z. 2014. Analysis of Rural Social Aspects in the Context of Land Consolidations and Land Use Planning, the Case Study, Czech Republic. Acta Universitatis Agriculturae et Silviculturae Mendelianae Brunensis, 62(3), 507–515.
4. Karásek, P., Tlapáková, L., Podhrázská, J. 2015. The location and extent of systematic drainage in relation to land use in the past and at present and in relation to soil vulnerability to accelerater infiltration in the protected landscape area Železné Hory. Acta Universitatis Agriculturae et Silviculturae Mendelianae Brunensis, 63(4), 1121–1131.

5. Konečná, J., Pražan, J. et. al. 2014. Assessment of economic aspects of erosion control of agricultural land. Certified methodology. Research Institute for Soil and Water Conservation, v.v.i. (in Czech).
6. Konečná J. et al. 2011. Soil and water conservation in the frame of the land consolidation process in the Hubenov cadastre (Czech Republic). Moravian Geographical Reports, (19)3, 8–11.
7. Krása, J. et al. 2013. Assessment of the risk of sediment deposition and eutrophication of water reservoirs caused by agricultural land erosion. Certified methodology. Czech Technical University in Prague.
8. Kulhavý, Z., et al. 2007. Management of agricultural drainage systems in the Czech Republic. Irrigation and drainage, 56, 141–149.
9. Kyselý J., Beranová R. 2009. Climate-change effects on extreme precipitation in central Europe: uncertainties of scenarios based on regional climate models. Theor. Appl. Climatol., 95(3–4), 361.
10. Langhammer, J., Vilímek, V. 2008. Landscape changes as a factor affecting the course and consequences of extreme floods in the Otava River basin, Czech Republic. Environmental monitoring and assessment, 144, 53–66.
11. Leń, P., Noga, K. 2018. Prioritization of Land Consolidation Interventions in the Villages of Central Poland. Journal of Ecological Engineering, 19(2):246–254.
12. Ministry of Agriculture of CR. 2015. Soil – situation and prospective report. Prague: Ministry of Agriculture of the Czech Republic. (in Czech).
13. Ministry of Agriculture of CR. 2016. Report on the state of water management in the Czech Republic. Prague: Ministry of Agriculture of the Czech Republic. (in Czech).
14. Miranda, D., Crecente, R., Alvarez, M.F. 2006. Land consolidation in inland rural Galicia, N.W. Spain, since 1950: An example of the formulation and use of questions, criteria and indicators for evaluation of rural development policies. Land Use Policy, 23, 511–520.
15. Nikodemus, O., Bell, S., Grine, I., Liepins, I. 2005. The impact of economic, social and political factors on the landscape structure of the Vidzeme Uplands in Latvia. Landscape and Urban Planning, 70, 57–67.
16. Novák, P., et al. 2003. Definition of agriculturally less favourable and vulnerable areas of the Czech Republic and proposals for land use including economic impacts. Project of the Ministry of Agriculture of the Czech Republic NAZV QC1293, Prague (in Czech).
17. Novák P., Batysta M., Havelková L. 2013. Impact of the intensive soil erosion for the water balance of landscape. In Water, Soil and Plants. Křtiny 29.-30.5.2013.
18. Pašakarnis, G., Maliene, V. 2010. Towards sustainable development in Central and Eastern Europe: applying land consolidation. Land Use Policy, 27(2), 545–549.
19. Pavelková R. et al. 2017. Historical ponds of the Czech Republic: an example of the interpretation of historic maps. Journal of Maps, 12(1), 551–559.
20. Podhrázská, J. et al. 2008. Optimization of windbreaker functions in agricultural land. Certified methodology. Research Institute for Soil and Water Conservation, v.v.i. (in Czech).
21. Podhrázská, J. et al. 2011. Evaluation of the effectiveness of permanent vegetation barriers in protection against wind erosion. Research Institute for Soil and Water Conservation, v.v.i. (in Czech).
22. Podhrázská, J., Kučera, J., Karásek, P., Konečná, J. 2015. Land degradation by erosion and its economic consequences for the region of Jihomoravský (Czech Republic). Soil and Water Research, 10(2), 105–113.
23. Podhrázská J., Kučera J., Středová H. 2015. The Methods of Locating Areas Exposed to Wind Erosion in the Jihomoravský Region. Acta Universitatis Agriculturae et Silviculturae Mendelianae Brunensis, 63(1), 113–121.
24. Podhrázská, J., Vaishar, A., Toman, F., Knotek, J., Ševelová, J., Stonawská, K., Vasylchenko, A., Karásek, P. 2015. Evaluation of Land Consolidation Process by Rural Stakeholders. European Countryside, 7(3), 144–155.
25. Rožnovský, J., Fukalová, P., Pokladníková, H. 2010. Prediction of climate in Southern Moravia. In Voda v krajině. Czech Hydro-meteorological Institute, Lednice, 31.5.-1.6.2010. (in Czech).
26. Sklenička, P. 2006. Applying evaluation criteria for the land consolidation effect to three contrasting study areas in the Czech Republic. Land Use Policy, 23(4), 502–510.
27. Sky, P. K. 2015. Land consolidation in Norway in an international perspective. Spanish Journal of Rural Development, 6(1–2), 81–90.
28. SLO. 2015. Concept of land consolidations for 2016–2020. Prague: State Land Office. (in Czech).
29. SLO. 2016. Methodological guidelines for application of land consolidations. Prague: State Land Office. (in Czech).
30. Thomas, J. 2006. Property rights, land fragmentation and the emerging structure of agriculture in Central and Eastern European countries. Journal of Agricultural and Development Economics, 3(2), 225–275.
31. Tlapáková, K., Stejskalová, D., Karásek, P., Podhrázská, J. 2013. Landscape metric as a tool for evaluation of landscape structure – case study Hustopeče. European Countryside, 5(1), 52–70.
32. Uhlířová J., Kaplická M., Kvítek T. 2009. Water erosion and characteristics of sediment load in the Kopaninský stream basin. Soil and Water Research, 4(1), 39–46.
33. Vitikainen, A. 2004. An overwiev of land consolidation in Europe. Nordic Journal of Surveying and Real Estate Research, 1(1), 25–44.

4

Investigation of Waste Properties of Subway Construction as a Potential Component of Soil Layer

Andrey Vladimirovich Ivanov[1*], Yuri Dmitrievich Smirnov[1], Georgiy Igorevich Petrov[1]

[1] Saint Petersburg Mining University, 21st Line of Vasilyevsky Island, 2, Saint-Petersburg, 199106, Russia

[*] Corresponding author's e-mail: andrey-racer@mail.ru

ABSTRACT

The construction of subway stations is inevitably accompanied by the formation of a large amount of waste that is extracted from the underground workings by rock. When the waste is transported, the air is polluted with the exhaust gases of the dump trucks, some of the waste can stay on the road surface, the use of dump trucks increases the load on the roadway, the disposal of waste at landfills harms the environment, and enterprises pay fees for the transportation as well as placement of the waste. The article deals with the waste from the construction of new stations in the Saint-Petersburg subway. In order to determine the possibility of using waste in the municipal economy, a comprehensive analysis of this waste was carried out, including determination of moisture content, acidity, total carbon content, as well as chemical and granulometric composition. On the basis of the study results, it was established that the waste in question belongs to the hazard class 5, which enables to use it as a component of the soil mixture. Afterwards, the hazard class 5 was confirmed by the method of bioassay based on the determination of water toxicity by changing the optical density of the Chlorella algae culture.

Keywords: subway, waste, soil, landfill, mine site, hazard class, total carbon, moisture, biotesting, Chlorella algae

INTRODUCTION

The subway of Saint-Petersburg ranks 4th in Europe and 19th in the world in terms of the total annual passenger traffic and is the deepest in the world in terms of the average depth of the stations. Currently in Saint-Petersburg, as in other major cities in Russia and Europe, there an active construction of new subway stations is underway. The article deals with the construction of the Nevsko-Vasileostrovskaya line of the Saint-Petersburg subway, namely the section from the station Primorskaya to the station Savushkina Street, including the station Novokrestovskaya-1. In the period of construction of subway stations, the inevitable generation of waste, the amount of which depends on the volume of building materials and seized masses of rock.

The resulting waste (rock) is stored on landfills. The landfills are located at a relatively large distance from the waste generation areas; the haulage of trucks is greater than 50 km.

The applied waste management system causes certain problems:

- Environmental damage is caused by the location of waste dumps on lands in the vicinity of Saint-Petersburg [Pashkevich, 2015, Danilov et. al., 2017];
- the company pays a fee for storage of waste;
- the question arises of the high cost of transporting waste from the construction site;
- As a result of the organization of long distances of dump trucks, additional exhaust gases are emitted into the atmosphere, such as carbon monoxide (CO), carbon dioxide (CO_2), nitrogen oxides (NO_x), hydrocarbons (C_xH_y), benzapyrene, soot, etc.;
- Part of the waste from loaded dump trucks falls to the roadway;
- There is an additional load on the roadway;
- traffic congestion of roads increases, the situation with traffic congestion is aggravated.

On the territory of Saint-Petersburg, there is no disposal of the waste being studied. In this regard, there is a need to study the properties pertaining to the waste construction of subway stations to determine the possibility of their use in the urban economy.

MATERIALS AND METHODS

The sampling of waste and their laboratory storage is in accordance with the state methodical recommendations [PND F 12.1:2:2.2:2.3:3.2–03].

The study of the physical properties of waste includes:

1. Thermogravimetric method of moisture analysis. The measurement of moisture content in the soil under study is necessary to take into account the moisture content of the samples for some of the analysis methods discussed below;
2. Ion metric method of pH determination. The values of the hydrogen index should be taken into account when selecting the composition of the soil mixture;
3. Determination of the granulometric composition by the method of laser radiation diffraction. It is carried out in order to determine the mass fraction of each fraction (clay, sand, dusty particles), according to which it will be possible to accurately classify the waste being studied.

The definition of the elemental chemical composition includes:

1. Determination of the total carbon (TC) content by the high-temperature oxidation method. The values of TC must be taken into account while determining the elemental composition of the waste;
2. Determination of the elemental composition of waste by the wave X-ray fluorescence spectrometry method. The knowledge of the elementary composition will allow the calculation of the hazard class.

Definition of hazard class included:

1. Calculation of the hazard class of waste in accordance with the technique of classifying hazardous wastes as hazard classes for the environment [Methodical manual… 2003];
2. Confirmation of the hazard class in accordance with the methodology for determining the toxicity of water by changing the optical density of the chlorella algae [PND F T 14.1:2:3:4.10–2004].

The study was conducted on the basis of the Center for the collective use of high-tech equipment of the Saint-Petersburg Mining University.

The measurements were carried out using the AND MX-50 moisture meter. The moisture determination method corresponds to [GOST 5180–84].

The measurement was performed using an automatic potentiometric ATP-02 titrator with a special attachment in accordance with [GOST 26423–85].

Determination of the total carbon content was performed using a Shimadzu TOC device using the SSM-5000A solid sample analysis attachment in accordance with [GOST R 54244–2010 (ISO 29541: 2010)].

The chemical composition of the samples was studied using a XRF-1800 Shimadzu wave X-ray fluorescence spectrometer.

The analysis of the particle size distribution was carried out by laser diffraction using a laser particle size analyzer LA-950V2 Horiba in accordance with [ISO 13320–1].

The calculation of the hazard class was carried out in accordance with the current methodical manual [Order of the Ministry… 2014]. After collecting all the necessary information, the calculation was carried out in the "Integral. Calculation of hazard class 4.0" program. The software fully implements the algorithm for calculating the hazard class of waste, prescribed in the order of the Ministry of Natural Resources and Ecology of the Russian Federation of December 4, 2014 # 536.

In order to determine the toxicity of waste, a methodology was used to identify the toxicity of water from the change in the optical density of the Chlorella algae [PND F T 14.1:2:3:4.10–2004]. Light exposure of samples was made in a KVM-05 multi-cultivator.

RESULTS AND DISCUSSION

The waste (rock) formed during excavation works is mostly clay (up to 50%), loam (up to 30%), and also sands (up to 20%). While choosing the method of utilization, special attention is paid to the properties of clays and studies the ways of their possible use. There are many directions for the use of clay, such as: the production of building materials, refractory materials, ceramic products, the production of drilling fluids, use as a sorbent, and also the production of soil [Chukhrov et al., 1970].

On the territory of Saint-Petersburg and the Leningrad region, there are areas for which the improvement works are required. As a rule, while carrying out such measures, the soil corresponding to the sanitary norms [SanPiN 2.1.7.1287–03], i.e. requirements for vegetative soil, ensuring optimal growth and healthy state of green plantations and being economically and logistically accessible, is necessary.

The secondary use of waste in close proximity to the places of their formation partially solves the problems associated with long hauling distances dump trucks to storage sites.

The main waste-generating production processes are connected with excavation. Excavation works include: development of soil for shafts and near-barreled workings, inclined (escalator) tunnels, stations with substation facilities, dead ends, exit chambers, distillation tunnels and dormitories, vestibules, upper ventilation units, built open and closed. The volumes of soil formed during these works for all 5 years of construction are given in Table 1.

In the work, waste is considered as a potential component of the soil mixture, which can be used to solve the problems arising in the city and the immediate area, such as:

- creation of a soil-vegetative layer (SVL) in the territories disturbed by construction;
- restoration of SVL on disturbed and polluted landscapes;
- increase in the power of the SVL, for example, in low-lying, flooded zones;
- subsidence of soil during the reorganization of the SVL;
- creation of SVL in alluvial landscapes;
- Reclamation of the disturbed and polluted areas of Saint-Petersburg and the region, for example, waste landfills.

Of course, any application of waste to create a soil-vegetation layer requires a preliminary study on the properties of this waste, on the basis of which the following tasks are formulated:

1. studying the physical properties of waste;
2. determination of the elemental chemical composition;
3. establishment of a hazard class and toxicity.

The mineral composition of the waste in question is mainly clay, fine-grained rock, consisting of particles smaller than 0.01 mm and containing about 30% of particles smaller than 0.002 mm. The material composition includes: clay minerals (kaolinite, montmorillonite, hydromica), quartz, mica, oxides and hydroxides of iron, aluminum, carbonates, sulfates, phosphates, organic matter.

Clay minerals are water aluminosilicates ($xAl_2O_3 \cdot ySiO_2 \cdot zH_2O$), where x, y, z have different values. Other secondary minerals include quartz (SiO_2), gypsum ($CaSO_4 \cdot 2H_2O$), dolomite ($MgO \cdot CaO \cdot CO_2$), glauconite ($K_2O \cdot Fe_2O_3 \cdot 4SiO_2 \cdot 10H_2O$), limonite ($Fe_2O_3 \cdot 3H_2O$), rutile ($TiO_2$), serpentine ($3MgO \cdot 2SiO_2 \cdot 2H_2O$) and others [Valkov, V.F. et al., 2004, Chukhrov F.V. et al., 1970].

The sampling of waste was carried out in the territories of four construction sites from temporary storage sites (Figure 1).

Temporary storage sites are the sites with a concrete pavement and a curb along the perimeter (Figure 2). In the places of temporary storage, the waste is stored for no more than 2 days; afterwards it is transported to the landfill.

Sequence of actions for sampling involved:

1. Choosing 5–7 points, located evenly over the surface of the temporary dump.
2. The point samples weighing 2 kg were selected with a scoop.

Table 1. The volumes of rock formed during the development of underground objects (by years)

Waste sources	In just five years	Years				
		2012	2013	2014	2015	2016
	m^3	m^3	m^3	m^3	m^3	m^3
Mine shafts and near-barrels	8710	1700	6120	-	890	-
Inclined (escalator) tunnels	-	-	-	-	-	-
Stations with station facilities. Dead ends and move out tunnels.	-	-	-	-	-	-
Subway tunnels and tunnel cameras	378 310		43 690	141 200	177 430	15 980
Subway tunnels, vestibules, upper ventumblers, constructed by open method	174 470	61 070	61 070	26 170	26 170	-
Total:	561 490	62 760	110 880	167 370	204 490	15 980

Figure 1. The layout of the construction sites of the subway: 1 – construction site of the mine #467, 2 – station "Novokrestovskaya", 3 – station "Savushkina Street", 4 – mine #462.

3. The point samples were combined on a clean flat section of the concrete platform to a mass of 5 kg combined sample.
4. The sample obtained was placed in a plastic bag.

Moisture was determined for the natural state and for air-dry conditions. The method of heating involves calculating the moisture content by weight of the water evaporated from a solid or liquid sample after heating for some time at a temperature equal to or higher than the evaporation temperature of the sample. The weight loss increases as the sample warms up and, in the end, tends to a constant value. The weight of the sample is 2 g.

When determining the natural moisture of the waste, the sample was crushed to a uniform particle size with a diameter of less than 1 mm, which provided a thorough heating and drying of each particle. The results of measuring the moisture content of samples in the natural state are presented in Table 2.

While determining the moisture of the waste in an air-dry state, the sample weights were pre-dried. In order to do this, a sample of 100 g was selected from the initial sample, shredded with a spatula and evenly dispersed on kraft paper. The weighed portions were dried for 5 days and mixed daily for uniform drying. Then, the dried samples were crushed manually in porcelain mortars, after which the moisture content was determined. The results of measuring the moisture of samples in the air-dry state are presented in Table 3.

In order to determine the pH, the pre-dried to air-dry state was weighed in a porcelain mortar, and the moisture was measured as described above. Afterwards, the sample was sieved through a sieve with a mesh size of 1–2 mm. The resulting sample had a mass of 100–150 g. Further, the conversion to dry matter was carried out according to the formula (1):

Figure 2. The site of temporary storage of rock on the mine site #462

Table 2. The results of measuring the moisture of samples in the natural state

Place of sampling	Station "Savushkina Street"	Station "Novokrestovskaya"	Mine #462	Mine #467
Moisture of sample, %	20.27	28.20	19.56	8.99
Moisture of the sample (parallel measurement), %	20.35	27.90	19.51	9.02
Average value, %	20.31	28.05	19.53	9.00

Table 3. Results of moisture measurement of samples in air-dry state

Place of sampling	Station "Savushkina Street"	Station "Novokrestovskaya"	Mine #467	Mine #462
Moisture of sample, %	1.74	2.37	2.03	2.10
Moisture of the sample (parallel measurement), %	1.77	2.31	2.20	2.15
Average value, %	1.75	2.34	2.11	2.12

$$M_{ad} = M_d/(1 - P) \quad (1)$$

where: M_{ad} is the mass of the substance in the air-dry state (g),
M_d is the mass in terms of dry matter (g),
P is the moisture content in the air-dry state (%).

The sifted sample was rolled on a flat surface and portioned to obtain a sample weighing 30 g in terms of absolutely dry matter by the formula (1). The resulting sample was poured into a glass beaker, mixed with distilled water in a ratio of 1:5, stirred on a magnetic stirrer for 10 minutes, then left to stand for 5–10 minutes and the pH was measured.

The measurement was carried out using an ATP-02 automatic potentiometric titrator with a special attachment. The measurement results were displayed in the software interface of the device in the digital and graphical form.

The pH was determined on October 17, 2017. The air temperature in the room was 23°C, the air humidity amounted to 32%. On the basis of the results presented in Table 4, it can be concluded that the waste in question is weakly acidic, and this property should be taken into account when developing the composition of the soil mixture.

Determination of the total carbon (TC) content was performed using a Shimadzu TOC device using the SSM-5000A solid-state assay attachment.

Principle of catalytic combustion method was as follows:

1. The analyzed sample is heated to 680°C;
2. As a result of combustion of organic substances of the sample, CO_2 was obtained;

Table 4. The results of determining the pH of the aqueous extract of waste

Place of waste generation	Measurement result, unit. pH	Result of parallel measurement, unit. pH	Average value, unit. pH	t, °C
Station "Novokrestovskaya"	6.01	6.04	6.02	21
Station "Savushkina Street"	5.53	5.50	5.51	21.5
Mine #467	5.65	5.67	5.66	21
Mine #462	5.91	5.92	5.91	21

3. The combustion products are carried away from the reaction tube by a carrier gas stream;
4. The gas mixture enters the electronic dehumidifier, where it is cooled and freed from water vapor. Then, the gas passes through a halogen trap, in which chlorine and other halogens are absorbed;
5. A mixture of carrier gas and combustion products is fed to the IR detector, in which the measurement of CO_2 absorption.

Sample preparation: drying to air-dry state, grinding in a porcelain mortar, determination of moisture.

The weight of the sample was 0.20 g in terms of dry matter according to formula (1). The weight of the sample was determined in accordance with the average carbon content in clay rocks (0.5–3%) [Basic Scales… 1982, Valkov et al., 2004] and the corresponding calibration curve. The TC contents were measured in two parallel samples for each sample. The determination of the contents of the vehicle was carried out on October 18, 2017. The room temperature was 24°C, air humidity 31%. Eight measurements were made.

The results obtained indicate a low content of carbon and organic matter in the waste and correspond to the content of carbon in the main types of clays. The results are shown in Table 5.

The chemical composition of the samples was studied using a XRF-1800 Shimadzu wave X-ray fluorescence spectrometer.

The method implemented in this device is based on the dependence of the intensity of X-ray fluorescence on the concentration of the element in the sample. When a powdered tablet of the sample is irradiated with a powerful radiation flux of the X-ray tube, characteristic fluorescent radiation of the atoms arises, which is proportional to their concentration in the sample.

Before the analysis, the sample was prepared as follows: drying to air-dry state, determination of moisture, grinding in a ball mill. A tablet is formed by means of the hydraulic press from the received powder. Afterwards, the tablet is placed in a special box of the spectrometer and an analysis is performed with a digital display of the results and plotting.

Calculation of the concentrations of elements is performed with standard software. When displaying the analysis results, the departing mineral composition consisting essentially of oxides of different elements is accounted for; therefore, the content of detectable conversion elements in oxide form has been set in the program.

During the analysis, the concentration of the already known total carbon and moisture of the sample was included in the instrument program for calculating the concentration.

The analysis was carried out on October 18, 2017. The room temperature was 21°C, whereas the air humidity equaled 32%. The results are shown in Table 6.

The measurement results were determined with the standard software for the XRF-1800 and by the fundamental parameters method with a normalization of 100% of the content of the detected elements. Averaging over the surface of the sample was provided by rotating the sample at a speed of 50 rpm during the recording of fluorescence radiation. The weight of the sample was 1 g. The values of the signal with an intensity exceeding 1/3 of the background value were taken into account.

Table 5. The content of total carbon in the waste

Place of waste generation	Measurement result, %	Result of parallel measurement, %	Average value, %
Station "Novokrestovskaya"	0.72	0.74	0.73
Station "Savushkina Street"	1.28	1.25	1.26
Mine #467	1.21	1.23	1.22
Mine #462	0.73	0.73	0.73

Table 6. Content of chemical elements in the tested waste (%)

Substance	Station "Savushkina Street"	Station "Novokrestovskaya"	Mine #462	Mine #467
SiO_2	71.20	62.71	61.86	62.03
Al_2O_3	12.58	15.57	19.01	19.93
K_2O	3.78	4.05	4.71	4.82
Fe_2O_3	3.46	6.74	7.30	6.39
CaO	2.30	3.20	0.67	0.20
MgO	1.13	2.06	1.70	1.69
TiO_2	0.70	0.95	1.13	1.13
Na_2O	0.66	0.73	0.14	0.08
SO_3	0.53	0.16	0.17	0.08
P_2O_5	0.25	0.26	0.12	0.05
BaO	0.16	0.15	0.14	0.12
Cr_2O_3	0.08	0.07	0.04	0.05
MnO	0.07	0.12	0.11	0.03
Nd_2O_3	0.00	0.03	0.00	0.00
ZrO_2	0.03	0.02	0.02	0.02
SrO	0.03	0.03	0.01	0.01
Cl	0.02	0.02	0.00	0.00
Ga_2O_3	0.00	0.02	0.00	0.00
ZnO	0.01	0.01	0.02	0.02
Rb_2O	0.01	0.01	0.01	0.01
Co_2O_3	0.00	0.00	0.00	0.00
NiO	0.00	0.01	0.01	0.01
CuO	0.00	0.01	0.00	0.00
C	1.26	0.73	0.73	1.22
Moisture	1.75	2.34	2.12	2.11

The granulometric analysis was carried out by laser diffraction using a laser particle size analyzer LA-950V2 Horiba. The principle of operation is based on scattering and detection of reflected / refracted laser light, red and blue spectra (650 and 405 nm). Very thin particles of the submicron range scatter light from the LEDs at large angles. These signals are recorded on 4-channel detectors with increased sensitivity and a frequency of 5000 measurements per second.

The determination of the granulometric composition is necessary for further research. While choosing the composition of the soil mixture, the results of the analysis will allow us to name and classify this soil, and also to establish some agrophysical properties. The sample preparation consisted in drying the sample to an air-dry state and grinding it in a porcelain mortar. The analysis was conducted on 19.10.2017. The air temperature in the room was 21°C, whereas the air humidity 32%. On the basis of the results obtained, it can be concluded that the wastes under study are mainly composed of physical clay consisting of clay and dust particles. The results of the analysis are shown in Table 7.

The hazard class of waste was calculated in the following way: in accordance with the Order of the Ministry of Natural Resources and Ecology of the Russian Federation #536 of December 4, 2014, the Methodological Handbook on the Application of Criteria for the Recognition of Hazardous Wastes to Hazardous Classes for

Table 7. Results of granulometric analysis of waste

Place of waste generation	Content of clay (d<0.01 mm), %	Content of sand (0.05<d<1.0 mm), %	The content of dust particles (0.001<d<0.05 mm), %
Station "Novokrestovskaya"	74.0	5.5	88.5
Station "Savushkina Street"	27.0	30.7	61.7
Mine #467	72.7	2.5	91.6
Mine #462	81.3	1.7	92.3

the Environment], a literature review was first conducted aimed at studying the concentrations of oxygen, nitrogen, carbon, phosphorus, sulfur, silicon, aluminum, iron, sodium, potassium, calcium, magnesium, titanium in basic types of soils [Basic scales… 1982, Valkov et al., 2004]. Then, information was collected on 19 primary hazard indicators for those substances that do not belong to the rock-forming, or if their content in the waste exceeds the concentrations in the main types of soils. For this purpose, a literature review of reliable literature sources containing the confirmed information on hazard indicators was made [SanPiN 2.3.2.1078–01, Lazarev, 1976, Grushko, 1982] and a review of relevant state standards in place in Russia [GOST 17.4.1.02–83, GN 2.1.7.2042–06, GN 2.1.5.1315–03, GN 2.1.5.2307–07, GN 2.1.6.1338–03, GN 2.1.6.2309–07, SanPiN 2.1.7.1287–03, SP 2.1 .7.1386–03, MU 2.1.7.730–99].

After collecting all the necessary information, the calculation was carried out in the program "Integral. Calculation of hazard class 4.0". The concentrations of certain substances were recalculated from the percentage content in mg/kg. For this purpose, formula (2) was used:

$$C_i = \frac{P_i}{100} * 1000000, \text{ mg/kg} \tag{2}$$

where: C_i is the desired concentration of the i-th substance (mg/kg),
P_i is the content of the i-th substance in the sample (%).

Further, the calculation of the content of atoms of certain elements from their oxide forms was calculated by formula (3):

$$C_{ei} = \frac{C_i * P_{ei}}{100}, \text{ mg/kg} \tag{3}$$

where: C_{ei} is the concentration of the i-th element in the sample (mg/kg),
P_{ei} is the content of the atoms of the i-th element in the molecule of the i-th substance (%) by the formula (4).

$$P_{ei} = \frac{M_e}{M_s} * 100\%, \% \tag{4}$$

where: M_e is the molecular mass of the element in the substance (g/mol),
M_s is the molecular mass of the i-th substance (g/mol).

After the calculations, the obtained concentrations of oxygen, nitrogen, carbon, phosphorus, sulfur, silicon, aluminum, iron, sodium, potassium, calcium, magnesium and titanium were compared in the main soil types with concentrations of the corresponding substances in the waste samples. As a result, it was found that the content of all listed elements is within the permissible ranges; therefore, they are related to non-hazardous rock-forming elements.

For the rest of the substances, the hazard class was calculated in accordance with the procedure [Order of the Ministry… 2014]. As a result of calculations, it was found that all four samples of waste correspond to the 5 hazard class.

In order to determine the toxicity of the waste, a methodology was used to specify the toxicity of water from the change in the optical density of the Chlorella algae [PND F T 14.1: 2: 3: 4.10–2004]. Chlorella algae are often used for biotesting of sewage and solid waste [Radziemska et. al., 2017; Śliwka, 2014]. The technique is based on recording the differences in the optical density of the test culture of Chlorella algae grown on a medium containing no toxic substances (control sample) and test water samples with appropriate dilution, in which these substances may be present. The algae culture is grown on a 50% Tamia nutrient medium, the composition of which is presented in Table 8.

Solutions A and B are prepared separately: solution A (H_3BO_3 – 2.86 g/dm³; $MnCl_2 \times 4H_2O$ – 1.81 g/dm³; $ZnSO_4 \times 5H_2O$ – 0.222 g/dm³) and solution B (MoO_3 – 17.64 mg/dm³; NH_4VO_3, – 22.96 mg/dm³, dissolved with heating).

Cultivation of algae was conducted in the KV-05 cultivator. A transparent bottle of colorless glass with a capacity of 400 cm³ was used as a reactor. Slurry of algae was poured into the reactor in a volume of 125 ± 10 cm3. The container with the suspension continuously rotated about its longitudinal axis to provide carbon dioxide. In the process of cultivation, a constant medium temperature of 36.0 ± 0.5°C was maintained.

Table 8. The composition of the nutrient medium Tamia for the cultivation of algae chlorella

Component	Concentration, g in 200 cm³
KNO_3	20
$MgSO_4 \times 7H_2O$	10
$KH_2PO_4 \times 3H_2O$	5
Iron citric acid	0.6
Trace elements	Solutions A and B

Before biotesting, the algae culture was filtered and diluted to an optical density of 0.125 ± 0.005 50% by Tamia medium. The registration of the optical density of the culture of the test object was carried out using the density meter of the suspension IPS-03.

The sample preparation was carried out by drying a sample weighing 120–150 g to an air-dry state. The sample was neatly ground with a spatula, eliminating the particles larger than 10 mm. Then, the sample were measured in 100 g in terms of dry matter and preparation of the extract of leaching was carried out afterwards. The resulting sample was poured into a glass beaker and 1000 ml of distilled water was added. The resulting extract was stirred on a magnetic stirrer for 7–8 hours. After mixing, 100 ml of the extract was purified from the suspensions by centrifugation for 10 minutes at 4000–4500 rpm. The resulting extract was ready for biotesting.

The water prepared in the biotesting in a volume of 50 cm^3 was transferred to a glass beaker with a capacity of 100 cm^3. In order to obtain a number of dilutions of the analyzed sample, divisible by three, 24 cm^3 of distilled water was added in 24 cups . Afterwards, 12 cm^3 of test water was introduced into the first of them, and in the second, third and fourth were carried by 12 cm^3, from the first, second and third glasses, respectively. The 12 cm^3 from the last glass is discarded. Along with the diluted test water, 24 cm^3 of initial test water and 24 cm^3 of control (distilled) water were added to separate beakers. Thus, we obtained the following six sample variants:

1. control sample, 0%;
2. initial (not diluted) water sample, 100%;
3. sample diluted 3 times, 33%;
4. sample diluted 9 times, 11%;
5. sample diluted 27 times, 3.7%;
6. sample diluted 81 times, 1.2%.

The prepared test-culture of algae was introduced in 1.5 cm^3 into all 6 cups. At the same time, as a result of 17-fold dilution of the planted culture, the content of the nutrients in the test water was required to ensure that the growth of algal cells would correspond to 0.6% Tamia medium, and the initial optical density of the alga test culture would be 0.030 ± 0.002. The contents of each glass were bottled in 5 cm^3 in vial bottles (4 bottles for each variant of the tested water, including a control sample). The light exposure of samples was made in a KVM-05 multi-cultivator.

Under the same and controlled conditions, the device allows 24 samples to be simultaneously cultured at the temperature, light intensity, CO_2 supply (0.03%) and mixing. The exposure time was 1 hour. Afterwards, the fluorescence intensity parameters were measured; for this purpose, all cuvettes with the samples were placed in a cassette of the Foton-10 fluorimeter for 10–15 minutes. After starting the device, the values of the registered parameters were displayed in the interface of the "Foton 10" program.

Processing of results was as follows:

1. The average value of optical density is calculated by the formula (5):

$$D = \sum D_i/n_i \tag{5}$$

where: D is the average value of the optical density;
D_i – the values of the optical density in the i-th parallel definition; n is the number of parallel definitions.

2. The relative difference in the optical density for each dilution as compared to control (I) was calculated by formula (6):

$$I = (Dk - Do)/Dk \times 100\ \%, \tag{6}$$

where: Dk and Do are the average values of the optical density in the control and in the experiment, respectively.

3. The criterion for the toxicity of a water sample is a decrease in the average optical density in comparison with the control variant by 20% or more in case of inhibiting the growth of the test culture or increasing it by 30% or more when stimulating growth processes.

The quality of the tested water was determined on the basis of its toxicological characteristics through the amount of toxic dilution of water. This required dilution of a water sample that is divisible by three. The dilution for which the deviation index I(calculated by formula (6)) exceeded the water toxicity criterion was selected.

The value of the toxic multiplicity of dilution (TMD) of water and water extracts, if the criterion of toxicity in the form of a 20 % growth suppression was exceeded, was calculated by the formula (7):

$$TMD = 10^{\frac{(\log P_l - \log P_s)\times(I_s - 0{,}2)}{I_s - I_l}} + \log P_s \tag{7}$$

If the criterion of toxicity in the form of 30% growth stimulation was exceeded, the calculation of *TMD* was carried out according to the formula (8):

$$TMD = 10^{\frac{(\log P_l - \log P_s)\times(I_s - 0,3)}{I_s - I_l}} + \log P_s \quad (8)$$

where: P_l is the dilution value (large) at which the deviation index was below the toxicity criterion;

P_s is the dilution value (smaller), at which the deviation index was higher than the toxicity criterion;

I_l and I_s are the values corresponding to these dilutions of the indices of the deviation in growth, expressed in fractions.

The date of water extract preparation was 21.10.2017. The biotesting was conducted on 22.10.2017 at room temperature 23°C and air humidity of 35%.

The results of biotesting on the example of the water extract of the waste station "Novokrestovskaya" were presented in Table 9.

The calculations were made in accordance with the procedure [PND F T 14.1: 2: 3: 4.10–2004]. As a result of biotesting, it was found that all four wastes considered do not have a toxic effect at a multiplication factor of k=1. This confirms the safety of the waste and the hazard class established by the calculation method.

CONCLUSION

A complex analysis of the waste generated at the subway construction sites was carried out. The purpose of this analysis was to determine the possibility of using waste in the urban economy, in particular, as a component of the soil mixture.

The chemical composition of the samples was established that they comprised SiO_2 (from 62.71% to 71.20%), Al_2O_3 (from 12.58% to 15.57%), K_2O (from 3.78% to 4.05%), Fe_2O_3 (from 3.46% to 6.74%), CaO (from 2.30% to 3.20%), and MgO (from 1.13% to 2.06%).

The determination of the granulometric composition is necessary for further research. While choosing the composition of the soil mixture, the results of the analysis will allow us to name and classify this soil, and also to establish agrophysical properties. As a result of calculations, it was found that all four samples of waste correspond to the 5th class of danger.

The technique to determine the toxicity of water was based on recording the differences in the optical density of the test culture of Chlorella algae grown on a medium containing no toxic substances (control sample) and test water samples with appropriate dilution, in which these substances may be present. As a result of biotesting, it was found that all four wastes considered do not have a toxic effect at a multiplication fac-

Table 9. The result of biotesting as an example of waste from station "Novokrestovskaya"

# Sample	Multiplicity of dilution	# Replicates	X	X_a	*I*, %	Have (does not have) acute toxic effect
Control	0	1 2 3 4	4.663 4.634 4.545 4.560	4.603	0	-
1	1	1 2 3 4	4.485 4.675 4.632 4.935	4.681	-1.70%	Not have
2	3	1 2 3 4	4.582 4.669 4.662 4.771	4.671	+-1.49	Not have
3	9	1 2 3 4	4.571 4.560 4.535 4.528	4.548	+1.18	Not have
4	27	1 2 3 4	4.651 4.594 4.699 4.787	4.683	-1.74%	Not have
5	81	1 2 3 4	4.796 4.774 4.870 4.675	4.772	-3.68%	Not have

tor of k=1. This confirms the safety of the waste and the hazard class established by the calculation method.

Acknowledgements

This scientific work was supported by the Government of Saint-Petersburg, a grant from the President of the Russian Federation and the Center for Collective Use of the Saint-Petersburg Mining University.

REFERENCES

1. Basic scales of properties of morphological elements of soils. VASKhNIL Publishing. Dokuchaev Soil Institute, Moscow 1982, pp. 55.
2. Chukhrov, F.V. Petrov, V.P., Nikitina, A.P. 1970. Clays, their mineralogy, properties and practical significance: Collection of articles. Academy of Sciences of the USSR, Institute of Geology of Ore Deposits, Petrography, Mineralogy and Geochemistry. Moscow: Science. 271 p.
3. Danilov, A., Smirnov, Y., Korelskiy, D. 2017. Effective methods for reclamation of area sources of dust emission. Journal of Ecological Engineering, 18(5), 1–7
4. GN 2.1.5.1315–03.Maximum permissible concentration (MPC) of chemical substances in water of water bodies of economic, drinking and cultural-household water use.
5. GN 2.1.5.2307–07.Approximate permissible levels (ODUs) of chemicals in water in water bodies for domestic and drinking and cultural and domestic water use.
6. GN 2.1.6.1338–03.Maximum permissible concentration (MPC) of pollutants in the atmospheric air of populated areas.
7. GN 2.1.6.2309–07.Indicative Safe Levels of Exposure (ISLE) of Pollutants in the Atmospheric Air of Populated Areas.
8. GN 2.1.7.2042–06.Approximate permissible concentrations (APC) of chemical substances in soil.
9. GOST 17.4.1.02–83.Protection of nature.Soil. Classification of chemicals for pollution control.
10. GOST 5180–84.Soils.Methods for laboratory determination of physical characteristics.
11. GOST 26423–85.Soils.Methods for determination of specific electric conductivity, pH and dense residue of aqueous extract.
12. GOST R 8.777–2011. State system for ensuring the uniformity of measurements (ICG). Disperse composition of aerosols and suspensions. Determination of particle size by diffraction of laser radiation. ISO 13320–1: 1999 Granulometric analysis. Methods of laser diffraction.Part 1.General principles.
13. Grushko, Ya.M. 1982. Harmful organic compounds in industrial wastewater. Leningrad: Chemistry, pp. 161.
14. Lazarev, N.V. 1976. Harmful substances in industry.A reference book for chemists, engineers and doctors. In 3 volumes. Leningrad: Chemistry, pp. 608.
15. Methodical manual on the application of "Criteria for classifying hazardous wastes as hazard classes for the environment". Ministry of Natural Resources of the Russian Federation. Moscow 2003, pp. 38.
16. MU 2.1.7.730–99.Hygienic assessment of soil quality in populated areas.
17. Order of the Ministry of Natural Resources and Ecology of the Russian Federation of December 4, 2014 #536. On the Approval of Criteria for classifying wastes as hazard class IV according to the degree of negative impact on the environment. http: //docs.cntd. com / document / 420240163
18. Pashkevich, M.A. 2015. Environmentally safe methods of technogenic deposits conservation. Notes of the Mining Institute. National Mineral Resources University "University of Mines" in Saint-Petersburg, Vol. 213, 86–93.
19. PND F 12.1: 2: 2.2: 2.3: 3.2–03. Methodical recommendations.Sampling of production and consumption wastes.
20. PND F T 14.1: 2: 3: 4.10–2004. Toxicological methods of analysis. Method for determining the toxicity of drinking, natural and waste water, water extracts from soils, sewage sludge and production and consumption wastes by changing the optical density of chlorella alga.
21. Radziemska, M., Mazur, Z. 2017. Chemical composition of spring rapeseed grown in copper contaminated soil amended with halloysite and Zeolite. Journal of Ecological Engineering, 18(2), 38–43
22. SanPiN 2.1.7.1287–03.Sanitary and epidemiological requirements to soil quality.
23. SanPiN 2.3.2.1078–01.Hygienic requirements for safety and nutritional value of food products.
24. SP 2.1.7.1386–03. Sanitary rules for determining the hazard class of toxic production and consumption wastes.
25. Valkov, V.F. Kazeev, K.Sh. Kolesnikov, S.I. 2004. Soil Science: Textbook for high schools.: Publishing Center "Mart". Moscow, pp. 496.

5

Legal Regulations and Methods Neutralising Expired Food Products

Joanna Kazimierowicz[1]

[1] Department of Environmental Engineering Technology and Systems, Faculty of Civil and Environmental Engineering, Bialystok University of Technology, ul. Wiejska 45A, 15-351 Białystok, Poland, e-mail: j.kazimierowicz@pb.edu.pl

ABSTRACT

In retail chains, an expired product is withdrawn from sales and becomes waste which should be managed properly. Expired products cannot be placed back on the market for sale, sold at a reduced price or in promotion. The losses in the food trade result mainly from an inefficient organisation and excessive stocks that cause food expiration and perishability. In this chain, the expired food products, which in majority go to landfill sites, constitute a significant part. This article presents the legal regulations regarding neutralisation of expired food products together with their utilisation possibilities. One of the basic documents is the Directive 2009/28/EC of the European Parliament and the Council of 23 April 2009 on the promotion of the use of energy from renewable sources. The second important document is the Directive of the Council 1999/31/EC on the landfill of waste. According to the Regulation No 852/2004 of 29 April 2004 on the hygiene of foodstuffs, every facility should have elaborated procedures describing the manner of handling foodstuffs not meeting the requirements of health quality as well as food waste. One of the most often applied methods to manage waste is its thermal treatment. Incineration is problematic in environmental and social aspects. It includes numerous requirements and legal restrictions. Pyrolysis is an alternative to incineration. Composting in aerated bed is a widely known utilisation method of biological waste from agriculture and agri-food industry. Mineralisation of organic substances in biogas plants is a method for controlled course of processes aiming at the production of greenhouse gases, the negative effect on meteorological conditions of which is unquestionable in many social circles. Biomass fermentation in biogas plants is prospective because it enables to limit the emission of methane during uncontrolled biochemical processes accompanying landfill of waste.

Keywords: expired food products, neutralising organic waste, legal regulations, methods neutralising

INTRODUCTION

According to the EU report (European Commission 2010), the average food losses in the EU countries amount to 11.7%. According to the FAO Report (Gustavsson et al. 2011) they are at least 30% and according to the Natural Resources Defence Council they are up to 50% in the case of easily perishable products (Gunders 2012). Currently, expired food, thrown away by the processing industry – trade and gastronomy, constitutes the biggest part of the waste. In Poland, about 500 thousand tons of food a year is wasted in this way (Byczyński 2012). The same material, energy and financial outlays are incurred to manufacture damped food as in the case of consumed products. Wasting and then throwing food away increases the use of mineral fertilizers which contributes to the environmental pollution.

LEGAL REGULATIONS

In order to ensure energy security and withstand the climate change, European Union conducts the policy oriented towards limiting energy requirements. One of the basic documents is the Directive 2009/28/EC of the European Parliament and the Council of 23 April 2009 on the promotion of the use of energy from renewable sources (O.J. EU L 09.140.16). This Directive presumes that the share of renewable energy in the total energy use in Poland till 2020 should be on the level of 15%. The second important document is the Directive of the Council 1999/31/EC on the landfill of waste (O.J. EU L 182 of 16 July 1999, page 1). According to the binding law, in 2020 the amount of biodegradable waste which will be allowed on landfill sites is to be reduced by 65% in relation to the base year 1995.

State Sanitary Inspection is obliged to control the facilities under its supervision, according to the binding regulations, including the Act of 14 March 1985 on State Sanitary Inspection (J of Laws 2006 No 122, item 851, later amended) and the Act of 25 August 2006 on food safety and nutrition (J of Laws of 2010, No 136, item 914, later amended). The agencies of the State Sanitary Inspection supervise the food waste management in controlled facilities and check documentation, including the documents identifying the food producers and receivers as well as the agreements with the companies obliged to collect waste. This guarantees that the food which does not fulfil requirements will be utilised. Depending on the type of waste and its purpose, proper agreement should be made between the entrepreneur and the recipient of the waste operating on the basis of the Act of 14 December 2012 on waste (J of Laws of 2013, item 21, later amended) or on the Regulation (EC) No 1096/2009 of the European Parliament and the Council of 21 October 2009 laying down health rules as regarded animal by-products and derived products not intended for human consumption. Additionally, in the Regulation of the Minister of Environment of 10 November 2015 on the list of types of waste that natural persons or non-business entities may recover for their own needs and acceptable recovery methods (J of Laws 2015, item 93) changing the Regulation of 21 April 2006, 12 items were crossed out, including the waste with the code 16 03 80, i.e. expired food products or unsuitable for consumption.

According to the Regulation No 852/2004 of 29 April 2004 on the hygiene of foodstuffs (O.J. EU L 139 of 30.04.2004, p. 1, later amended) every facility should have elaborated procedures describing the manner of handling foodstuffs not meeting the requirements of health quality as well as food waste, having regard to eliminate potential contamination of other raw materials and finished products. In accordance with this Regulation, the food waste and non-edible by-products in production and food trading facilities are to be stored in closed containers which need to be properly constructed and maintained in good condition, easy to clean and disinfect. This is a basic way for a hygienic and environmentally-friendly removal of food waste from such facilities.

The position of the Ministry of Environment clearly determines that the use of colloidal grinders is contrary to the binding regulations from the environmental protection. According to the Act of 18 July 2001 Water law (J of Laws of 2005, No 239, item 2019, later amended) it is prohibited to dispose of waste– within the meaning of the Act on waste – as well as liquid animal excrement, into water. The provisions of the Act of 7 June 2001 on collective water supply and sewage removal (J of Laws of 2006, No 123, item 858, later amended) implemented the prohibition of waste input into sewage. According to this Act, it is forbidden to input solid waste into the sewage system, which could decrease the sewage pipes capacity, especially including gravel, sand, ash, glass, pomace, yeast, bristles, cuttings of furskin, textiles, fibres, even if shredded. The food waste should be neutralised or recycled with the use of different methods.

Expired food of animal origin or containing products of animal origin which are not hazardous is regarded as raw material category 3 (Regulation (EC) No 1069/2009 of the European Parliament and of the Council of 21 October 2009). In a disposal plant, the expired food products are subjected to the thermal processing and combustion is the most often used method. Products may also be processed in composting plants or in digestion installation where pasteurisation and hygienization processes need to be included.

METHODS NEUTRALISING EXPIRED FOOD PRODUCTS

One of the most often applied methods of waste management is its thermal treatment. Thermal waste utilisation means a set of consistent technological segments realising waste combustion, recovery and use of heat generated by burning, treatment of exhaust gases and other combustion products as well as preparation of combustion products and exhaust gases for economic use or storage. Such set of installation technological segments fulfils the legal requirements (Skowron 2003). Thermal combustion may take place in waste incineration as well as in utility boilers. Implementation of the projects aiming at the construction of waste thermal treatment units requires applying reliable and tested technologies.

Currently, the only such technology is grate combustion. This technology, created in the first half of the 20th century, is systematically developed and updated and is applied to both lower calorific value waste (4–6 MJ/kg) as well as to high calorific value waste (12–18 MJ/kg). In the case of lower calorific value waste, air-cooled grates are used with special system of mixing remains subjected to burning, whereas in the case of high calorific waste, grates are water-cooled. Such grates operate in a few hundred installations worldwide. No other technological solution allows to combust such lower calorific value waste as grate incineration. Rotary kiln incinerators to autothermal work require waste with calorific values of minimum 15–18 JM/kg and chamber incineration plants – of minimum 16–19 MJ/kg (Marchwińska and Budka 2014). Regarding technical parameters, universality and reliability, the only type of incineration plant which may be compared to grate incineration plant is fluidized bed incineration. However, it requires initial waste shredding which decreases the amount of generated net useful electric power. The application of these technologies is limited by high costs which are generated by the most technologically advanced incineration plants as well as the difficulties connected with achieving complete combustion.

As a result of thermal disposal and utilisation of organic waste, the problem of air pollution produced by incineration plants arises. The exhaust fumes contain highly toxic chemical compounds such as polychorinated dibenzo-para-dioxins (PCDDs) and polychlorinated dibenzofurans (PCDFs), commonly called dioxins and furans, which – if they infiltrate into the environment – pose a serious threat (Grochowalski et al. 1993, Sokołowski 1994, Yoshimura and Masuda 1994 and 1996). Dangerous pollution comprises nitrogen oxides which, together with exhaust fumes, connect with ozone and then transform into nitric acid (V). It reacts with metals and drops with rain in the form of salt. As a result of reaction with hydrocarbons toxic organic nitrate, perioxides are created. Significant and common exhaust gases pollution is sulphur dioxide. If the commonness of organic waste combustion is assumed, the problem of excessive air pollution with sulphur dioxide and carbon dioxide (increases the existing greenhouse effect) as well as carbon monoxide (a very toxic gas created by incomplete combustion) will return. In the exhaust gases after thermal disposal of waste the hydrogen chloride appears as well, which is water-soluble and forming hydrochloric acid and hydrogen fluoride – corrosive gas that is also easily soluble in water. Exhaust fumes may also contain heavy metals. Removal of those substances from waste gases is usually difficult and expensive (Marchwińska and Budka 2014). Every company building incineration plants has their own checked grates and heat-recovery systems together with boiler configuration. There is no preferred type of grate (besides currently unadvisable – cylindrical-shaped grate) as well as structural form of the boiler or required parameters of produced steam. The only but significant limitation is achieving the required energy efficiency index above 0.65. Then, the combustion process is regarded as a recovery operation (Wielgosiński 2012).

Incineration is problematic in the environmental and social aspects. It includes numerous requirements and legal restrictions (Gawłowski 2011, Piecuch 2006, 1999). There is a need to select such combustion technology which will be accepted in economic and ecological aspect and will not have social resistance (Nowak 2013, Mokrzycki and Uliasz-Bocheńczyk 2005, Piecuch 2006, Poskrobko B. and Poskrobko T. 2012).

Pyrolysis is an alternative to incineration. It is a set of physicochemical processes which need to be initiated and conducted to obtain effects in the form of thermal decomposition of solid, liquid or gas fuel of different types of hydrocarbon making organic substance in waste without oxygen. In the process of pyrolysis, due to the lack of oxygen, no toxic compounds such as PCDDs and PCDFs are created (Piecuch 2006). This method is used for particular groups of waste which must be selected from the whole mass of waste delivered to the processing plant. The group of waste whose utilisation by pyrolysis is effective includes mainly organic chemical products such as: plastics, tires, paints, varnishes, cosmetics and other plastics. The biggest merit of this process and its main advantage over combustion is the fact that it is safe for air cleanness. However, the process of bonding post-pyrolytic secondary waste and secondary waste after combustion (slags) is similar (Piecuch 2006).

Composting in aerated bed is widely known as a utilisation method of biological waste from agriculture and agri-food industry (Błaszczyk 2007, Matcalf and Eddy Inc. 2003). It is based on the biochemical processes consisting in decomposition of organic substances. Composting is generally conducted in two stages (Manczarski 2007):

I. intensive composting (thermophilic stage) – in this stage fresh compost is obtained from organic waste. The material is hygienised, easily degradable substances are decomposed, potential odour emission is decreased;
II. maturing stage (mesophilic stage) – during this stage mature compost is obtained from fresh compost. Hard degradable substances are decomposed and stable humus structures are created, resistant to the external factors and rich in nutrients.

The time of each phase depends on the composition of the composted biomass and the applied technology. Aerobic mineralisation is an exothermal process and the intensity of decomposition depends on the compounds susceptibility to it. Fats, majority of proteins and sugars including starch are easily decomposed, whereas hemicellulose and cellulose are harder to decompose. Lignin and proteins from scleroproteins group such as keratin are very resistant to decomposition (Manczarski 2007). The most significant parameter of composting is temperature which may even exceed 70°C (Yamada and Kawase 2005). Such high temperature is detrimental to the majority of thermophilic microorganisms taking part in the decomposition of organic matter in the thermophilic phase of composting. A temperature decrease in the composting bed may be achieved by increasing the degree of aeration; however, the results are not always as expected. It is also important to maintain moisture in the composted material between 50 and 70% through the whole period of microorganisms' activity. On the other hand, excessive aeration may cause drying of pile which may result in decreasing the activity of microorganisms or even hinder the whole process (Sołowiej et al. 2010).

There are many composting systems, depending on the adopted criteria (Staszczyk 2003). The simplest way involves the process conducted in piles with periodic waste mixing. It is conducive to the homogenisation of substrate mix subjected to composting and provides the oxygen necessary to the proper course of the process (Aniszewska 2007, Baeta-Hall et al. 2005). The advantages of this composting method are mainly low investments costs and simple technology (Puyuelo 2010). The greatest disadvantages include huge space requirements and susceptibility of piles to weather conditions, especially if there is no roofing (Czekała et al. 2013). The second group of composting methods constitute closed technologies. They are characterised by the fact that the processes take place in reactors, usually chambers or containers (Olszewski et al. 2005). During operation, they have a large energy demand. They require an installation with precise aeration and chamber leachates seizing systems (Boniecki et al. 2012). The construction and exploitation of such installations are connected with large financial outlays but are compensated by many advantages. The most significant ones include: precise process control, independence from the weather conditions and the possibility to recover and use heat from the installation (Czekała et al. 2013).

The greatest advantage due to which this method is popular in many countries is low financial expenditures. Maintenance of optimal process parameters such as proper temperature and moisture as well as access of oxygen necessary for microorganism development may be achieved with the use of specialised measuring equipment. Additionally, appropriate fraction of previously sorted waste is selected (Kucharczak et al. 2010). More and more problems are caused by the compost management – the main product of composting – due to the protection of agricultural products against soil pollution. Very strict restrictions have been implemented regarding the amount of harmful substances in all materials applied into the soil. This mainly concerns the amount of heavy metals and some organic pollutants as well as the sanitary-epidemiological safety (Manczarski 2012).

Mineralisation of organic substances in biogas plants is a method for controlled course of processes aiming at the production of greenhouse gases, the negative effect on meteorological conditions of which is unquestionable in many social circles (Węglarzy et al. 2011). The biomass fermentation in biogas plants is prospective because it enables to limit the emission of methane during uncontrolled biochemical processes accompanying a waste landfill (Szlachta 2008, Szlachta and Fugol 2009). Fermentation is a biochemical process occurring with the participation of microorganisms and their enzymes. As a result of organic matter conversion, methane and carbon dioxide are obtained (Pilarski and Adamski 2009). Fermentation fulfils three functions, it enables to (Ledakowicz and Krzystek 2005):

- convert the energy from waste into a useful fuel (biogas) which may be stored and transported,
- recycle organic waste into stable soil improvers, liquid fertilizer and energy,

- make the waste inert, aiming at decreasing the negative impact on the environment.

The organic waste from such industries as: agriculture, meat industry, dairies, distillers, breweries and fruit and vegetables processing may be successfully used in biogas plants (Curkowski et al. 2013).

The applied technological process as well as used substrates influence the composition of obtained biogas. Biogas mainly consists of methane, carbon dioxide and hydrogen oxide (Cebula 2012) and the amount of methane decides about biogas quality and usefulness. Table 1 presents the chemical composition of biogas according to different sources (Mollera et al. 2004, Arvanitoyannis and Kassaveti 2008, Luostarinen et al. 2009, Lansing et al. 2008). Table 2 (Lewandowski 2012, Rosik-Dulewska 2006, Romaniuk et al. 2010) presents content of biogas components and pollution as well as their effects. The potential of biogas production mainly depends on the availability of raw materials (Dinuccio et al. 2010).

Biogas is produced constantly and does not depend on weather conditions. Its application has the following advantages (Lewandowski 2012):

- it decreases the use of non-renewable resources and emission of compounds produced during their combustion as well as lowers the greenhouse gasses emission,
- it reduced odours by over 80%,
- it eliminates pathogens in the process of hygienisation,
- it enhances the conditions of soil fertilizers and eliminates synthetic fertilizers in agricultural crops,
- it decreases the risk of surface and ground waters contamination and enables to improve the water efficiency,
- it allows the developing countries to increase the level of civilisation by providing light, electricity and water.

The production costs are similar to the costs of electric energy from the power grid and with higher interest rates, they may even be lower.

The production of energy from biogas has its weaknesses. The most significant ones include high investments connected with the construction of tanks, fermenters, acquisition of engine, generator, control-measurement instruments, but also the necessity to strictly comply with the re-

Table 1. Biogas chemical composition

Source	Biogas composition [%]					
	CH_4	CO_2	N_2	H_2	H_2S	O_2
Mollera et al. 2004	55–75	25–45	0–0.3	0.1–0.5	0–3	0.1–0.5
Arvanitoyannis and Kassaveti 2008, Luostarinen et al. 2009	50–75	25–50	0–2	0–1	0–2	0–1
Lansing et al. 2008	52–85	14–18	0.6–7.5	0–5	0.08–5.5	0.1–0.2

Table 2. The content of biogas components and pollution as well as their effects (Lewandowski 2012)

Component	Content	Effect
Methane	45–70%	• Determines calorific value. • With the content >45% biogas is combustible. • With the content 5–15% biogas and air mixture is highly explosive
Carbon dioxide	25–50%	• Decreases calorific value • Causes corrosion (low concentration carbonic acid) when the gas is moist. • Damages alkaline fuel cells
Hydrogen sulphide	0–0.5%	• Causes corrosion of appliances and pipes (many engine manufacturers defines maximal level of 0.05%) and with the content of 40 mgm^{-3} gas desulphurisation is conducted. • SO_2 emission behind burner, H_2S emission with incomplete combustion. • Decreases the effect of catalysts.
Ammonia	0–0.05%	• NO_x emission behind the burners.
Water vapour	1–5%	• Causes corrosion of appliances • Condensate may damage appliances and installations • Risk of freezing pipes and nozzles system
Dust	>5 pm	• Blocks nozzles and fuel cells
Nitrogen	0–5%	• Decreases biogas calorific value • With high temperature of burning it creates dioxides and trioxides (NO_2, NO_3)
Siloxanes	0–50 mgm^{-3}	• Are abrasive and damage engines

gimes of fermentation processes such as: temperature, pH, balanced hydraulic retention time and constant level of organic compounds wasteload, homogeneity of substrate mix and air-tightness (Lewandwski 2012).

The post-fermentation sludge, rich in nutrients, is a byproduct of anaerobic decomposition of organic matter. Although huge amounts of post-fermentation sludge require special utilisation method, it should not be treated as waste but rather as a valuable source of minerals and energy. It may be used in agriculture or in natural areas, thermally processed or ultimately deposited at a landfill site. Each of the described methods has both advantages and disadvantages and affects the environment in a different way. Therefore, the methods should be selected depending on the composition of digestate, characteristic features and geographical location of the area of its designation so that they are safe and cost-effective. In Poland, the post-fermentation sludge is usually thermally utilised or used in agriculture. Large part of sludge is landfilled; however, this method should not be used because it does not bring any benefits. Fertilisation or energy potential generated in the substance is lost. Application in agriculture, in contrast landfilling, is very beneficial due to appropriate fertilisation properties of the post-fermentation sludge. The thermal use enables producing additional energy as well as valuable ash, which may be used as fertiliser owing to high amount of minerals.

CONCLUSION

1. About 1 to 3% of food production includes perished or expired foods, i.e. thousands tons every year. Shops still rarely use the services of disposal companies which means that waste is dumped. Hypermarkets and supermarkets lower prices and make discounts; with huge turnover, they constantly have products requiring utilisation. Therefore, they sign agreements for collecting the expired food. Improperly managed waste poses a threat to human health and pollutes the environment.
2. Realisation of the sustainable development concept is achieved among others by proper waste management. The newest research indicates that food waste does not have to be deposited on landfill sites. They may be a valuable material in processing because microelements might be recycled from them (Galanakis 2012, Kazimierowicz 2014,2017, Śmiechowska 2015).
3. According to the author, the applicable legal regulations and methods for neutralizing this type of waste are sufficient.

Acknowledgements

This article has been written within the work S/WBiIŚ/02/2014, funded by the Ministry of Science and Higher Education.

REFERENCES

1. Aniszewska M. 2007. Modern self-propelled machines for processing compost prisms. Agricultural, Horticultual and Forest Engineering, No. 4.
2. Arvanitoyannis I.S., Kassaveti A. 2008. Fish industry waste: treatments, environmental impacts, current and potential uses. International Journal of Food Science and Technology, vol. 43, 726–745.
3. Baeta-Hall L., CeuSaagua M., Lourdes Bartolomeu M., Anselmo A.M. 2005. Bio-degradation of olive oil husks in composting aerated piles. Bioresource Technology, 96, 69–78.
4. Błaszczyk J.K. 2007. Microorganisms in environmental protection. PWN, Warszawa.
5. Boniecki P., Dach J., Pilarski K., Piekarska-Boniecka H. 2012. Artificial neural networks for modeling ammonia emissions released from sewage sludge composting. Atmospheric Environment, 57, 49–54.
6. Byczyński J. 2012. A new era of utilization. Gazeta Polska. No. 35, http://www.gazetapolska.pl [access 13.10.2016].
7. Cebula J. 2012. Selected methods of treatment of agricultural and landfill biogas. Monograph, Wydawnictwo Politechniki Śląskiej, Gliwice.
8. Curkowski A., Oniszk-Popławka A., Haładyj A. 2013. Biogas plant – a deliberate choice. Foundation Institute for Sustainable Development. Warszawa.
9. Czekała W., Witaszek K., Rodriguez Carmona P. C., Grzelak M. 2013. Installations for industrial bio-waste composting: advantages and disadvantages. Agricultural, Horticultual and Forest Engineering, 2, 23–25.
10. Dinuccio E., Balsari P., Gioelli F., Menardo S. 2010. Evaluation of the biogas productivity potential of some Italian agro-industrial biomasses. Bioresource Technology, vol. 101, 3780–3783.
11. Directive 2009/28/EC of the European Parliament and of the Council of 23 April 2009 on the promotion of the use of energy from renewable sources

(OJ L EU L 09.140.16).

12. Council Directive 1999/31/EC on the landfill of waste (Journal of Laws L182 of 16.7.1999, p. 1).

13. European Commission (DG ENV) 2010. Preparatory study on food waste across EU 27. Final Report ss. 213 Directorate C – Industry.

14. Galanakis C.M. 2012. Recovery of high added-value components from food wastes: Conventional, emerging technologies and commercialized applications. Trends in Food Science & Technolog. 26(2), August, 68–87.

15. Gawłowski S. 2011. Management of EU funds acquired in Poland in the area of environmental protection. Rocznik Ochrony Środowiska, vol. 13, 269–282.

16. Grochowalski A., Wybraniec S., Górski L., Sokołowski M. 1993. Solid phase extraction and capillary GC-ECD analysis of polychlorinated dibenzo-p-dioxins in chlorinated phenols. Chem. Anal., 38, 279–286.

17. Gunders D. 2012. Wasted: How America is losing up to 40 percent of its food from farm to fork to landfill. The Natural Resources Defense Council (NRDC). Issue paper, August 2012. iP:12–06-B. http://www.nrdc.org/food/files/wastedfood-IP.pdf.

18. Gustavsson J., Cederberg CH., Van Otterdijk U. S. R., Meybeck A. 2011. Global food losses and food waste. Study conducted for the International Congress SAVE FOOD! at Interpack 2011. Food and Agriculture Organization of the United Nation, Düsseldorf, pp. 28.

19. Kazimierowicz J. 2014. Organic waste used in agricultural biogas plants, Journal of Ecological Engineering, 15(2), 88–92.

20. Kazimierowicz J. 2017. Neutralisation of expired food products in the process of methane fermentation. Doctorate. Bialystok University of Technology. Faculty of Civil and Environmental Engineering, 28/09/2017.

21. Kucharczak K., Stępień W., Gworek B. 2010. Composting of municipal waste as a method of organic substance recovery. Protection of the Environment and Natural Resources, No. 42, 240–254.

22. Lansing S., Botero R., Martin J.F. 2008. Waste treatment and biogas quality In small-scale agricultural digesters. Bioresource Technology, vol. 99, 5881–5890,

23. Ledakowicz S., Krzystek L. 2005. The use of methane fermentation in the utilization of waste from the agri-food industry. Biotechnology 3(70), 165–183.

24. Lewandowski W.M. 2012. Pro-ecological renewable energy sources, WNT Publisher, Warsaw, 360–374.

25. Luostarinen S., Luste S., Sillanpää M. 2009. Increased biogas production at wastewater treatment plants through co-digestion of sewage sludge with grease trap sludge from a meat processing plant. Bioresource Technology, vol. 100, 79–85.

26. Manczarski P. 2007. Composting of municipal waste. A paper on the Forum of Environmental Protection Technology POLEKO

27. Manczarski P. 2012. Mechanical and biological treatment and storage of waste in the light of new regulations. Zarządzanie Gospodarką Odpadami, 117–144.

28. Marchwińska E., Budka D.: The problem of waste in the aspect of public health. http://www.srodowiskoazdrowie.pl/wpr/Aktualnosci/Czestochowa/Referaty/Marchwinska.pdf?f27ba39e183cc4811d3754669e5fce7a=96a08867e409ac927ff0b619a555c326

29. Matcalf and Eddy Inc. 2003. Treatment and Reuse, fourth ed. McGraw-Hill. Wastewater Egineering. New York, pp. 1546–1554.

30. Mokrzycki E., Uliasz-Bocheńczyk A. 2005. Alternative fuels from waste for the power industry. Energy Policy Journal, Vol. 8, pp. 508.

31. Mollera H.B., Sommer S.G., Ahring B.K. 2004. Methane productivity of manure, straw and solid fractions of manure. Biomass and Bioenergy, Vol. 26, 485–495.

32. Nowak B. 2013. Dilemmas of economic efficiency of projects for the thermal treatment of municipal waste. Energy Policy Journal, 16(4), 201–216.

33. Olszewski T., Dach J., Jędruś A. 2005. Modeling of the composting process of natural fertilizers in the aspect of heat generation. Journal of Research and Applications in Agricultural Engineering, 50(2), 40–42,

34. Piecuch T. 1999. The Pyrolitic Convective Waste Utilizer. Environmental Science Research, Volume 58, Kluwer Academic (Plenum Publisher – New York, Boston, Dordrecht, London, Moscow).

35. Piecuch T. 2006. Outline of thermal waste utilization methods. Handbook of the Koszalin University of Technology, pp. 396.

36. Pilarski K., Adamski M.: Perspectives of biogas production with taking into consideration reaction mechanism in the range of quantitative and qualitative analyses of fermentation processes. Journal of Research and Application in Agriculture Engineering, 54(2), 81–86.

37. Poskrobko B., Poskrobko T. 2012. Environmental management in Poland. Polskie Wydawnictwo Ekonomiczne Warsaw, pp. 273.

38. Puyuelo B., Gea T., Sánchez A. 2010. A new control strategy for the composting process based on the oxygen uptake rate. Chemical Engineering Journal, 165, 161–169.

39. Romaniuk W., Łukaszuk M., Karbowy A. 2010.

Potential for development of biogas plants on farms in Poland. Problemy Inżynierii Rolniczej No. 4, 129–139.

40. Rosik-Dulewska C. 2006. Basics of waste management. PWN, Warsaw, pp. 342.
41. Regulation (EC) No 1069/2009 of the European Parliament and of the Council of 21 October 2009 specifying sanitary rules for animal by-products not intended for human consumption.
42. Regulation of the Minister of the Environment of November 10, 2015 on the list of waste types that natural persons or organizational units that are not entrepreneurs may recycle for their own needs, and acceptable methods of their recovery (Journal of Laws 2015, item 93).
43. Skowron H. 2003. Thermal treatment of waste – news and comments continued. Przegląd Komunalny, August.
44. Śmiechowska M. 2015. Sustainable consumption and food waste. Ann. Acad. Med. Gedan. 45, 89–97.
45. Sokołowski M. 1994. Dioxins – assessment of environmental hazards and methods of their detection. Military Institute of Chemistry and Radiometry. Brochure issued by PIOŚ – WARJNTECH, Warsaw.
46. Sołowiej P., Neugebauer M., Piechocki J. 2010. The effect of additives and aeration on the dynamics of the composting process. Agricultural Engineering 5(123), 259–265.
47. Staszczyk J. 2003. Composting of organic waste and sewage sludge – machinery, equipment, technologies, legal conditions and sources of financing. International Scientific and Technical Conference "Practical Ecology". Ustka.
48. Szlachta J., Fugol M. 2009. Analysis of biogas production possibilities based on slurry and maize silage. Agricultural Engineering 5(114), 275–280.
49. Szlachta J. 2008. Possibilities of biogas production from agricultural products. A paper at the International Conference at IBMER, Warsaw, 223–229.
50. The Act of 14 December 2012 on waste, Dz.U. 2013 item 21.
51. The Act of 14 March 1985 on the State Sanitary Inspection, Dz. U. of 2006 No. 122, item 851 with later d.
52. The Act of 18 July 2001 on Water Law (Journal of Laws of 2005 No. 239, item 2019, as amended).
53. The Act of 25 August 2006 on food and nutrition safety, Dz. U. of 2010, No. 136, item 914 with later d.
54. The Act of June 7, 2001 on collective water supply and collective sewage disposal (Journal of Laws of 2006, No 123, item 858, as amended).
55. Węglarzy K., Skrzyżala I., Pellar A. 2011. Agricultural biogas plant in kostkowice. First experiences. Journal of Research and Applications in Agricultural Engineering, 56(4), 189–192.
56. Wielgosiński G.: The choice of technology for the thermal transformation of municipal waste. Nowa Energia, No. 1, 2012.
57. Yamada Y., Kawase Y. 2006. Aerobic composting of waste activated sludge: Kinetic analysis for microbiological re action and oxygen consumption. Waste Management 26, 49–61,
58. Yoshimura H., Masuda Y., Hori Y., Kuratsune M., Okumara M. Yusho A. 1996. Human Disasters Caused by PCBs and Related Compounds. Kynshu University Press, Fukuaka, pp. 361.
59. Yoshimura H., Masuda Y. 1994. Approach to Risk Assessment of Chlorinated Dioxins from Yusho PCB Poisoning. Organohalogen Compounds 21, pp. 11.

6

Surface and Subsurface Water Runoff and Selected Matter Components from the Forested Loess Slope

Andrzej Mazur[1]

[1] University of Life Sciences in Lublin, Department of Environmental Engineering and Geodesy, ul. Leszczyńskiego 7, 20-069 Lublin, Poland, e-mail: amazur70@op.pl

ABSTRACT

In the years 2008–2011, the study on the surface and subsurface water runoff from the forested loess slope was carried out to determine the concentrations of selected chemical indicators of water quality, soil suspension and loss of the pure matter component. The maximum tested concentrations of water quality indicators were low and amounted to: 1.841 mg·dm^{-3} N-Nog, 0.943 mg·dm^{-3} N-NH_4, 0.478 mg·dm^{-3} N-NO_3, 0.213 mg·dm^{-3} N-NO_2, 0.423 mg·dm^{-3} P, 1.621 mg·dm^{-3} K. The masses of the eroded matter constituents were low and amounted to: 0.808 kg·ha^{-1} N, 0.157 kg·ha^{-1} P, 0.142 kg·ha^{-1} K and 2.989 kg·ha^{-1} soil. The parameters of erosive precipitation and water outflow were statistically significantly correlated with the concentration of soil suspended matter and losses of the analyzed components of matter, as well as negligible concentration of chemical indicators of water quality. Afforestation of the loess slopes threatened by erosion is a treatment that effectively protects the soil against water erosion.

Keywords: soil erosion, forest, surface and subsurface outflow, water quality, loess soil

INTRODUCTION

The soils developed from loess belong to the most fertile soils in the world [Catt 2001], but they are highly susceptible to soil erosion [Józefaciuk, Józefaciuk 1999]. Water erosion is a destructive process. Currently, it is considered one of the major soil degradation factors in the world. As a result of its operation, nutrients for plants and humus are leached from the soil [Lal 2005, Mazur 2018, Olson 2007], the physicochemical properties of soils deteriorate [Hladký et al. 2016, Ijaz et al. 2006], which leads to a decrease in their fertility [Duan et al. 2011, Lobo et al. 2005] and in the obtained crop yields [Arriaga, Lowery 2003, Papiernik et al. 2005]. The plant nutrients and soil material eroded from the fields [Mazur 2018], contribute to water eutrophication [Dupas et al. 2015, Gabryszewska et al. 2016, Grzywna et al. 2015] and silting of technical infrastructure devices [Asiedu 2018].

For many years, interdisciplinary research on soil erosion has been carried out. However, the studies covering the quantitative and qualitative surface and subsurface water runoff in the zone of eroded slopes, especially on forested slopes, are still poorly recognized and significantly less accentuated in the scientific literature [Mazur 2018, Święchowicz 2012, Żmuda 2006]. It has long been believed that the forest provides best protection for the soil against water erosion [Tyszka 2008], which does not mean that there is no soil erosion at all [Robichaud et al. 2010]. Therefore, it is necessary to strive for a better identification of the functioning of fluvial transport on the erosion-threatened forested areas, in order to rationally manage soil resources, in accordance with the principle of sustainable development.

The paper presents the results of research from 2008–2011, carried out under the conditions of a forested loess slope in the Lublin Upland, regarding the quantity and quality of surface and subsurface water runoff as well as selected components of matter.

MATERIAL AND METHODS

The research was carried out in the area of loess catchment with a periodic outflow of water in Wielkopole village, located in the eastern part of the Wyniosłość Giełczewska mesoregion (Lublin Upland) [Kondracki 2000]. The measurements of surface and subsurface outflows were carried out on a forested slope of 0.5 ha, with an average slope of about 11% and an NE exhibition. The slope was overgrown by a 13-year-old young tress pine (Pinus sylvestris L.) planted in a 1.5 x 1.0 m rectangular spacing. The surface coverage was 100%. The brushwood was not found, but few herbaceous plants grew in the undergrowth.

At the beginning of April 2008, a catcher of the surface runoff and subsurface outflows was installed in the lower part of the slope. The volume of outflowing water was measured and samples were taken for a laboratory analysis to determine: suspension, total nitrogen, ammonium nitrogen, nitrate nitrogen and nitrite nitrogen as well as phosphorus and potassium.

In April 2008, a soil research was carried out. A description of the soil profile was made, the soil samples were taken from the distinguished diagnostic horizon and the following parameters were determined: granulometric composition, density of the solid phase of the soil and soil bulk density, total porosity and water permeability coefficient.

The atmospheric precipitation in the basin was measured using a daily recording pluvograph and the Hellmann rain gauge. The meteorological station was established within the catchment area, about 400 m from the site of research. On the basis of pluvographs, the unit kinetic energy of rainfall and the erosive indicator of rain and surface runoff were calculated – EI_{30} index.

On the basis of the obtained research results, the dependences pertaining to the quality indicators of outflowing waters and losses of matter components on the parameters of precipitation and outflows were determined, based on the Pearson's linear $r_{x,y}$ correlation index, at a significance level of $\alpha = 0.05$. A detailed description of the research methodology was given in the work of Mazur [2018].

RESULTS AND DISCUSSION

On the basis of the stratification of the distinguished diagnostic horizons made in the soil outcrop, in this study area the soils developed from loess and are of the lessive type (Haplic Luvisols). These are medium-eroded soils with an O-A-B2t-BC-Cca. The dominant fraction in the granulometric composition was dust, constituting about 62% (Table 1). The alluvial parts (<0.02 mm) also constituted a high percentage (about 34%), and a thick dusty loam constituted the largest share of these.

The average specific soil density was 2.6 Mg m^{-3} and increased with depth (Table 2), as well as the bulk density, which ranged from 1.34 to 1.48 $Mg \cdot m^{-3}$. In contrast, the total porosity decreased with depth and fluctuated between 44.2 and 47.5%. The water permeability coefficient also decreased with depth. Its highest value ($26.245 \cdot 10^{-6}$ $m \cdot s^{-1}$) was recorded in the humus horizon.

The meteorological conditions have a very significant impact on the intensity of soil erosion [Żmuda 2006, Święchowicz 2012]. Their characteristics during the research period are presented in the publication of Mazur [2018]. The study

Table 1. Soil granulometric composition

Horizon	Depth [cm]	Percent of fraction of diameter [mm]						Sum of floatable parts <0.02
		1 – 0.1	0.1 – 0.05	0.05 – 0.02	0.02 – 0.006	0.006 – 0.002	<0.002	
O	0-4	-	-	-	-	-	-	-
A	4-17	4.0	15.5	46.2	16.9	11.1	6.3	34.3
Bt2	17-48	3.7	15.7	46.8	16.2	10.2	7.4	33.8
BC	48-65	2.9	16.5	45.9	16.5	9.8	8.4	34.7
Cca	>65	3.3	15.5	47.6	17.6	8.8	7.2	33.6

Table 2. Selected physical properties of the studied soils

Horizon	Depth [cm]	Specific density [$Mg \cdot m^{-3}$]	Bulk density [$Mg \cdot m^{-3}$]	Total porosity [%]	Water permeability [$x10^{-6}$ $m \cdot s^{-1}$]
A	4-17	2.55	1.34	47.5	26.245
Bt2	17-48	2.61	1.39	46.7	17.326
BC	48-65	2.62	1.42	45.8	14.547
Cca	>65	2.65	1.48	44.2	12.258

of surface runoff on a forested slope was carried out in 9 periods, while the subsurface outflow in 7 measurement periods. Rainfall heights then varied from 31.5 to 63.6 mm (Table 3). The calculated coefficient of variation was 24.21% (Table 4) – an average variation [Mucha 1994]. Out of all the recorded instances of erosive precipitation, they occurred most frequently in July (55.6%). In August, their incidence was 33.3%, and in June 11.1%.

The EI_{30} precipitation and surface runoff intensity index values ranged from 145.3 to 450.2 $MJ \cdot mm \cdot ha^{-1} \cdot h^{-1}$ (Table 4). The variation of the EI_{30} index amounted to 41.77% – high variation [Mucha 1994]. Dependence between the EI_{30} index and the total rainfall is shown in Figure 1. High precipitation erosion potential, expressed by the EI_{30} index, does not necessarily depend on the sum of the rainfall. The calculated value of the correlation coefficient between the EI_{30} index and the sum of rainfall r = 80, indicates a very high correlation [Stanisz 1998]. The R^2 determination index was 0.64. The high dependence ($R^2 = 0.54$) between the EI_{30} index and the amount of precipitation was also confirmed by the results obtained by Rejman [2006]. The correlation coefficient between EI_{30} and the precipitation height calculated by Święchowicz [2012] amounted to 0.51.

The variation of surface water runoff from the slope was 79.75%, indicating high variability [Mucha 1994]. The surface water runoff level on the slope was from 0.11 to 2.84 mm (Table 4), which constituted from 0.3 to 4.5% of the precipitation causing the erosive event. These are low rates in relation to the maximum surface outflows on arable lands, which can account for even more than 40% of precipitation causing the ero-

Table 3. Date of occurrence and amount of erosive precipitation: a – surface runoff, b – subsurface runoff

Month	2008		2009		2010		2011	
	Date	Rainfall [mm]	Date	Rainfall [mm]	Date	Rainfall [mm]	Date	Rainfall [mm]
Jun	-	-	25 (a, b)	50.8	-	-	-	-
Jul	7 (a)	31.5	-	-	24 (a, b)	59.8	3-4 (a, b) 5 (a, b) 20 (a)	63.5 57.2 32.1
Aug	-	-	-	-	6-7 (a) 9 (a, b) 30 (a, b)	44.2 63.6 41.5	-	-

Table 4. Indicators characterizing rainfall, surface and subsurface runoff: a – surface runoff, b – subsurface runoff

Indicator		Medium	Standard Deviation	Variation	Minimum	Maximum
Rainfall [mm]		49.4	11.9	24.21	31.5	63.6
Kinetic energy [$J \cdot h^{-1}$]		491.7	153.4	31.19	245.3	699.3
Index EI_{30} [$MJ \cdot mm \cdot ha^{-1} \cdot h^{-1}$]		277.7	115.9	41.77	145.3	450.2
Outflow [mm]	a	1.27	1.01	79.75	0.11	2.84
	b	0.05	0.02	30.38	0.03	0.08
Soil suspension [$g \cdot dm^{-3}$]	a	0.033	0.02	53.77	0.012	0.059
	b	0.018	0.01	30.39	0.011	0.025
N-Nog [$mg \cdot dm^{-3}$]	a	1.139	0.22	19.15	0.895	1.464
	b	1.485	0.30	20.31	1.025	1.841
$N-NH_4$ [$mg \cdot dm^{-3}$]	a	0.591	0.12	20.66	0.442	0.819
	b	0.795	0.15	20.35	0.548	0.943
$N-NO_3$ [$mg \cdot dm^{-3}$]	a	0.301	0.07	22.53	0.218	0.386
	b	0.382	0.09	23.62	0.254	0.478
$N-NO_2$ [$mg \cdot dm^{-3}$]	a	0.146	0.03	18.96	0.104	0.185
	b	0.178	0.03	18.92	0.162	0.213
P [$mg \cdot dm^{-3}$]	a	0.257	0.05	19.72	0.194	0.379
	b	0.331	0.07	22.25	0.212	0.423
K [$mg \cdot dm^{-3}$]	a	1.123	0.16	14.36	0.946	1.421
	b	1.335	0.18	13.12	1.311	1.621

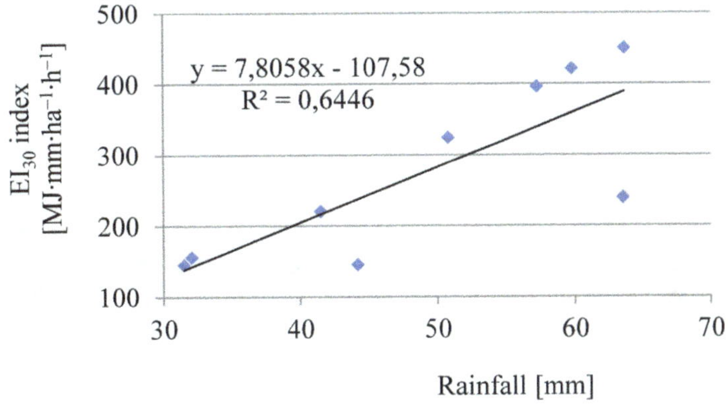

Fig. 1. Rainfall erosivity index EI_{30} in dependence on rainfall amount

sive event [Rejman 2006]. The subsurface runoff during the research period occurred only from the confined layer up to 0.25 m. Their variation amounted to 30.38% – average variability [Mucha 1994].

The subsurface water outflows ranged from 0.03 to 0.08 mm (Table 4), which accounted for 0.07 and 0.13%, respectively, of the precipitation causing the erosive event. In total, the largest outflows were recorded in 2010 (a wet year with precipitation of 867.7 mm). The surface runoff was then 5.74 mm, and the subsurface outflow was 0.17 mm, which accounted for 0.66% and 0.02%, of annual precipitation, respectively. Very low runoff rates are the result of the high retention capacity of forest ecosystems. According to a study conducted by Tyszka [2008], interception, retention of forest litter and a relatively high permeability of forest soils enable the retention of a large part of the precipitation and conversion of surface runoff into a vertical ground drain. Thus, the forest increases the resources of groundwater.

The concentration of the soil suspension contained in the surface runoff waters was characterized by high variability (53.77%) and ranged from 0.012 to 0.059 $g \cdot dm^{-3}$, and in the subsurface runoff waters 0.011-0.025 $g \cdot dm^{-3}$ (Table 4). With respect to the concentrations of soil suspended in waters flowing from arable land reaching the maximum values of 170-240 $g \cdot dm^{-3}$ [Mazur 2018, Żmuda 2006], they should be recognized as very low. The mass of the eroded soil on the slope during individual events was also very low, from 0.01 to 1.67 $kg \cdot ha^{-1}$ during the surface runoff and 0.003 to 0.020 $kg \cdot ha^{-1}$ during the subsurface outflow. In the field of sugar beet cultivation, the maximum surface soil wash during a single event was estimated by Żmuda [2006] at 60.132 $Mg \cdot ha^{-1}$ and Rejman [2006] at 20.640 $Mg \cdot ha^{-1}$ (cultivation of root crops). The surface wash of the soil from the forested slope during the four years of research was 5.390 $kg \cdot ha^{-1}$. The largest amount of the eroded soil (54.8%) flowed in 2010 (2.953 $kg \cdot ha^{-1}$), and the lowest in 2008 (0.013 $kg \cdot ha^{-1}$ (Figure 2a). During the subsurface outflow, the mass of the eroded soil during the four-year study period was 0.066 $kg \cdot ha^{-1}$. The highest soil losses (0.036 $kg \cdot ha^{-1}$, which is 54.52% of the total eroded soil mass) were recorded in 2010, while the lowest (0.008 $kg \cdot ha^{-1}$) in 2009 (Figure 2b). In comparison with arable land, soil loss from the forested slope, should be considered as very low. According to the studies by Mazur [2018] and Rejman [2006], the annual soil losses on arable lands can even exceed 160 $Mg \cdot ha^{-1}$.

The chemistry of waters flowing from the slope was characterized by low diversity. This is evidenced by the calculated values of variation coefficients ranging from 19.15 to 23.62% (Table 4) – small variability [Mucha 1994]. The concentration of total nitrogen in the surface flowing waters ranged from 0.895 to 1.464 $mg \cdot dm^{-3}$ N-Nog. On the other hand, the concentrations of other forms of nitrogen were at the level of: ammonium 0.442-0.819 $mg \cdot dm^{-3}$ $N\text{-}NH_4$, nitrate 0.28-0.386 $mg \cdot dm^{-3}$ $N\text{-}NO_3$, nitrite 0.104-0.185 $mg \cdot dm^{-3}$ $N\text{-}NO_2$. The concentration of phosphorus flowing away in the dissolved form was in the range of 0.194-0.379 $mg \cdot dm^{-3}$ P, and potassium 0.946-1.421 $mg \cdot dm^{-3}$ K. In the subsurface flowing waters, total nitrogen concentrations ranged from 1.025 to 1.841 $mg \cdot dm^{-3}$ N-Nog. The concentrations of other forms of nitrogen were at the level of: ammonium 0.548-0.943 $mg \cdot dm^{-3}$ $N\text{-}NH_4$, nitrate 0.254-0.478 $mg \cdot dm^{-3}$ $N\text{-}NO_3$, nitrite 0.162-0.213 $mg \cdot dm^{-3}$ $N\text{-}NO_2$. The concentration of phosphorus was on the level of 0.212-0.423 $mg \cdot dm^{-3}$ P, and potas-

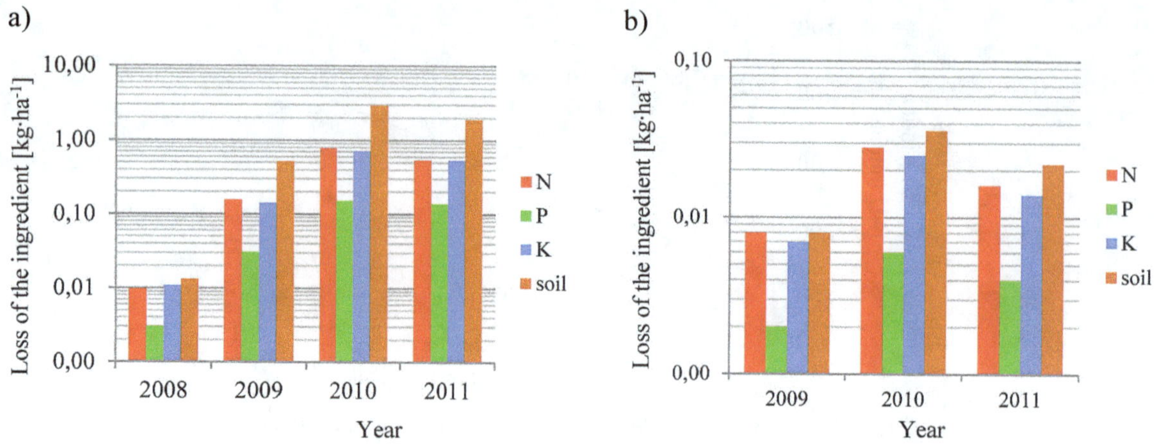

Fig. 2. Losses of N, P, K and soil as a result of surface (a) and subsurface (b) runoff from the slope

sium 1.311-1.621 mg·dm^{-3} K. It is worth noting that only in the case of ammonium nitrogen and phosphorus, the maximum concentration exceeded the limit values for the I class of water purity established in the Regulation of the Minister of Environment of November 9, 2011 [Reg... 2011]. The maximum concentrations of the remaining water quality indicators tested were at a low level and did not exceed the limit values of indicators for the first class of water purity.

The average annual concentration of the tested water quality index was varied in individual years of research, both in the surface (Figure 3a) and in subsurface water runoff (Figure 3b). In the surface outflow, the difference between the minimum, maximum and average annual concentrations of the tested indicator was at the level of 14 to 308%. The least diverse (14-22%) was the average annual concentration of N-NO_2, P and K. For N-Nog and N-NH_4, the difference was 40 and 41%, respectively. However, for N-NO_3 it was 77%, and for soil suspension as much as 308%. A smaller variation in average annual concentrations occurred in the subsurface outflow waters (12-40%). For N-Nog, K and N-NO_3 the difference was from 12 to 14%. For the soil suspension, N-NH_4 and N-NO_2 it amounted to 19, 23, 25%, respectively. However, for P it was 40%.

Together with water during surface runoff of waters, in a form dissolved from the forested slope, in the following year flowed: from 0.010 to 0.780 kg·ha^{-1} N-Nog, 0.003-0.151 kg·ha^{-1} P, 0.011-0.117 kg·ha^{-1} K (Figure 2a). During the subsurface outflow of water, in the dissolved form, it flowed annually: from 0.008 to 0.028 kg·ha^{-1} N-Nog, 0.002-0.006 kg·ha^{-1} P, 0.007-0.025 kg·ha^{-1} K (Figure 2b). These are low losses in relation to N, P, K losses occurring on agriculturally used lands. Kim et al. [2018] determined that the nitrogen losses range from 5.63-13.97 kg·ha^{-1}, and the phosphorus losses from 0.96-11.00 kg·ha^{-1}. However, according to Mazur's research [2018], the annual nitrogen losses range from 5.762 to 23.448 kg·ha^{-1}, phosphorus from 0.696 to 4.644 kg·ha^{-1}, and potassium from 5.937 to 19.216 kg·ha^{-1}.

The calculated low values of Pearson's correlation coefficients enable to conclude that the concentration of the tested chemical indicator of water quality cannot be inferred from the parameters of erosive rainfall (its height, kinetic energy and EI_{30} index) or the size of water outflow. The analysis of the dependence, at the significance level of $\alpha = 0.05$, did not show any statistically significant correlations between the analyzed factors (Table 5), both for surface and subsurface flowing waters – correlations from faint to weak [Stanisz 1998]. The statistically significant dependencies were obtained between the precipitation parameters as well as the surface and subsurface outflows (correlations from high to almost certain [Stanisz 1998]). Additionally, the soil suspension concentrations, as well as soil losses, N, P and K were in most cases significantly statistically correlated with precipitation erosive rainfall and the size of the water outflow (correlations from high to almost certain [Stanisz 1998]).

CONCLUSIONS

The research carried out on the forested loess slope, on the outflow of surface and subsurface water and selected components of matter, prove that this is an area threatened by erosion, and the flowing waters are laden with chemical and me-

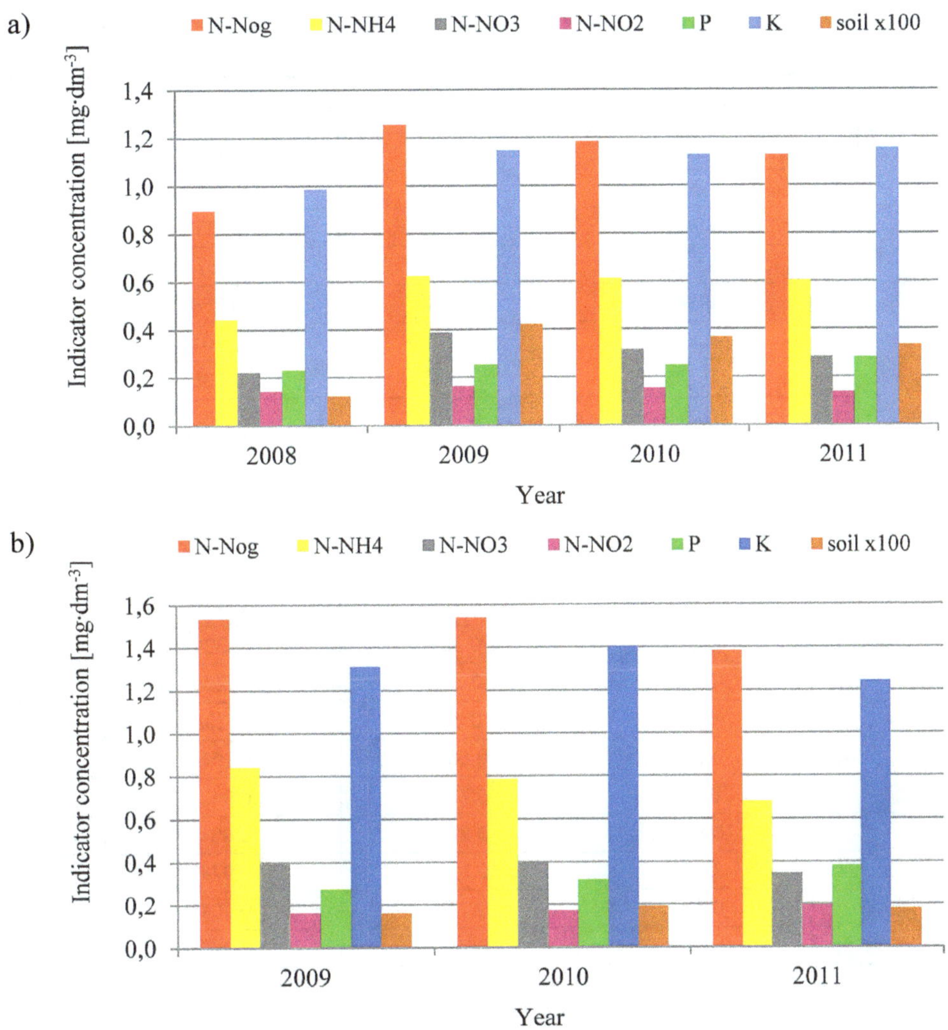

Fig. 3. Average annual concentration of tested indicators of water quality in surface (a) and subsurface (b) runoff from forested slope

chanical pollution. This is the result of the denuding processes. However, the losses of N, P, K and soil are low in relation to the losses recorded on arable lands. The following conclusions can be drawn on the basis of the research:

The surface runoffs from the slopes were at a low level of 0.11 to 2.84 mm, representing 0.3 to 4.5% of the precipitation causing the erosive event. The subsurface outflows occurred only from the layer up to 0.25 m and ranged from 0.03 to 0.08 mm, which accounted for 0.07 and 0.13%, of erosive precipitation, respectively.

The maximum concentrations of ammonium nitrogen and phosphorus in the waters flowing from the slope exceeded the limit values for the class I water purity and were: 0.943 mg·dm^{-3} N-NH$_4$, 0.423 mg·dm^{-3} P, respectively. In contrast, the maximum concentrations of the other quality indicators tested water (total nitrogen, ammonium and nitrite, potassium and slurry) were at a low level and did not exceed the limit values for class I water purity.

The maximum annual losses of plant nutrients and soil were: 0.808 kg·ha^{-1} N, 0.157 kg·ha^{-1} P, 0.142 kg·ha^{-1} K and 2.989 kg·ha^{-1} soil.

There were no statistically significant relationships between the concentration of the tested chemical indicator of water quality and erosive precipitation and its parameters. water outflow, a concentration of soil suspension, soil losses and N, P, K were significantly correlated with erosive precipitation and its parameters. High correlation was also obtained between water outflow and soil suspension concentration as well as soil losses, N, P and K .

Afforestation of the loess slopes threatened by erosion is a treatment that effectively protects the soil against water erosion, as well as surface waters against eutrophication, due to the low level of leaching of biogenic elements.

Table 5. Correlation between the quality indicators of outflowing waters and losses of constituents of matter, precipitation parameters and outflow

Indicator		Rainfall [mm]	Kinetic energy [J·h⁻¹]	Index EI_{30} [MJ·mm·ha⁻¹·h⁻¹]	Outflow [mm]
Outflow [mm]	a	0.898*	0.875*	0.948*	1
	b	0.678*	0.812*	0.951*	1
Soil suspension [g·dm⁻³]	a	0.878*	0.913*	0.983*	0.972*
	b	0.535	0.739*	0.967*	0.981*
N-Nog [mg·dm⁻³]	a	-0.148	-0.195	-0.254	-0.142
	b	-0.194	-0.243	-0.301	-0.199
$N\text{-}NH_4$ [mg·dm⁻³]	a	-0.189	-0.252	-0.235	0.195
	b	-0.275	-0.297	-0.352	-0.295
$N\text{-}NO_3$ [mg·dm⁻³]	a	-0.254	-0.289	-0.292	-0.164
	b	-0.312	-0.398	-0.324	-0.253
$N\text{-}NO_2$ [mg·dm⁻³]	a	-0.325	-0.381	-0.415	-0.299
	b	-0.392	-0.465	-0.451	-0.354
P [mg·dm⁻³]	a	-0.256	-0.251	-0.245	-0.189
	b	-0.323	-0.359	-0.321	-0.256
K [mg·dm⁻³]	a	-0.199	-0.342	-0.235	-0.195
	b	-0.232	-0.389	-0.298	-0.248
Soil [kg·ha⁻¹]	a	0.835*	0.861*	0.955*	0.983*
	b	0.566	0.731*	0.942*	0.991*
N-Nog [kg·ha⁻¹]	a	0.864*	0.875*	0.964*	0.993*
	b	0594	0.813*	0.974*	0.992*
P [kg·ha⁻¹]	a	0.803*	0.788*	0.907*	0.956*
	b	0.665	0.743*	0.790*	0.927*
K [kg·ha⁻¹]	a	0.841*	0.845*	0.945*	0.990*
	b	0.599	0.790*	0.954*	0.987*

Comments:
* statistically significant.
a – surface runoff, b – subsurface runoff.

Acknowledgements

Publication supported by the Polish Ministry of Science and Higher Education as a part of the program of activities disseminating science from the project „Organization of the First International Science Conference – Ecological and Environmental Engineering", 26-29 June 2018, Kraków.

REFERENCES

1. Arriaga F.J., Lowery B. 2003. Corn production on an eroded soil: effects of total rainfall and soil water storage. Soil and Till. Res., 71, 87–93.
2. Asiedu J.K. 2018. Assessing the Threat of Erosion to Nature-Based Interventions for Stormwater Management and Flood Control in the Greater Accra Metropolitan Area, Ghana. Journal of Ecological Engineering, 19(1), 1–13.
3. Catt J.A. 2001. The agricultural importance of loess. Earth-Sci. Reviews, 54, 213–229.
4. Duan X., Xie Y., Ou T., Lu H. 2011. Effects of soil erosion on long-term soil productivity in the black soil region of northeastern China. Catena, 87, 268–275.
5. Dupas R., Delmas M., Dorioz J.M., Garnier J., Moatar F., Gascuel-Odoux C. 2015. Assessing the impact of agricultural pressures on N and P loads and eutrophication risk. Ecol. Indic., 48, 396–407.
6. Gabryszewska M., Kucharczak K., Maciaszek D., Gworek B., Kijeńska M., Tokarz L. 2016. The computational models for determining predicted substance concentrations in surface waters. Przem. Chem., 95/3, 609–612 (in Polish).
7. Grzywna A., Tarkowska-Kukuryk M., Bochniak A., Marczuk A., Jóźwiakowski K., Marzec M., Mazur A., Obroślak R., Nieścioruk K., Zarajczyk J. 2015. Application of chemical and biological indicators for assessment of an ecological potential

of artificial watercourses. Przem. Chem., 94/11, 1954–1957 (in Polish).

8. Hladký J., Novotná J., Elbl J., Kynický J., Juřička D., Novotná J., Brtnický M., 2016. Impacts of Water Erosion on Soil Physical Properties Acta Univ. Agric. Silvic. Mendelianae Brun., 64, 1523–1527.
9. Ijaz A., Khan F., Bhatti A.U. 2006. Some physico-chemical properties of soil as influenced by surface erosion under different cropping systems on upland-sloping soil. Soil and Environ., 25(1), 28–34.
10. Józefaciuk A., Józefaciuk Cz. 1999. Mechanism and methodical guidelines for the study of erosion processes. PIOŚ, Bibl. Monit. Środ. (in Polish).
11. Kondracki J. 2000. Regional geography of Poland. Wyd. Nauk. PWN (in Polish).
12. Kim K., Kim B., Eum J., Seo B., Christopher L., Shope C.L., Peiffer S. 2018. Impacts of land use change and summer monsoon on nutrients and sediment exports from an agricultural catchment. Water, 10, 544, doi:10.3390/w10050544.
13. Lal R. 2005. Soil erosion and carbon dynamics. Soil Till. Res., 81,137–142.
14. Lobo D., Lozano Z., Delgado F. 2005. Water erosion risk assessment and impact on productivity of a Venezuelan soil. Catena, 64, 297–306.
15. Mazur A. 2018. Quantity and Quality of Surface and Subsurface Runoff from an Eroded Loess Slope Used for Agricultural Purposes. Water, 10, 1132, doi:10.3390/w10091132.
16. Mucha J. 1994. Geostatistical methods in documenting deposits. Skrypt Katedry Geologii Kopalnianej AGH (in Polish).
17. Olson K.R. 2007. Soil organic carbon storage in southern Illinois woodland and cropland. Soil Sci., 172, 623–630.
18. Papiernik S.K., Lindstrom M.J., Schumacher J.A., Farenhorst A., Stephens K.D., Schumacher T.E., Lobb D.A. 2005. Variation in soil properties and crop yield across an eroded prairie landscape. J. Soil Water Conserv., 60, 388–395.
19. Regulation of the Minister of Environment of November 9, 2011 on the method of classification of uniform state surface water bodies and environmental quality standards for priority substances. Dz. U. 2011. nr 257 poz. 1545 (in Polish).
20. Rejman J. 2006. Effect of water and tillage erosion on transformation of soils and loess slopes. Acta Agroph., 136 (in Polish).
21. Robichaud P. R., Wagenbrenner J. W., Brown R. E. 2010. Rill erosion in natural and disturbed forests: 1. Measurements. Water Resources Research, 46, W10506.
22. Stanisz A. 1998. An affordable statistics course. Vol. 1. StatSoft Polska Sp. z o.o. Kraków (in Polish).
23. Święchowicz J. 2012. Rainfall thresholds for erosion processes in agricultural catchments. Wyd. IGiGP UJ, Kraków (in Polish).
24. Tyszka J. 2008. Hydrological functions of forests in small lowland catchments. Prace IBL Rozprawy i Monografie, 10 (in Polish).
25. Żmuda R. 2006. Fluvial transport system functioning in small catchment threatened by soil water erosion. Zesz. Nauk. AR we Wrocławiu, 544 (in Polish).

7

Monitoring of Road Noise in the Urban Environment of Lublin

Agnieszka Malec[1*], Gabriel Borowski[1]

[1] Department of Environmental Engineering, Lublin University of Technology, ul. Nadbystrzycka 40B, 20-618 Lublin, Poland

* Corresponding author's e-mail: aga_malec@interia.pl

ABSTRACT

In this article we presented the issues concerning the measurement of road noise, measurement results, and the discussion pertaining to the methods of limiting the excessive noise in the city of Lublin. Mobile systems, equipped with software which enables editing the measurement data, analysis and presentation of the data in the form of an acoustic map, are used to measure the noise level. Exemplary streets of the city are shown, where the road noise level has been significantly exceeded, and the effects of acoustic screens installation is presented. Moreover, the preventive and educational actions which are carried out in Lublin, meant to popularise alternative methods of transport, are discussed. It was concluded that an effective reduction of the traffic noise in the city can be achieved already at the stage of designing housing estates, along with the planning of access roads and noise barriers.

Keywords: road noise, means of transport, urban environment

INTRODUCTION

The noise originating from traffic is a factor which has a negative impact on the human health and on the natural environment. The noise levels on roads are most often between 55 dB and 65 dB. It was found that a prolonged exposure to the noise of such intensity is detrimental to the human health and results in stress reactions [Okokon et al. 2015]. The studies on road noise carried out in Stockholm have shown its influence on increasingly greater sleeping problems of its citizens. Sleeping disorders were much more often found in the persons living in the apartments with windows facing the street, as opposed to the persons whose bedroom windows were facing the "calm" side [Bluhm et al. 2004]. Apart from insomnia, the effects of long-term exposure to noise among people are: irritation, decreased efficiency, aggressive behaviour and an increased risk of a heart attack [Moudon 2009]. It was also concluded that people do not grow accustomed to the constant noise which can cause such ailments as asthma, headaches, depression and increased blood pressure [Ouis 2001].

The noise coming from the means of transport has been a major problem since the 1980s – more than a half of the respondents living in American cities had problems with the excessive noise, while 80% of them all thought that its main source is traffic or, in a broader sense, transport (road vehicles, rail vehicles, planes, water vehicles) [Brown and Lee 1987]. In Beijing, the main roads are congested with traffic during the day, and the noise level along these roads exceeds the appropriate environment protection standards by 5 dB. The unbalanced development of the infrastructure caused the diversification of traffic in this city. In the northern part, the traffic intensity is greater and the road noise level is higher than in the southern part of Beijing [Bengang Li et al. 2002]. The results of the tests concerning the exposure to noise of the citizens of Tainana (Taiwan) showed that more than 90% of the population is exposed to onerous acoustic conditions. In particular, the level of noise in commercial areas on summer evenings was 23 dB higher than the permitted level [Tsai et al. 2009]. Similar results of road traffic noise exposure can be found in Karachi (Pakistan). It was concluded that noise level in the morning and in the evening was higher due to the citizens commuting to and from work. The average noise level was over 66 dB and the maximum level was more than 101 dB [Mehdi et al. 2011].

The road noise also influences the habitats of many animals, their walking and communication trails – distorting, i.a., the possibility of birds' communicating [Polak 2014].

It was concluded that a proper design plan of a modern city, created by urban planners, taking into account the specificity of the acoustic environment, has an effect on the decrease of the noise level or even its elimination [Raimbault and Dubois 2005]. Currently, the urban planners are able to foresee the influence of noise in the city already at the stage of designing a street or an intersection, using special, prognostic software for acoustic optimisation [Quartieri et al. 2015].

In Poland, in ca. 20% of living areas, the level of noise is greater than 60 dB. Overall, ca. 33% of the country's population is at risk of being exposed to excessive noise caused, above all, by road traffic [Łukasik et al. 2013]. The permitted levels of environmental noise in Poland are shown in Table 1 [Announcement… 2014].

For the past few years, the municipal authorities of Lublin have been trying to resolve the problem of traffic noise, among others by developing and implementing the programme for the protection of the environment against noise [Environmental protection program…, 2009]. This programme has defined the direction and scope of the actions necessary to restore the permitted environmental levels of noise created due to the use of roads, railway lines and industrial plants.

The purpose of this paper is to discuss the system of road traffic noise measurement, present the results of these measurements and indicate the methods of limiting the excessive noise in Lublin.

ROAD TRAFFIC NOISE MEASUREMENTS

The road traffic noise measurements constitute an important piece of information on the condition of the environment and are useful in the process of creating acoustic maps as part of the strategic plans for environmental protection. The effects of noise monitoring show the direction of implementing remedial actions. First and foremost, they allow to be indication of places where the noise level is significantly exceeded, enabling to introduce, i.a., noise barriers, in order to improve the comfort of the city's population [Boczkowski 2016].

The Voivodship Inspectorate for Environmental Protection is responsible for the environmental noise measurements in Lublin. The tests of road traffic, railway and industrial noise in Lublin were conducted in 2012. They should have been repeated in 2017, but were abandoned due to a number of new road investments in the city.

The measurements are carried out using the referential method, in accordance with the requirements of the Regulation of the Minister for the Environment of 16th June 2011 [Regulation… 2011]. This method is used for measuring the value of noise levels introduced into the environment expressed with L_{DWN} (for the day-evening-night time) and L_N (for the night time) indices, where:

- L_{DWN} – long-term, average A sound level expressed in decibels (dB), determined during all days of the year, taking into account the time of day (from 6 a.m. to 6 p.m.), time of the evening (from 6 p.m. to 10 p.m.) and the time of night (from 10 p.m. to 6 a.m.)

Table 1. The permitted levels of environmental noise in Poland [Announcement… 2014]

Item No.	Area type	Permitted noise level in [dB]			
		Roads and railroads		Other structures and activities being the source of noise	
		during the day (16 h)	at night (8 h)	during the day (8 h)	at night (1 h)
1	a) Resort protection zone b) Hospital areas outside the city	50	45	45	40
2	a) Single-family residential areas b) Residential areas connected with constant or temporary stay of children and youth c) Areas of public nursing homes d) Areas of hospitals in the city	61	56	50	40
3	a) Multi-family and collective residential areas b) Smallholding areas c) Recreational and leisure areas d) Residential and service areas	65	56	55	45
4	Downtown areas of cities with population above 100,000	68	60	55	45

- L_N – long-term, average A sound level expressed in decibels (dB), determined during all times of the night in the year (from 10 p.m. to 6 a.m.).

An example of a measurement device is the AZ8921 sonometer with the sound intensity measurement range from 30 to 130 dB, work frequency from 0.0315 to 8 kHz, resolution of sound intensity measurement of 0.1 dB, and the precision of measurement of 1dB. The measurements are done in designated measurement points, in day- and night-time.

Mobile measurement systems, which are fully automated and consist of a transport vehicle and measurement station, were used to measure the noise intensity in Lublin. The systems are also equipped with the software responsible for the control of the station, collecting measurement data, analysis and presentation of the measurement results, and for creating acoustic maps using specialised software. The acoustic map is used in planning the remedial actions which make the adjustment of the noise level to the required environmental quality standards possible.

The acoustic map of Lublin was elaborated with the calculation and measurement methods, using specialised LimA software by Bruel & Kjaer, with the initial data in the form of a 3D model of the structures, a numeric terrain model, and a satellite orthophotomap. The model obtained in this way was then calibrated on the basis of the acoustic data which came from the measurements of the sound level determined using the noise indices measured in 22 measurement points relating to the vehicle traffic noise emission, 5 points relating to rail traffic, as well as the noise emitted by 12 industrial plants in the city. Two maps with day-evening-night time (L_{DWN}) and long-term night (L_N) indices were elaborated for the noise coming from all the afore-mentioned sources, i.e. vehicle, rail and industrial.

The utilised mobile noise measurement systems, along with the software, were purchased by the Chief Inspectorate for Environmental Protection and co-financed by the European Union funds under the Infrastructure and Environment Operational Programme [www.wios.lublin.pl].

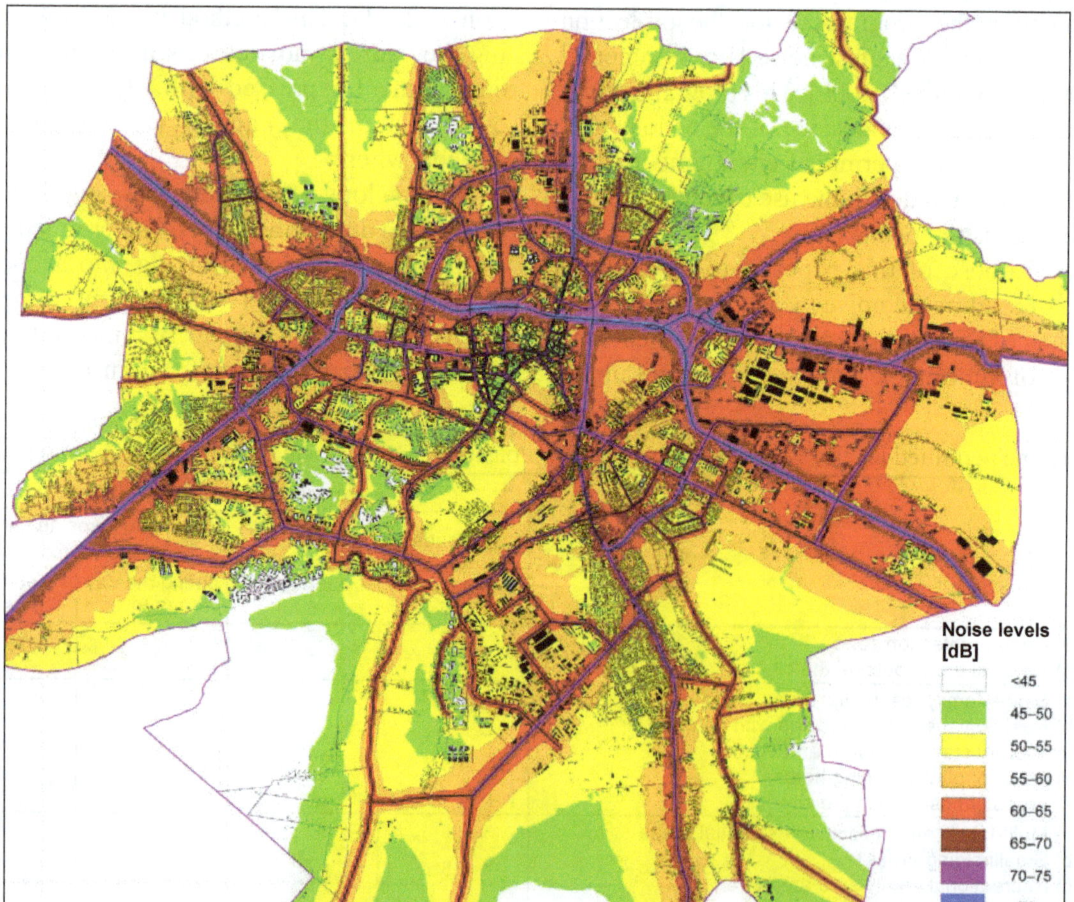

Fig. 1. Day-time acoustic map of Lublin [Acoustic map..., 2012]

NOISE MEASUREMENT RESULTS AND DISCUSSION

The results of the noise measurement in Lublin show that its greatest emitter is the road transport during the day (Fig. 1), rail transport and industrial plants [Acoustic map…, 2012].

The measurements of road traffic noise showed that the permitted standards were exceeded in some measurement points in Lublin. For example, at ul. Krańcowa, the standards were exceeded by 3.8 dB during the day and by 6.9 at night. The highest instantaneous values of the noise emitted to the environment reached up to 90 dB.

In accordance with the environmental protection programme, the *M* index was determined which is the function of the number of citizens and the noise level excessiveness degree [Environmental protection program…, 2009]. This index pertains to the residential housing areas and has the value of zero, when the permitted noise levels are exceeded for the residential structure in question, located in the city. The greater the *M* index, the greater the need to implement the noise prevention actions. The acoustic map of the city contains the information on the *M* index value for each of the residential buildings. On the basis of this information, two zones were differentiated in Lublin, where the noise coming from the identified sources exceeded the permitted level. The 27 zones characterised by excessive impact of road traffic and 5 zones encompassing the impact of the railway transport were distinguished (Fig. 3).

On the basis of the calculated values of the *M* index, it was determined that the most urgent need is the noise protection of the entire city centre (M = 5124) and residential buildings at Al. Sikorskiego (M= 1319), ul. Szeligowskiego (M = 1258) and Al. Racławickie (M = 1003). In the city centre of Lublin, the permitted noise levels were found to be exceeded by up to the value of 12 dB, while the proposed remedial actions include the installation of wood joincry with high acoustic insulation (> 40 dB). At Al. Sikorskiego, the permitted noise level was found to be exceeded by up to 8 dB (Fig. 4a), thus it is proposed to install 700 metres long acoustic screens (Fig. 4b). On ul. Szeligowskiego, the permitted noise level was exceeded by up to 11

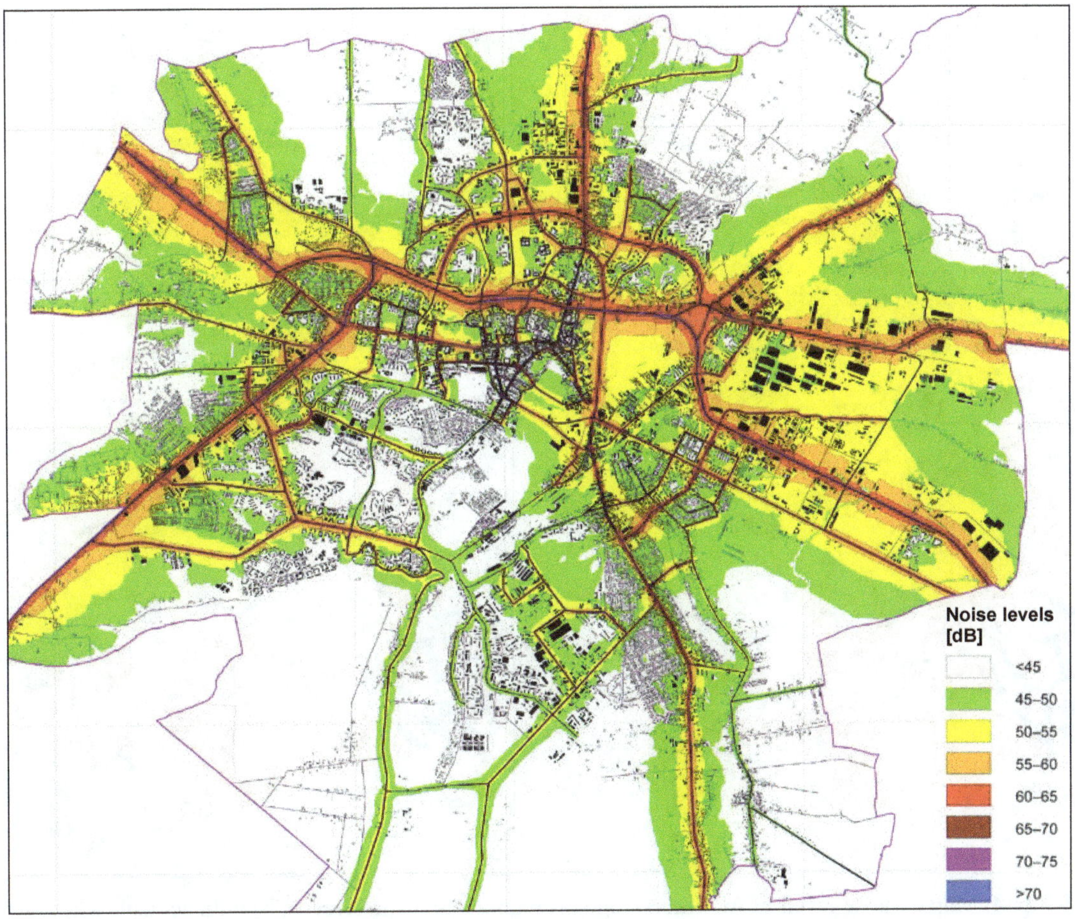

Fig. 2. Night-time acoustic map of Lublin [Acoustic map..., 2012]

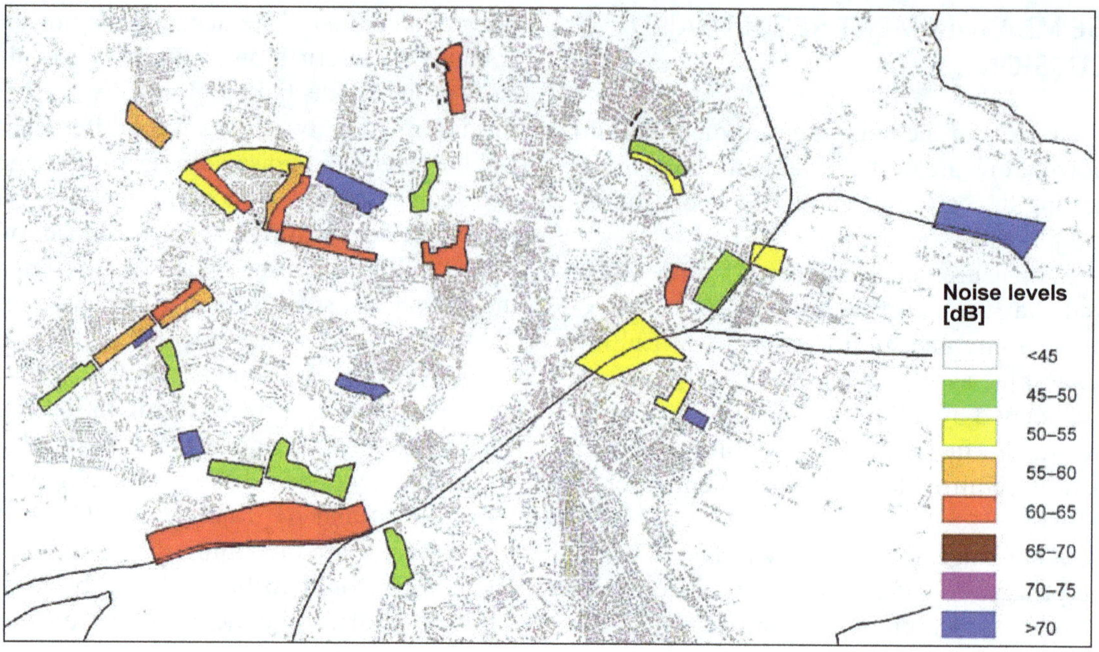

Fig. 3. Residential housing areas in Lublin exposed to road traffic and railway noise to the greatest degree [Environmental protection program…, 2009]

a) b)

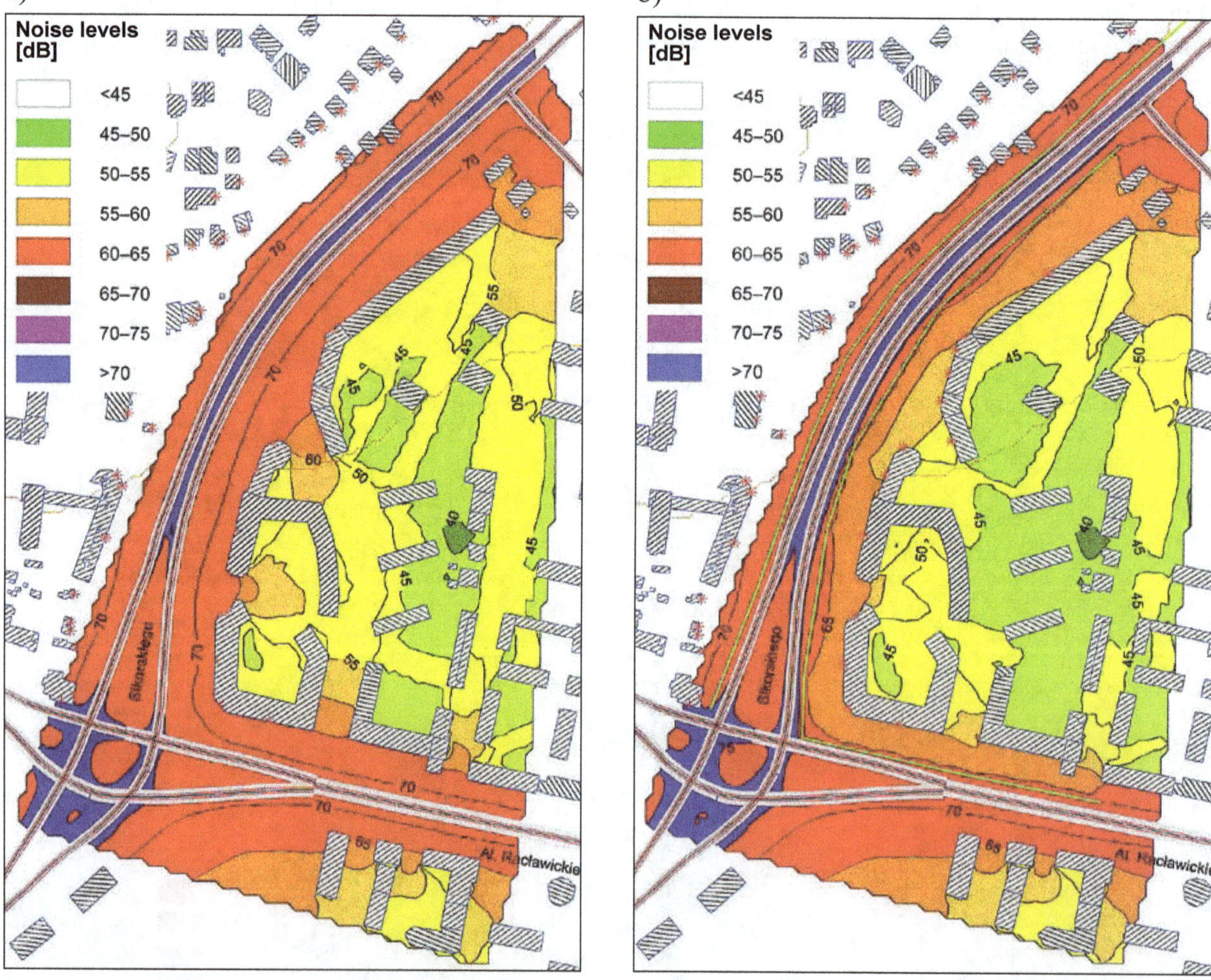

Fig. 4. Map of noise levels at Al. Sikorskiego: a) without acoustic screens, b) with acoustic screens [Environmental protection program…, 2009]

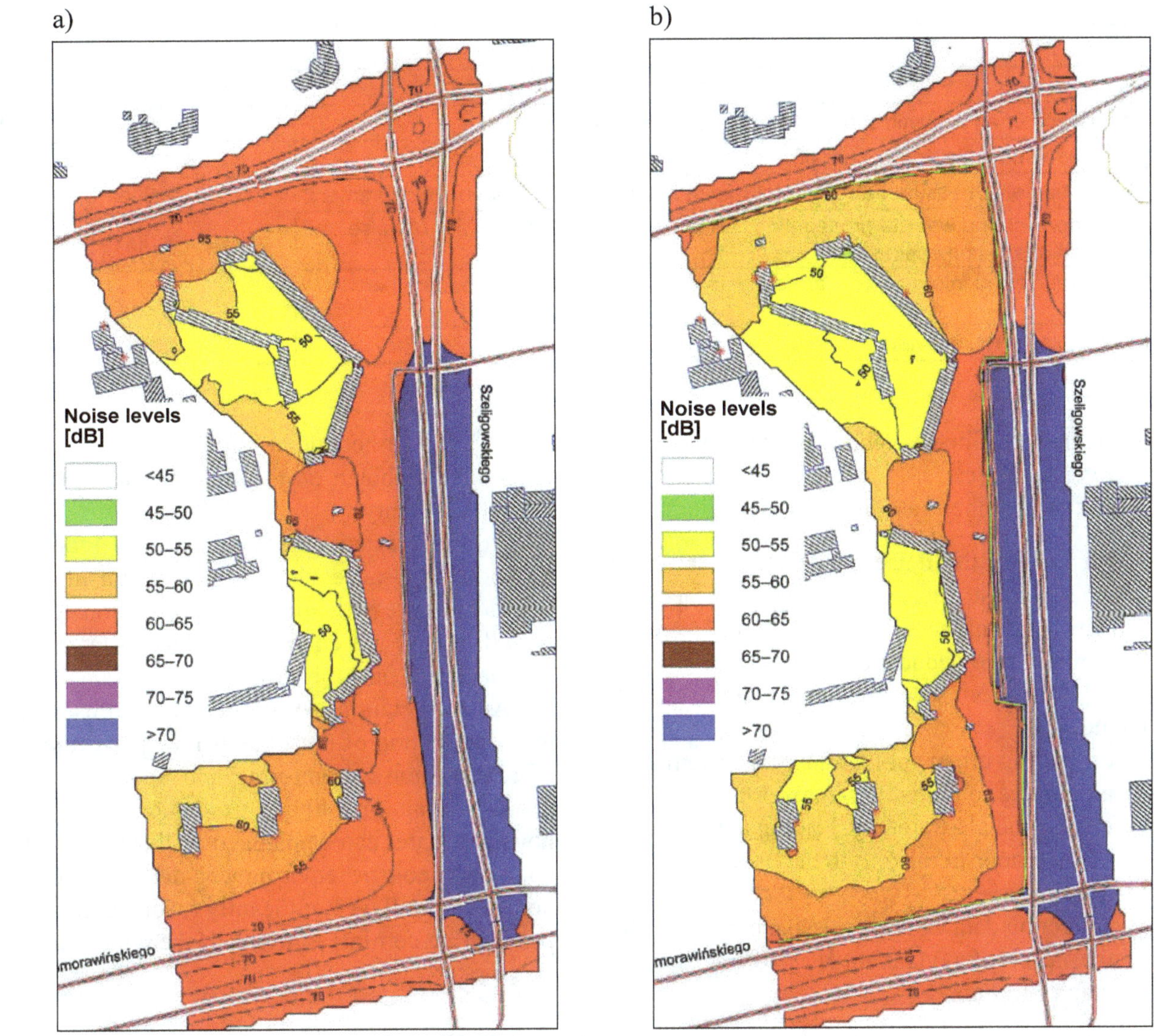

Fig. 5. Map of noise levels on ul. Szeligowskiego: a) without acoustic screens, b) with acoustic screens [Environmental protection program…, 2009]

dB (Fig. 5a), and its reduction will take place after the installation of 1130 metres long acoustic screens (Fig. 5b). At Al. Racławickie, the permitted noise level was exceeded by 7 dB and in order to reduce it is necessary to replace the wood joinery due to the lack of possibility to install acoustic screens.

The M index values indicating the risk of exposure to road traffic noise in the remaining residential housing zones in Lublin were much lower, compared to the above-mentioned examples. The second significant source of noise in Lublin is the presence of a functioning railway, especially on the Warsaw-Lublin line. The calculated M index value was the highest on ul. Wyżynna – M = 1552. The installation of screens with a sound-absorption surface is proposed in the first line of the multi-family residential buildings.

NOISE PREVENTION METHODS AND ACTIONS

The reduction of noise in the city is possible in two ways: by installing the noise protection means and by taking the actions involving proper management of the traffic, its structure or the speed of vehicles. Good results can be achieved by taking into account the proper location of road investments in relation to the residential buildings during the design stage. The most common solutions reducing the noise in urban agglomerations are [Boczkowski 2016]:

- installation of road acoustic screens,
- use of silent surfaces,
- relocation of the traffic from the city to the bypass routes,
- reduction of speed of motor vehicles,

Table 2. Actions for the reduction of noise in Lublin, and the estimated costs [Environmental protection program…, 2009]

Task	Gross cost, PLN/pc
Installation of an acoustic screen (average price level for 1 m^2)	1 000
Purchase and installation of a traffic enforcement camera	200 000
Installation of a measurement point for the 24 h industrial noise level measurement	2 500
Development of a website propagating content connected with the risk of noise and the methods of combatting it	8 000
Installation of an information board cooperating with sound level measurement devices	20 000
Wood joinery replacement	2 500

- reduction of heavy vehicle participation in the overall traffic,
- use of greenbelts around the roads.

In Lublin, the noise protection installations which are most often invested into are acoustic screens, embankments and soundproof screens. Preventive and educational actions are proposed, apart from specific remedial actions. They include:

- expanding urban public transport system as an alternative to individual transport;
- raising social awareness, leading to the change of behaviour in relation to the use of means of transport;
- collecting data on the excessive noise level for the needs of regular updates of the acoustic map of Lublin.

The calculated costs of implementing specific actions for the reduction of noise in the city are shown in Table 2.

An important investment, limiting the noise in Lublin, was the commissioning of the city bypass in December 2016. Its task was to divert the transit traffic outside the city centre, which traffic was calculated to generate 20% of the urban traffic, half of which was arriving from and leaving for Warsaw. Favourable conditions were created in Lublin to develop the bicycle traffic (city bike system and new bicycle paths), as an alternative to the automotive transport.

CONCLUSION

To sum up the discussion, it must be concluded that the environmental noise pollution in Lublin is caused mainly by road traffic, and, to a lesser degree, by railway. It was determined that acoustic screens can be installed by the busiest roads, and the beneficial effects of these installations were shown on the example of two busy streets in the city.

Additional actions in the scope of reducing the noise nuisance should include raising the social awareness of the city's population which would lead to the change of habits in favour of alternative means of transport.

An important action to be taken with regard to the reduction of noise is also the performance of regular noise monitoring and updates of the acoustic map of Lublin. Construction of new housing estates can be planned on the basis of the collected acoustic environment data, where the noise level will be significantly reduced, thus improving the comfort of living of the population.

REFERENCES

1. Bengang Li, Shu Tao, Dawson R.W. 2002. Evaluation and analysis of traffic noise from the main urban roads in Beijing. Applied Acoustics, 63, 1137-1142.
2. Bluhm G., Nordling E., Berglind N. 2004. Road traffic noise and annoyance – an increasing environmental health problem. Noise & Health, 6(24), 43-49.
3. Boczkowski A. 2016. Analysis of the possibilities of noise reduction in the urban environment (in Polish). In: R. Knosala (Ed.) Innowacje w zarządzaniu i inżynierii produkcji. Oficyna Wydawnicza Polskiego Towarzystwa Zarządzania Produkcją, 343-352.
4. Brown A.L., Lam K.C., 1987. Urban noise surveys. Applied Acoustics, 20(1), 23–39, doi: https://doi.org/10.1016/0003-682X(87)90081-8.
5. Łukasik Z., Bril J., Bril D. 2013. Risks associated with road transport (in Polish). Autobusy, 3, 45-57.
6. Acoustic map of the city of Lublin (in Polish), 2012. https://lublin.eu/mieszkancy/srodowisko/mapa-akustyczna/ (access: 10.03.2018).

7. Mehdi M.R., Kim M., Seong J.C., Arsalan M.H. 2011. Spatio-temporal patterns of road traffic noise pollution in Karachi, Pakistan. Environment International, 37, 97–104.
8. Moudon A.V., 2009. Real noise from the urban environment. American Journal of Preventive Medicine, 37(2), 167–171, doi: https://doi.org/10.1016/j.amepre.2009.03.019.
9. Announcement of the Minister of the Environment of January 22, 2014 regarding the publication of a uniform text of the ordinance of the Minister of the Environment regarding permissible noise levels in the environment (in Polish) (Dz.U. 2014, poz. 112).
10. Okokon E., Turunen A.W., Ung-Lanki S., Vartiainen A-K., Tiittanen P., Lanki T., 2015. Road-traffic noise: Annoyance, risk perception, and noise sensitivity in the Finnish adult population. International Journal of Environmental Research and Public Health, 12(6), 5712–5734, doi: 10.3390/ijerph120605712.
11. Osmólska A., Łazarz B., Czech P. 2011. Assessment of communication noise danger on a section of a national road No. 94 running through Dąbrowa Górnicza (in Polish). Zeszyty Naukowe Politechniki Śląskiej, Seria: Transport, 72, 75-82.
12. Ouis D. 2001. Annoyance from road traffic noise: A review. Journal of Environmental Psychology, 21, 101–120, doi: 10.1006/jevp.2000.0187.
13. Environmental protection program against noise for the city of Lublin (in Polish), 2009. Załącznik do uchwały nr 594/XXIX/2009 Rady Miasta Lublin z dnia 19 lutego 2009 r.
14. Polak M., 2014. Relationship between traffic noise levels and song perch height in a common passerine bird. Transportation Research Part D: Transport and Environment, 30, 72–75, doi: https://doi.org/10.1016/j.trd.2014.05.004.
15. Quartieri J., Mastorakis N.E., Guarnaccia C., Troisi A., D'Ambrosio S., Iannone G. 2015. Road intersections noise impact on urban environment quality. In: Proc. of the 11th WSEAS International Conference "Recent Advances in Applied and Theoretical Mechanics", 23-25 April 2015, Kuala Lumpur, Malaysia, 162-171.
16. Raimbault M., Dubois D. 2005. Urban soundscapes: Experiences and knowledge. Cities, 22(5), 339–350, doi: 10.1016/j.cities.2005.05.003.
17. Regulation of the Minister of the Environment of 16.06.2011 on the requirements for conducting measurements of levels of substances or energy in the environment by the manager of a road, railway line, tram line, airport or port (in Polish) (Dz.U. 2011 nr 140, poz. 824).
18. Tsai K-T., Lin M-D., Chen Y-H. 2009. Noise mapping in urban environments: A Taiwan study. Applied Acoustics, 70, 964–972.
19. www.wios.lublin.pl – Provincial Inspectorate for Environmental Protection in Lublin (in Polish) (access: 10 March 2018).

8

Poznań Air Pollution Analysis for 2015–2017

Anna Smurzyńska[1*], Wojciech Czekała[1], Piotr Hektus[2], Stanisław Marks[1], Jakub Mazurkiewicz[1], Michał Brzoski[1], Dawid Chełkowski[1], Kamil Kozłowski[1]

[1] Poznań University of Live Sciences, Faculty of Biosystem Engineering, ul. Wojska Polskiego 50, 60-637 Poznań, Poland

[2] Adam Mickiewicz University in Poznań, Faculty of Socio-economic Geography and Spatial Economy, Krygowskiego 10, 61-680 Poznań, Poland

[*] Corresponding author's e-mail: a.smurzynska@gmail.com

ABSTRACT

Air pollution is the result of natural processes and intense urban development. The undesired emission of volatile substances causes environmental threats such as acid rains, aggravated greenhouse effect or the ozone depletion. Moreover, the pollution released into the air is harmful to the human respiratory system, eyes and skin. This paper presents the body of analyses conducted in Poznań between 2015–2017 on the changes in the local emission concentration of PM10 and PM2.5. The data concerning the emission of suspended particulates were provided by the meteorological station on Polanka St. in Poznań. The research included a correlation analysis. The results point to a steady decrease in the amount of produced particulates. It was also noted that the emissions of PM10 and PM2.5 change seasonally, with the highest levels in the autumn and winter. Furthermore, the amount of emitted suspended particulates is correlated with the temperature; hence, it is supposed that the main source of air pollution in Poznań involves low-efficiency heaters and boilers.

Keywords: Poznań, air pollution, PM10, PM2.5

INTRODUCTION

As the industry develops and the urbanization continues, we note the tremendous influence of advanced civilization on the natural environment. Further development results in both technological advancement and the degradation of the natural environment. The main threat is the air pollution, which is a global problem, prevalent especially in the developed and developing countries [Cembrzyńska et al. 2012]. The air is polluted by gas, liquid, and solid substances in the form of suspended particulates (PM10, PM2.5). Table 1 presents the maximum accepted levels, the information threshold and the alarm levels for air pollution [Regulation of the Minister of the Environment ... 2012].

The sources of air pollution are divided into natural and anthropogenic. The natural sources include biological processes, volcanic eruptions, forest fires and lightning discharges. The anthropogenic pollution, on the other hand, results from the natural population growth and from the development of technology and industry. These sources can be divided into the pollution generated by the industry, traffic, energy sector and municipal waste. In the case of the industry, one should especially consider the consequences of the heavy, chemical and food industries [Dreiseidler et al. 2000]. The transportation is also an important branch of the industry affecting the air pollution [Roy et al. 2013]. Large cities are mostly affected by the car exhaust; therefore, the highest pollution is most often observed near the busy streets [McDonald et al. 2013]. The vehicles with diesel engines produce most nitrogen oxide emissions. The energy sector is the next significant source of air pollution, resulting from, among others, the usage of low-efficiency heaters in the households. Coal boilers and burning of waste and dry plant

Table 1. Maximum acceptable levels, the information threshold and the alarm levels for air pollution, considering the protection of humans and plants [μg/m³]

Specification	NO_2	SO_2	O_3	C_6H_6	PM2.5	PM10
Maximum accepted level of the annual average concentration	40	-	-	5	25	40
Maximum accepted level of the daily average concentration	-	125[2]	-	-	-	50[4]
Maximum accepted level of the hourly concentration	200[1]	350[3]	-	-	-	50
Information threshold	-	-	180	-	-	200
Alarm level for the hourly concentration	400	500	240	-	-	-
Alarm level for the daily concentration	-	-	-	-	-	300

waste is another significant source of pollution [Lelek et al. 2016]. Therefore, the air pollution is most frequently observed in large cities and urban areas. The municipal pollution, in turn, includes the emissions resulting from waste management in households and on farms.

The air pollution increases the risk of many diseases, especially the respiratory system diseases, such as lung cancer [Gładka and Zatoński 2016; Sosa et al. 2017]. It causes irritation of the respiratory tract as well as increases the risk of respiratory infections and asthmatic diseases [Kampa and Castanas 2008]. The air pollution may also result in the conjunctiva and skin irritation. Increased heart rate and lower fitness levels are also common adverse effects. The particulates with a diameter over 2.5 micrometers are absorbed in the upper and lower respiratory tracts and they can penetrate into the bloodstream. Similarly to the PM10 particulates, they can cause coughing, breathing difficulties and breathlessness, especially during the physical exercise.

Apart from the direct effect on the human health, another considerable threat related to the air pollution are acid rains, which are caused by nitrogen and sulphur oxides. Moreover, the air pollutants, after they settle on the ground, cause a wide range of negative effects, such as acidification and heavy metals contamination of the soil, waters, and vegetation [Du et al. 2017]. These effects also directly impact the human life and health. The acidic pollution of waters is another example. Low pH levels of the water leads to the release of toxic heavy metals, most of them bioaccumulate in fish tissues. Besides the degradation of flora and fauna, we also observe the damage to the structure of buildings [Maeda et al. 2001].

Smog is particularly dangerous for the human health [Mishra 2017]. It is produced as a result of high levels of air pollution under specific weather conditions [Ma et al. 2012]. The acidic smog (London type) is created in high humidity in the air contaminated mainly by sulphur oxides, carbon monoxide, and coal dust. It is observed in the winter at the temperatures between -3 and -5°C. The pollutants bind with the mist and create drops of sulphuric acid suspended in the air. This type of smog is dangerous to humans and animals, as well as the soil and water. It also causes damage to the structure of buildings. The London smog significantly limits the visibility (to several dozen meters). The photochemical smog (Los Angeles type) is produced in the period of increased emissions of peroxides and ozone, as well as nitrogen oxides, aldehydes, and aromatic hydrocarbons. It is usually observed in summer, at the temperatures over +26°C. In the sunlight, certain air pollutants react photochemically, creating carcinogenic toxic compounds which may also cause genetic mutations, for example, a poisonous gas created in the reaction of nitrogen oxides with hydrocarbons, i.e. peroxyacetyl nitrate (PAN). It is produced mainly in highly urbanized and densely populated areas [Wang et al. 2014]. The concentration of pollutants in the photochemical smog is usually extremely high and it significantly exceeds the accepted levels of air pollution recommended by WHO.

The ozone hole is the next danger linked to the air pollution. The increasing emissions of ozone and chlorofluorocarbons damage the protective ozone layer [Son et al. 2009]. As a result, the UV radiation permeates through the layer and it poses a serious threat to human and animal health, and even to the proper development of plants.

The air pollution aggravates the greenhouse effect, which is reported in the scientific research [Ramanathan and Feng 2009]. The climate change is a global problem affecting the proper functioning of flora and fauna.

In the face of the impending danger, multiple actions are taken to lower the emissions of gas and dust that are harmful to the environment. Biogas plants are one of the options that will not only allow to dispose of the harmful waste, but also produce energy (heating and/or electricity), which is crucial to the functioning of the urban areas [Cieślik et al. 2016; Czekała et al. 2016]. There are also other proposed methods of dangerous waste disposal for both the industrial and agricultural waste that produces harmful emissions [Smurzyńska et al. 2015; Czekała et al. 2017; Dzikuć 2017].

In consideration of the risks to the environment and the local community, linked to the pollutant emissions, an air quality analysis was conducted in Poznań. This is a large agglomeration, so it is at risk of the aforementioned industrial, traffic, energy and municipal air pollution. An analysis of changes in the emissions of PM10 and PM2.5 was conducted, as these particulates demonstrate a substantial risk to the human health. The research included the air quality measurements for the chosen agglomeration from 2015 to 2017.

RESEARCH METHODOLOGY

The data used consisted of the daily emissions of the suspended particulates: PM10 and PM2.5. The analyses included also the differences in the air temperatures. The research involved the measurements from 2015 to 2017. The data were provided by the meteorological station on Polanka St. in Poznań.

The influence of the meteorological elements on the concentration of the pollutants was identified through the analysis of Pearson's correlation (r). The coefficient of variation (V) was used to specify the changes in the concentration of the suspended particulates PM10 and PM2.5.

The research concerned the air pollution produced in Poznań, a city in west-central Poland with the surface of 262 km^2 and a population of 540 thousand [SVS 2017].

RESULTS AND DISCUSSION

Yearly concentration changes of PM10 and PM2.5

In 2015, the daily maximum accepted level of the suspended particulate PM10 was exceeded 54 times. The median for this year was 27.5 μg/m^3 (see Figure 1); 50% of all the cases fit in the concentration range between 18–41.5 μg/m^3. The lowest recorded value was 7 μg/m^3, whereas the highest was 168 μg/m^3. The values for the year 2016 were similar to the previous year. The daily median of the concentration of the particulate PM10 was 29 μg/m^3. The range for 50% of all

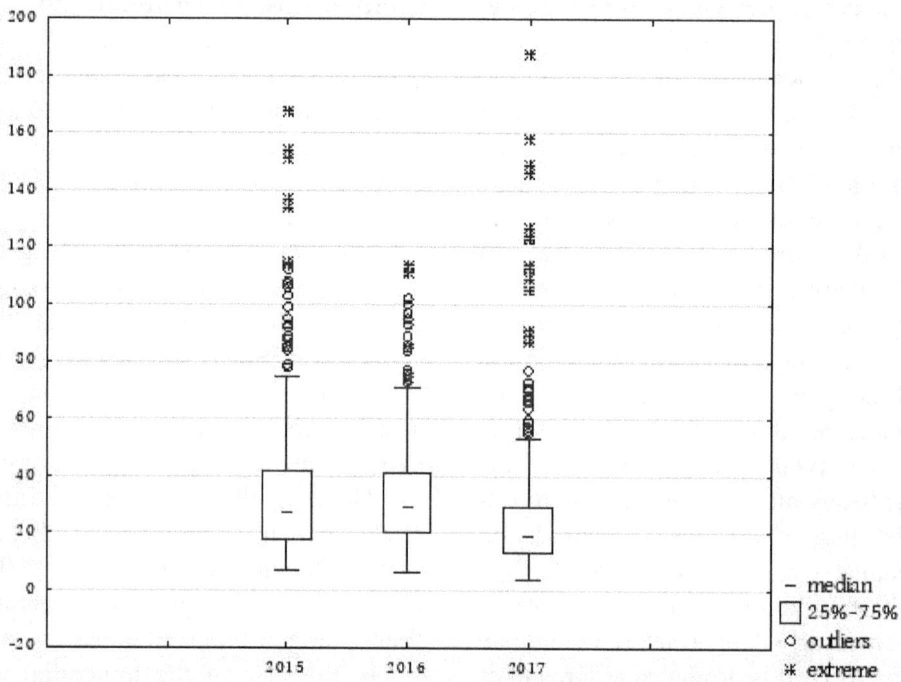

Figure 1. The concentration of the suspended particulate PM10 in the years 2015–2017 [μg/m^3] – daily measurements

(based on the data from the meteorological station on Polanka St. in Poznań) the cases was 20–41 μg/m³. In 2016, less extreme values were recorded, with the highest daily concentration of 114 μg/m³, while the norm was exceeded 49 times. In 2017, a decrease of the daily concentration was noted for PM10 in comparison to the 2015–2016 period. The median for 2017 was 19 μg/m³. The value of the third quartile was close to the median for the previous years and it was 29 μg/m³. In 2017, the accepted norm was exceeded 35 times.

The changes in the daily concentrations of the suspended particulate PM2.5 for the years 2015–2017 are shown in Figure 2. It has been noted that in the analyzed years, the yearly concentration of PM2.5 did not show significant changes. The median for the analyzed years was respectively 14, 16, 15 μg/m³. The values for the first quartile were identical – 10 μg/m³. On this basis, it is concluded that the lower limit of the daily concentration of the particulate is relatively constant. The differences in the concentration of PM2.5 for the years 2015–2017 concern the third quartile and the outliers. The conclusion is that every year, there are days when the concentration value of the particulate PM2.5 is frequently exceeded, but the numbers of the outliers for the analysed years differ.

The sources of suspended particulate emissions are not only the volcano eruptions or forest fires, but in case of the Poznań agglomeration, it mainly involves the production processes. Poznań is an important centre of the car, food, pharmaceutical, chemical and machine industry. The leading producers on the national level are: Volkswagen Poznań Sp. z o.o. [Ltd.], Bridgestone Poland Sp. z o.o. [Ltd.], Exide Technologies SA [JSC], Kompania Piwowarska SA [JSC], Unilever Polska Sp. z o.o. [Ltd.], Wrigley Poland Sp. z o. o. [Ltd.], Wyborowa SA [JSC], Jutrzenka Colian Sp. z o.o. [Ltd.] facility in Poznań, Uniq Lisner Sp. z o.o. [Ltd.], Terravita Sp. z o.o. [Ltd.], Poznańska Palarnia Kawy Astra® Sp. z o.o. [Ltd.], Spółdzielnia Piekarsko-Ciastkarska "Fawor", GlaxoSmithKline Pharmaceuticals SA [JSC], Nivea Polska Sp. z o. o. [Ltd.], Przedsiębiorstwo Farmaceutyczne "Ziołolek" Sp. z o. o. [Ltd.], Poznańskie Zakłady Farmaceutyczne "Herbapol" SA [JSC], Grupa Kapitałowa H. Cegielski – Poznań SA [JSC], SKF Poznań SA [JSC]. The greatest amounts of the suspended particulates are produced by the industries such as energy, chemical, mining, metal producing and construction, as confirmed by the research on the coal energy industry by Wang et al. [2018]. In the case of Poznań, the air pollution with PM10 and PM2.5 is certainly heavily influenced by the energy production and the heating sector, as many installations require considerable investments to improve their condition [Mayer 1999].

The collected data suggest a gradual decrease in the emissions of PM10 and PM2.5 to the atmosphere, especially in 2017. It may be the result of the actions of the local authorities who promoted the idea of replacing private transport with the

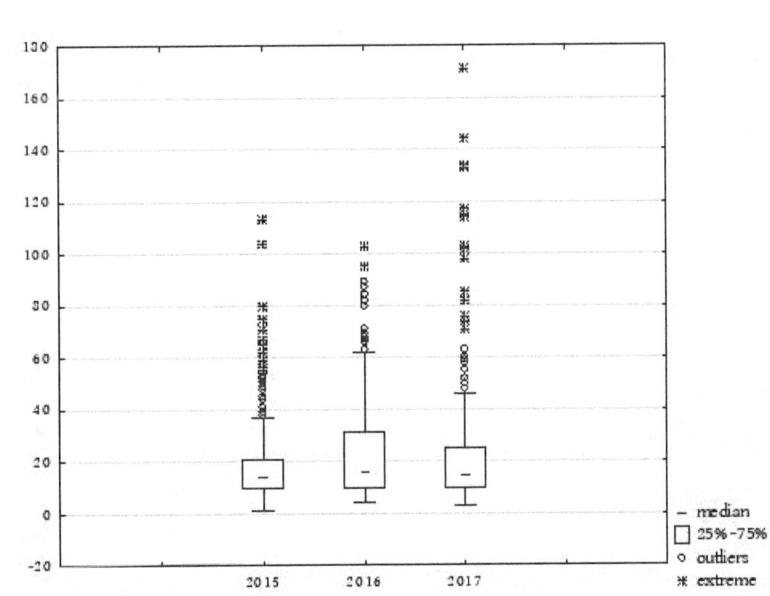

Figure 2. The concentration of the suspended particulate PM2.5 in the years 2015–2017 [μg/m³] – daily measurements

(based on data from the meteorological station on Polanka St. in Poznań) public transport and introduced a number of improvements (bus lanes) that made the public transport a more attractive means of transportation [Fajczak-Kowalska et al. 2017]. Free tram passes were offered to the inhabitants during the periods of heavy air pollution. Moreover, Poznań offers a well-developed train and tram network. A widespread use of the public transport can efficiently lower the air pollution with PM10 and PM2.5. On the other hand, the air and car transportation cause the PM10 and PM2.5 emissions [Penn et al. 2017]. In the researched urban area, there is an international airport, Poznań-Ławica, which generates massive emissions of air pollutants. According to the statistics for Poznań, the number of passenger cars is growing. In 2014, there were 601 vehicles per 1000 inhabitants, while in 2016, the number grew to 660 vehicles [SVS 2017]. Therefore, there are a number of proposed investments in the traffic infrastructure which may lower the air pollution in the long term. Some of them were previously implemented in China [Sun et al. 2018].

Within the KAWKA program, Poznań introduced subsidies for the replacement of the old heaters and coal boilers with the system heat, gas or electric heaters. It allowed to modernize many old, low-efficiency installations. Lower energy usage related to heating is also possible through thermo-modernisation of the public buildings, tenement houses and private houses.

Another significant factor which contributes to the regular decrease of the suspended particulates emissions is raising the public awareness of the issue [Environmental Protection Plan... 2013]. The quality of emissions has been raised through numerous educational programs, as well as promoting proper segregation of waste instead of burning it in heaters and boilers.

Monthly concentration changes of PM10 and PM2.5

In 2015, the maximum accepted norm of the concentration of PM10 (50 μg/m3) was most often exceeded in February (14 days), March (9 days), October (10 days) and November (10 days) (see Figure 3). In those four months, the accepted concentration was exceeded 43 times, that is 80% of all the days in the year with the exceeded norm. Additionally, the average concentration was the highest in the above-mentioned months: February – 55 μg/m³, March – 45 μg/m³, October – 53 μg/m³, November – 49 μg/m³. The months with the lowest average concentration of the harmful substances were: April (27 μg/m³), May (22 μg/m³), June (20 μg/m³), July (27 μg/m³), December (26 μg/m³). In April, the daily accepted norm of the concentration was exceeded 4 times, and in the other months, it was not exceeded at all. The level of concentration of the suspended particulates PM10 and PM2.5 is characterised by a very high degree of interdependence. The value of correlation was: r = 0.93.

In 2016, the daily norms were exceeded less frequently than in the previous year (see Figure 4). The daily concentration of PM10 was decidedly exceeded most often in January (14 days), which constitutes 24% of the days in the year when the norm was exceeded. The highest average concentrations were noted in the autumn and winter months: January (52 μg/m³), February (34 μg/m³), March (36 μg/m³), April (33 μg/m³), September (41 μg/m³), November (38 μg/m³), December (40 μg/m³). The daily concentration level of PM10 in 2016 was less variable than in 2015, the coefficient of variation was V = 57%, while in the previous year it was V = 73%. The

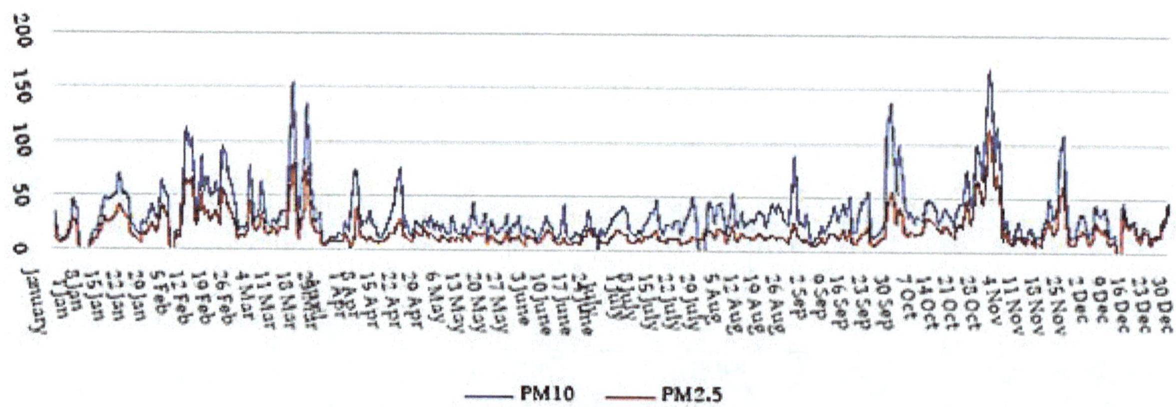

Figure 3. The level of concentration of the suspended particulates PM10 and PM2.5 in 2015 [μg/m³] (based on data from the meteorological station on Polanka St. in Poznań)

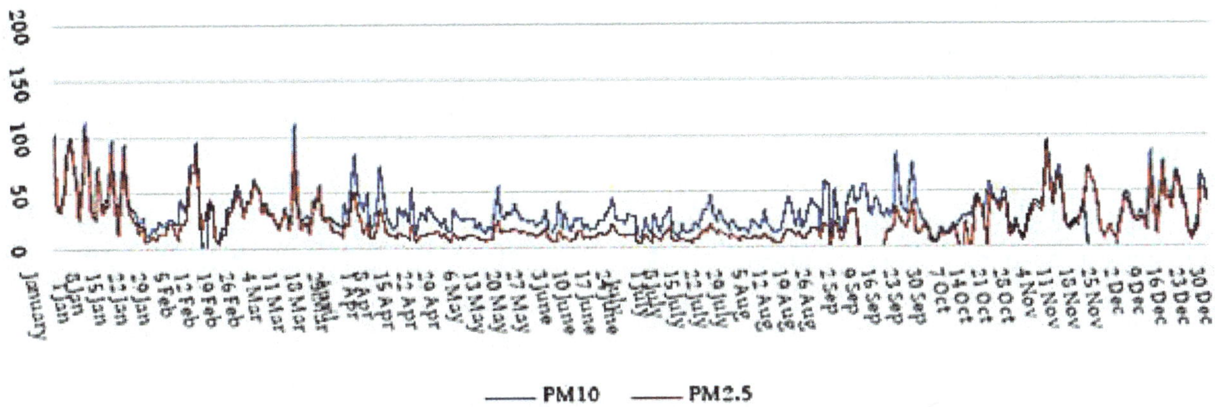

Figure 4. The level of concentration of the suspended particulates PM10 and PM2.5 in 2016 [μg/m^3] (based on data from the meteorological station on Polanka St. in Poznań)

concentration level of PM2.5 was again strongly correlated with the concentration level of PM10, with the value of correlation r = 0.92.

The year 2017 was marked by the highest level of variation of the concentration of PM10, with the coefficient of variation V = 101%. It was mainly due to the extremely high concentration values in January and February (see Figure 5). In those months the daily accepted norm was most frequently exceeded; in total, on 30 days during the two months, which constitutes 85% of all the days in the year when the norm was exceeded. The average daily concentration of PM10 in January was 61 μg/m^3, while in February it was 59 μg/m^3. In the remaining months, the level of concentration was relatively low, with average monthly values not exceeding 28 μg/m^3, and there were 5 days with a higher daily concentration (over 50 μg/m^3). The value of correlation between the concentration of PM10 and PM2.5 was nearly linear with r = 1.00.

The presented data indicate that the acceptable limits of suspended particulates emission in Poznań were exceeded in the autumn and winter period in 2015–2017. Similarly to the heating season, the produced pollution is most likely causing the highest emissions of the suspended particulates. It is probably the effect of the increased usage of old, low-efficiency heaters, resulting in the high emission of PM10 and PM2.5. The strong correlation between the emission of the suspended particulates and the temperature confirms this assumption. The value of correlation coefficient for PM10 and the temperature was r = -0.311, and for PM2.5 and the temperature, it was r = -0.48. The air humidity also influences the amount of the suspended particulates. It is noted that the higher the humidity, the higher the concentration of PM10 and PM2.5; however, it is much more visible in the case of PM2.5 (r = 0.29), than PM10 (r = 0.15). In light of the above, it is observed that the urban development have changed

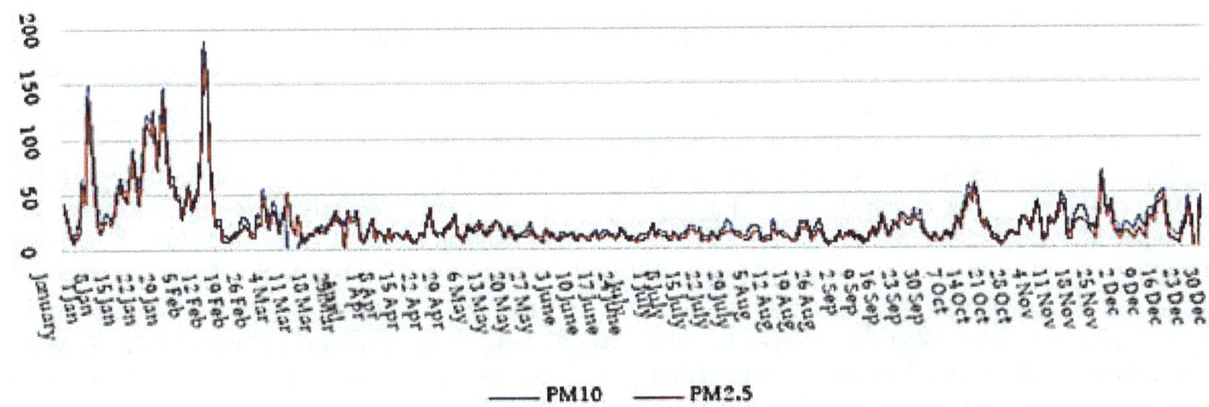

Figure 5. The level of concentration of the suspended particulates PM10 and PM2.5 in 2017 [μg/m^3] (based on data from the meteorological station on Polanka St. in Poznań)

the temperature, wind direction (due to the buildings) and humidity. The weather-influencing factors, together with the traffic and improper heating systems have caused a certain deterioration of the urban living conditions [de Lieto Vollaro et al. 2014; Battista 2017].

The sources of the suspended particulates emissions (transport, energy) in Poznań produce both PM10 and PM2.5, as is indicated by the calculated high correlation coefficient of the emissions.

CONCLUSIONS

The presented research shows a gradual decrease in the suspended particulates emissions, resulting from the programs that improve the effectiveness and efficiency of the heating installations and the thermomodernisation of the buildings. The public awareness has also increased when it comes to waste management. The extensive train and tram network in Poznań allows for lowering the emissions of PM10 and PM2.5. The annual characteristics of the suspended particulates emissions indicate that the emissions are the highest in the autumn and winter period, which is linked to the heating necessities. The correlation of the emissions and the temperature also confirms that the main sources of the emissions are the inefficient heaters and boilers. Moreover, high humidity is also an adverse weather factor. The water vapor binds with the suspended particulates and it causes them to settle down on the surfaces quicker.

REFERENCES

1. Battista G. 2017. Analysis of the Air Pollution Sources in the city of Rome (Italy). Energy Procedia, 126, 392–397.
2. Cembrzyńska, J., Krakowiak, E., Brewczyński, P.Z. 2012. Particulate pollution of PM10 and PM2.5 due to strong anthropopressure in Sosnowiec city (in Polish). Medycyna Środowiskowa – Environmental Medicine, 15(4), 31–38.
3. Cieślik, M., Dach, J., Lewicki, A., Smurzyńska, A., Janczak, D., Pawlicka-Kaczorowska, J., Boniecki, P., Cyplik, P., Czekała, W., Jożwiakowski, K. 2016. Methane fermentation of the maize straw silage under meso- and thermophilic conditions. Energy, 115, 1495–1502.
4. Czekała, W., Smurzyńska, A., Cieślik, M., Boniecki, P., Kozłowski, K. 2016. Biogas efficiency of selected fresh fruit covered by the russian embargo. Energy and Clean Technologies Conference Proceedings, SGEM 2016, VOL III, Book Series: International Multidisciplinary Scientific GeoConference-SGEM, 227–233.
5. Czekała, W., Dach, J., Dong, R., Janczak, D., Malińska, K., Jożwiakowski, K., Smurzyńska, A., Cieślik, M. 2017. Composting potential of the solid fraction of digested pulp produced by a biogas plant. Biosystems Engineering, 160, 25–29.
6. de Lieto Vollaro, A., de Simone, G., Romagnoli, R., Vallati, A., Bottillo, S. 2014. Numerical study of urban canyon microclimate related to geometrical parameters. Sustainability, 6, 7894–7905.
7. Dreiseidler, A.M., Baumbach, G., Straub, D. 2000. Results of fine particulate matter emission measurements at different industrial plants. Journal of Aerosol Science, Volume 31, Supplement 1, 242–243.
8. Dzikuć, M. 2017. Problems associated with the low emission limitation in Zielona Góra (Poland): Prospects and challenges. Journal of Cleaner Production, 166, 81–87.
9. Du, E., Dong, D., Zeng, X., Sun, Z., Jiang, X., Vriescd, W. 2017. Direct effect of acid rain on leaf chlorophyll content of terrestrial plants in China. Science of The Total Environment, Vol. 605–606, 764–769.
10. Fajczak-Kowalska, A., Ziemiński, P., Rudowska, P. 2017. Sustainable public transportation as a cornerstone of citizen-friendly urban logistics (in Polish). Gospodarka w praktyce i teorii, 1(46), 27–43.
11. Gładka, A., Zatoński, T. 2016. Impact of air pollution on respiratory diseases (in Polish). Kosmos Problemy Nauk Biologicznych, Vol. 65, 4(313), 573–582.
12. SVS 2017. Statistical Vademecum of Local Government (in Polish). Urząd statystyczny w Poznaniu.
13. Kampa, M., Castanas, E. 2008. Human health effects of air pollution. Environmental Pollution, 151(2), 362–367.
14. Lelek, Ł., Kulczycka, J., Lewandowska, A., Zarębska, J. 2016. Life cycle assessment of energy generation in Poland. Int. J. Life Cycle Assess., 21(1), 1–14.
15. Ma, J., Xu, X., Zhao, C., Yan, P. 2012. A review of atmospheric chemistry research in China: photochemical smog, haze pollution, and gas-aerosol interactions. Adv. Atmos. Sci., 29, 1006–1026.
16. Maeda, Y., Morioka, J., Tsujino, Y., Satoh, Y., Zhang, X., Mizoguchi, T., Hatakeyama, S. Material damage caused by acidic air pollution in East Asia. Water Air Soil Poll., 130, 141–150.
17. Mayer H. 1999. Air pollution in cities. Atmospheric Environment, 33, 4029–4037.

18. McDonald, D.R., Gentner, A.H., Goldstein, R.A. 2013. Harley Long-term trends in motor vehicle emissions in U.S. urban areas. Environ. Sci. Technol., 47, 10022–10031.
19. Mishra, S. 2017. Is smog innocuous? Air pollution and cardiovascular disease. Indian Heart Journal, 69(4), 425–429.
20. Penn, S.L., Boone, S.T., Harvey, B.C., Heiger-Bernays, W., Tripodis, Y., Arunachalam, S., Levy, J.I. 2017. Modeling variability in air pollution-related health damages from individual airport emissions. Environmental Research, 156, 791–800.
21. Environmental Protection Program for the city of Poznań for 2013–2016 with a perspective for 2020 (in Polish). 2013. Poznań, 53–70.
22. Ramanathan, V., Feng, Y. 2009. Air pollution, greenhouse gases and climate change: Global and regional perspectives. Atmospheric Environment, 43(1), 37–50.
23. Roy, J., Ghosh, D., Ghosh, A., Dasgupta, S. 2013. Fiscal instruments: crucial role in financing low carbon transition in energy systems. Curr. Opin. Environ. Sustain., 5, 261–269.
24. Regulation of the Minister of the Environment of 24 August 2012 on certain substances in the air (in Polish).
25. Smurzyńska, A., Dach, J., Szulc, R. 2015. Influence of different slurry treatment technologies for methane emissions after application of slurry to the soil. Acta Scientiarum Polonorum-Formatio Circumiectus, 14(3), 165–174.
26. Son, S-W., Tandon, N.F., Polvani, L.M., Waugh, D.W. 2009. Ozone hole and Southern Hemisphere climate change. Geophys. Res. Lett., 36, L15705.
27. Sosa, B.S., Porta, A., Lerner, J.E.C., Noriega, R.B., Massolo, L. 2017. Human health risk due to variations in PM10-PM2.5 and associated PAHs levels. Atmospheric Environment, 160, 27–35.
28. Sun, Ch., Luo, Y., Li, J. 2018. Urban traffic infrastructure investment and air pollution: Evidence from the 83 cities in China. Journal of Cleaner Production, 172, 488–496.
29. Wang, X., Chen, J., Cheng, T., Zhang, R., Wang, X. 2014. Particle number concentration, size distribution and chemical composition during haze and photochemical smog episodes in Shanghai. Journal of Environmental Sciences, 26(9), 1894–1902.
30. Wang, Ch., Li, Y., Liu, Y. 2018. Investigation of water-energy-emission nexus of air pollution control of the coal-fired power industry: A case study of Beijing-Tianjin-Hebei region, China. Energy Policy, 115, 291–301.

9

Reduction of Odour Nuisance of Industrial Plants – Biofiltering in Composting Plants

Monika Gałwa-Widera[1*], Anna Kwarciak-Kozłowska[1], January Bień[1]

[1] Czestochowa University of Technology, Faculty of Infrastructure and Environment, Institute of Environmental Engineering, ul. Brzeźnicka 60a, 42-200 Czestochowa, Poland

[*] Corresponding author's e-mail: mwidera@is.pcz.czest.pl

ABSTRACT

Composting produces several hundred different substances. Among them, active fragrance compounds constitute a considerable number. Composting technology and the method of conducting the process are important elements. Due to the different ways of conducting the composting process, there are also areas of odour emissions as well as the volume of emissions. Depending on whether the process is conducted statically or dynamically, in a natural or artificial way and also in an open (e.g. field prism) or closed system (reactors, containers, encapsulated plots), the amount of odourant emitted will be uneven. One of the methods is the use of biofiltration. The biological filter for air purification is designed to remove volatile air pollutants leaving various installations. By using a suitable filter bed, total organic and inorganic carbon, sulfur and nitrogen compounds such as ammonia and hydrogen sulfide can be reduced. They are part of the odours that emerge, for example, from a composting plant. The use of biofilter eliminates polluting emissions, which in practice means reducing the negative impact on the environment and neutralizing odours. The article presents the results of biogas analysis generated in the composting process and ecological methods of removing odorants in the biofiltration process. Two types of filling were used to fill the biofilter. Both are fully biodegradable. It has been shown that these fillings have a high impact on the development of biofilm, and thus on the reduction of pollution.

Keywords: biomass, composting, industry, sewage sludge, odour

INTRODUCTION

The development of civilization is associated with the production of waste. A large part of waste is subjected to heat treatment, recycling, and biological processes. Unfortunately, all these processes also cause the formation of other types of pollution. The main reason for the complaints of the population on health and environmental quality are usually arising from odours. They are not only unpleasant, but also contribute to the development of anxiety about the state of health. Unfortunately, despite the high-performance technology of removing odour, the complaints of the population are not unfounded. All kinds of waste reprocessing plants still represent an environmental problem. The provisions concerning the emission limit do not specify the obligation to reduce the emission of odourants [Kośmider 2002]. Odour pollution, however, shows the presence of substances referred to in the Regulation of the Minister of Environment of 26th January 2010 [Rozporządzenie Ministra Środowiska 2012].

Of course, not only the waste reprocessing plants, such as composting plants, sewage treatment plants or landfills, generate odour. This group also includes other branches such as animal breeding farms or processing plants, refineries [Rozporządzenie Ministra Środowiska 2012]. The compounds that are characterized by odour nuisance are mainly hydrogen sulfide, ammonia, sulfides and volatile organic compounds. According to the literature data, there are more than 100 of these compounds [Kwarciak-Kozłowska 2014].

The aim of the research is to optimize the process. The research described in the article included the stage of selection of biofilter filling. The authors focused on two fillings selected from among several analyzed in the laboratory taking into account porosity, nutrient content, availability, price and degradability. The aim of the research was first of all to check the effectiveness of the biofiltration process on a technical scale [Van Groenestjin 2005].

METHODS USED FOR DEODOURIZATION OF GASES

Total elimination of odourant emissions into the atmosphere seems impossible. This is due to the fact those are released during many technological processes.

Various stages such as unloading, transshipment or transport also form part of the technology, and it is not always possible to resign from them. For many gases, which are characterized by unpleasant odour, the sensing threshold is very low. Therefore, their complete removal from waste gases, should be taken into account at the design stage [Twardowski 2000, Warych 1998].

Deodourization methods can be basically divided into two groups. A group of chemical methods includes oxidation, chlorination, disposal of thermal as well as the methods using catalytic and sorptive processes. The second group includes, greener biological methods, such as biological MUDs called scrubbers or the methods based on natural materials and micro-organisms. The biological MUDs use activated sludge as a filtering layer, which constitutes a suspension of microorganisms in the water. On the other hand, filtration uses heterotrophic activity aerobic bacteria. Microorganisms have the ability to conduct the process of purifying the sorbent in an aerated chamber. After such treatment, a stream returns to the scrubber. Packed filters are characterized by slightly different design . Here, the main element is a layer filter material with high porosity. Living micro-organisms, capable of biological decomposition of impurities contained in the air, deposit themselves on such filters. The process is run by the transmission method (blowing out) of contaminated gases through the deposit, followed by two types of processes: sorption filtration and biological pollutant distribution in which the active micro-organism flourish. This method, as in the case of the scrubber, is is susceptible to self-healing. Last year shows that biological methods are effective for the disposal of volatile contaminants, both those associated with the formation of odours and volatile organic compounds such as chemical or physical methods [Dicks 1991].

When a contaminated gas flows through a layer of filter material, a diffusion of impurities from the gas phase to an active biological layer, surrounding the particle filter material, occurs [Warych 1998]. Consequently, the growth of microorganisms on the biomass "handles" a greater amount of pollutants.

The process of mineralization of pollutants in biofilter is possible thanks to the bacteria, fungi and thermophilic Actinomycetes forming on the surface of the sorbent active biofilm.

A biofilter, depending on the process parameters can be of varying size and shape. The way of supplying air also differs, as it is dependent on the spatial capabilities and scale. Large containers with height of up to several meters are used most often. Examples of such biofilters are illustrated in Figure 1.

The air supply to the layer of the deposit may also be diverse. One of the methods is to push air through the slotted chambers directly from shots or gas through the humidifier, as presented in Figures 2 and 3 [van Groenesjin 2005].

In the first case, a system of perforated pipes, placed in a layer of gravel is implemented – its task is to facilitate even distribution of gas in the biofilter. In the second case slotted Chambers may be used, as they are a durable design.

In the case of a large load of dirt, one can apply a storied building, *i.e.* a couple of biofilters placed one above the other. In addition, the Strip subjected to biofiltration process should be moistened for a relative humidity of 95–100%. In this situation, it is perfect to bring water vapour (steam with a humidifier) to the stream of gases.

The phenomenon of condensation often occurs in the biofilter. The condensate can be used to spray deposits on the filter. This is an effective solution to ensure that the installation will work on waste without wastewater. This includes the sealed biofilters. In the case of biofilters and open drains, the humidity of the biomass depends on various atmospheric conditions.

MATERIAL AND METHODS

The study had a trial basis. The analysis focuses mainly on the elimination of odours of inorganic origin such as hydrogen sulphide and ammonia. The scope of the research included the

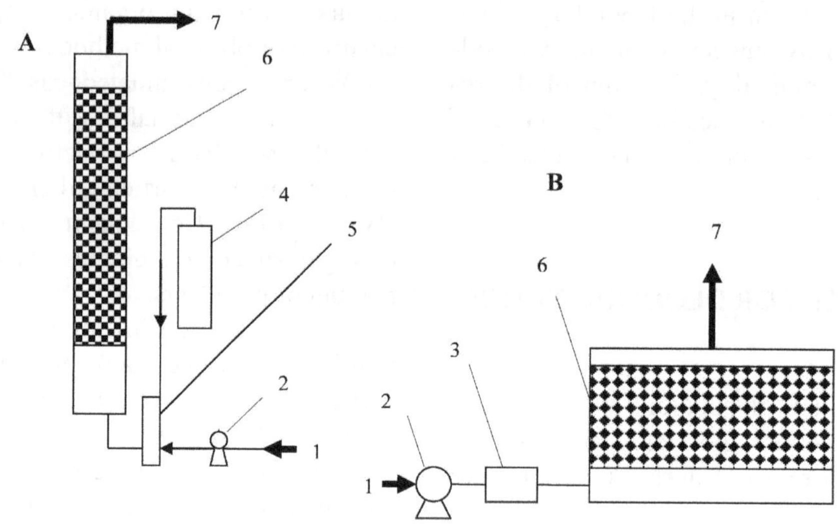

Fig. 1. Wiring diagrams for deodourization of gases: A – installation of the column, B – installation of the container, 1 – supply of contaminated gas, 2 – pressure-suction, 3 – humidification chamber, 4 – steam tank, 5 – mixing, 6 –biologically active deposits, 7 – the outflow of the purified gas) [Wierzbińska 2015]

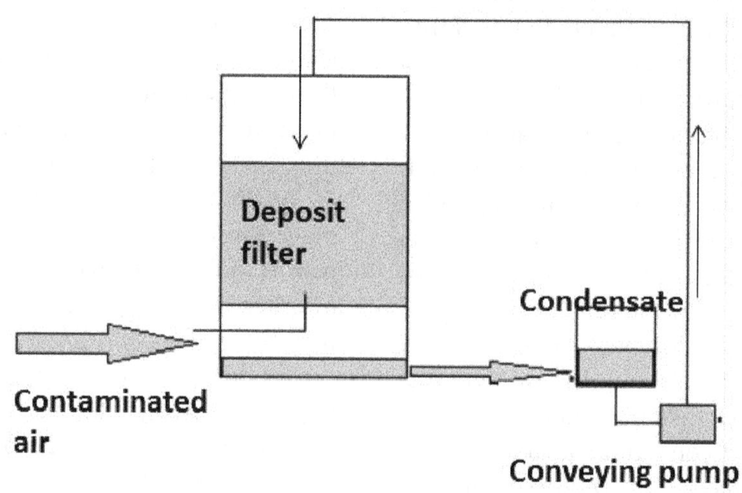

Figure 2. Diagram of biogas supply to the deposits without humidification

analysis of biogas suction from a reactor that carried out the composting process. The scale of the research was defined as half-technical. The experiment was carried out using two types of filling, i.e. peat and sawdust.

The composting process itself occurred in a commercial composting plant running on "MUT", container, consisting of 6 containers with one control unit. Each container was filled in ¾. The capacity of a single unit was 25 m^3. The system was capable of monitoring temperature, humidity, pH, and forced aeration.

A mixture of compost kitchen waste and the organic fraction of municipal waste were investigated. The process was carried out for 5 weeks. The parameters of composting were determined: moisture content of the mixture – 50–60%, pH – 6–7, the level of aeration – 4 dm^3/min.

All parameters of the composting process were maintained in accordance with guidelines on the processes of composting, fermentation and mechanical-biological treatment of waste by Ministry of Environment. A biofilter was installed on the bioreactor.

The research was carried out on the microorganisms adapted to the distribution of the contaminants on the fill layer inside the biofilter casing. The volume of the biofilter was 2 m^3. The biofilter has to be active for a certain amount of time before the biofilm with microorganisms flourishes on it. The experiment uses the settling of the deposits commercially.

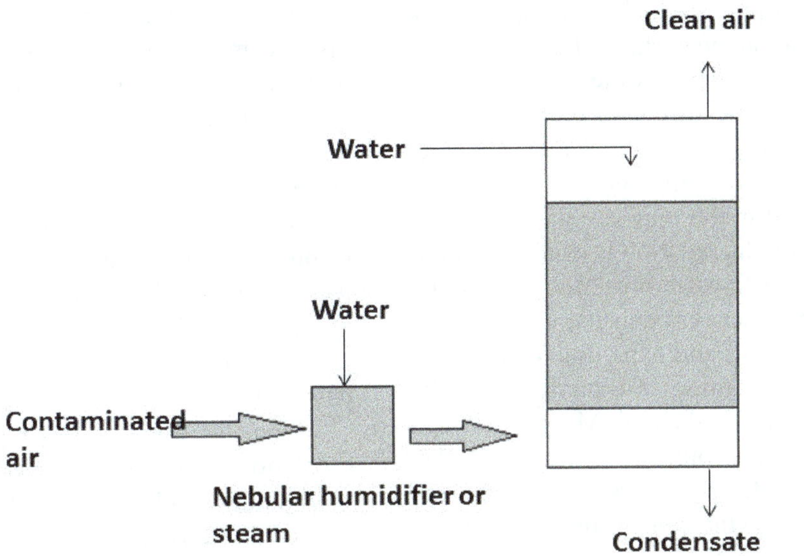

Figure 3. Diagram of biogas supply to the deposits by the installation of the steam humidification

The contaminated gases were carried out laterally in accordance with the circuit diagram (Figure 4).

The experiment uses a layout without humidification. In the solution, there is a possibility that the condensate accumulated on the walls of the biofilter returns. It may be re-used for spraying. This makes the installation work without waste water discharge, and is followed by a sprinkling of filling and microorganisms that colonize in the leachate in addition to deposit.

The analysis of power consumption of biogas is made directly from the reactor and after biofiltration.

The analysis uses the portable biogas Analyzer series 5000 GA certified ATEX II 2 g Ex ib IIA T1 Gb (Ta =-10°C to + 50°C), IECEx, CSA.

RESULTS AND CONCLUSIONS

Biofiltration represents a highly complex process that is carried out in the liquid phase and gas and its effect is the formation of biodegradable products: CO_2, H_2O and other inorganic compounds. The odour pollution contained in flowing air becomes liquid adsorption on the surface of the biofilm covering the particles of the filter material. The odour particles adsorbed on the surface of the biolayer, pass to the liquid phase and undergo diffusion towards the micro-organisms, induced by biofilm.

The process that takes place on the surface of the biofilter particles consists in the distribution of contaminants and is called “biodegradable”. Biodegradation takes place in two stages. The

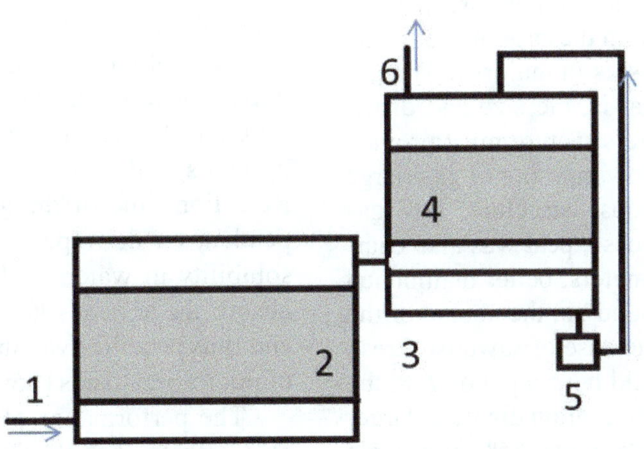

Figure 4. Layout scheme used in experiment 1 – air inlet, 2 – reactor-reaction containers set, 3 – biofilter, 4 – deposit , 5 – effluent of the condensate, 6 – clean air

first step is biosorption, which consists of stopping the pollution on the surface of cells.

The surface of the cell is the site of the Exchange, subject to the diffusion components to the Interior of the cell as well as the excretion of waste products and enzymes decomposing matter outside the cell in the opposite direction. The second stage of biodegradation is mineralisation.

It involves assisting the breakdown of organic compounds by micro-organisms, using energy and biogenic elements and expelling simple mineral products. Distribution of organic molecules occurs generally within the cell. Macromolecular compounds, such as proteins, cellulose, or other natural and synthetic polymers, are hydrolyzed outside the cell. The Interior of the system products of this diffusion were subject to oxidation. These changes occur under aerobic conditions [Pawlaczyk-Szpilowa 1980]. The fundamental mechanisms in the process of biofiltration, consisting in the kinetics of the removal process of impurities from gases are-sorption and their biodegradation [Adler 2001].

The result of the distribution of biological biolayer is the absorption of pollutants on the surface of the biofilm. The self-cleaning of gas treatment using biofilters and sorbent technology and is virtually waste-free [Szklarczyk 1991, Szynkowska 2010].

The filter material should be regarded as the most favourable condition for the development of micro-organisms in order to obtain the maximum of biodegradable pollutants. In this case, the two types of material are applied. Both of these materials have proven to be effective in the filtration process. However, better results in the formation of biofilms were obtained in the case of peat. It was found on the basis of random microbial analysis. Plating method was executed on enriched agar incubated for 48 h at temperatures of 20° and 37°; it was stated that in the case of peat plate, moved the micro-organisms in majority. The porosity of the deposit, that is, the size and distribution of grains and pore structure must provide high surface activity at low pressure of gas (large specific surface area, loose structure, low gas flow resistance). In this case, peat was also characterized by better parameters; better maintained moisture levels, not "settled" in the course of the process as much as in the case of sawdust. Fresh (raw) filter material should have a pH of 7–8, the volume of pores above 90%, grain diameter larger than 4 mm. It should contain over 55% of organic material [Szklarczyk 1991]. These conditions are definitely better with peat filling.

In accordance with the research and literature, organic materials good biological filters. These include: fertile soil of loosened structure, peat and peat waste composts made from the bark of trees, wood waste, and composts from municipal waste [Jorio 2000].

In nature, there is a great variety of heterotrophic bacteria, mostly located in the soil and water, having the ability of absorbing organic matter; these bacteria are involved in important processes of purification and regeneration of the environment.

Many of these bacteria have the ability to break down organic substances, uncommon in the wild nature, such as industrial waste [Szklarczyk 1996, Ulfik 2010].

Micro-organisms are very easy to adapt to the distribution of pollutants containing alcohols, ketones, and esters. Compost usually contains sufficient number and variety of micro-organisms capable of complex biochemical decomposition of pollutants.

At the same time, the micro-organisms show a great ability of adapting to specific pollutants contained in the waste gases. In the experiment, the waste gases ran through the entire technological line; therefore, there is a high probability the partial purification of gases occurred during blowing by the compost mass. Hence, a significant drop in pollution along with the progress of the humification process was observed.

The experiment involved the intentional introduction of seeding consisting of a mixture of micro-organisms and nutrient solution. The aim was to speed up the creation of the biofilm and increase the number of micro-organisms. Owing to such treatment, very good results related to the removal of most hydrocarbons from the industrial gases are achieved [Szklarczyk 1996, Wierzbińska 2015].

Technological reasons for the correct operation of the biofilter are important parameters such as: contact time gas deposit, deposit load, porosity and specific surface area, the flow resistance. Retention time of the gas in the biofilter is dependent on the type of contamination and their solubility in water. It should be long enough to absorb the aqueous layer of biofilter's pollution, and thus become available for the biodegradation of micro-organisms [Sówka 2011, Kapusta 2007].

The performance of the biofiltration process also affects the physical and chemical conditions found in a fluid. They have an impact on the health and the quantity of the micro-organisms responsible for the process gas. Therefore, it is important to provide the optimal environment for their development. The effects of the impact of a given factor may be different depending on its type and intensity, as well as the quality of the body. The most important factors that affect growth, reproduction, survival and metabolic activity of microorganisms include: temperature, concentration of hydrogen ions, oxy-reduction potential, water content in the environment, the availability of food and harmful chemicals. Temperature has a direct impact on the life processes of micro-organisms. Therefore, it is one of the most important factors determining the growth rate, enzyme activity or the chemical composition of cells and nutritional requirements. It can also indirectly influence the solubility of the compound molecules, intracellular transport of ions and the diffusion of the chemicals as well as the characteristics of osmotic membranes. The optimum temperature is 25–40°C. In the case of different micro-organisms, the quality of the treated gas, the type of compounds should be taken into account so as to optimize the temperature of the deposit and obtain the highest efficiency. Due to the microbiological activity, large amounts of heat may be released, raising the temperature in the filter deposits.

In winter, the temperature inside the biofilter may be higher than the ambient temperature 10–20°C [Brugess 2001, Adamiak 2012, Wierzbińska 2010]. An equally important factor, not only on the viability of organisms, but also absorption from the gas phase contaminants, is humidity. The bacteria are generally extremely resistant to desiccation, but it can cause deterioration of the effects of biofiltration by gradually slowing down, or even halting the processes of metabolism. Furthermore, in a dry deposit, the area of direct contact is decreased, and this in turn reduces the efficiency of the biofilter operation. Therefore, the gas can be pre-humidified before entering the filter deposit. Not every process requires such a procedure, but it is beneficial to prepare a system so that this option can be used. Of course, the excess water is also not advisable. The microorganisms inhabiting the deposit are heterotrophic aerobic organisms, and the excess moisture may lead to the formation of anaerobic areas, causing decay, which also deteriorates the deodourization effects [Adamiak 2012, Gałwa-Widera 2016]. The activity of microbial life is heavily dependent on the concentration of hydrogen ions. Their level is determined by the numerical value (pH). The optimal pH is neutral, but the biofilter process gas containing, for example hydrogen sulfide may cause acidification of the substrate, which leading to a reduction of bacterial activity. The reaction of the deposit should be adjusted, for example, by means of calcium carbonate in neutralizing acids, or by using microscopic fungi in the biofilter, because they are more resistant to the changes in pH than bacteria [Ramirez 2012, Adamiak 2012, McNevin 2000, Ulfik 2010]. As is clear from the analyses, the pH prevailing in the deposit was maintained as neutral. Important factors in the inhibition of vital micro-organisms are different poison types. The detrimental effect for the organism depends on the type of toxins, as well as the dose. The substances harmful to microorganisms are: heavy metals, detergents, cyanides, pesticides, strong oxidants. They damage the cell membranes of bacteria and result in denaturation of proteins or enzymes inhibition. However, all bacteria and living organisms are resistant to environmental factors, and they are able to increase the resistance through adaptive processes. The parameters of the composting process while using biofiltration for deposits of solid odourants are specified for both the mixture of compost and filter deposit (Table 1).

In the biomass rich in nutrients, microbial growth occurs, which in effect, "converts" a larger amount of impurities. Bacteria and actinomycetes, as well as fungi and myco-trophic algae such as green algae are primarily involved in the mineralization of organic pollutants,. The result of biodegradation of absorbed impurities in the biolayer is self-healing filter material, covered with a biofilm. This self-cleaning sorbent makes gas treatment using a biofilter a virtually waste-free process.

Each bacterial species grows in a specified temperature range. It should be assumed that the optimal temperature range in the biofilter is 25–33°C. In winter, the temperature inside the biofilter is higher than the ambient temperature,

Table 1. Parameters of the process biofiltration air containing odourants, created during composting

Parameter	Value
pH compost mixture	5,8 – 7,5
pH filter media	6–7
Relative humidity mixture of compost	45%-60%
Relative humidity filter media	50%-60%
C/N (compost mixture)	2–30:1

ranging from 10 to 20°C. This is because the large amounts of heat are released in the microbiological processes. Moreover, a process of composting is involved, which generates heat, especially in the first 14 days. Through this process, the air discharged from the reactor goes to the biofilter already in the heated form. A significant increase in the temperature in the dynamic phase of the composting process starts relatively early (Figure 4).

For most bacteria, the optimum pH value is 7.0–7.5. A drop in pH below 4 and an increase over 11, results in a significant reduction of the bacterial activity, as well as frequent and extreme changes in the environmental pH (Table 1). It is therefore important to maintain proper humidity and pH in the reservoir. The wetting process was conducted by recycling the condensate and putting it in an aerosol form. The air flow was regulated. Some species have the ability to actively influence the external pH and change them to favourable values.

The use of biofilters in composting of municipal waste allows the removal of almost all of specific pollutants characterized by odour nuisance.

The filtering material used in the process included sawdust and peat. The gas analysis conducted before and after filtration demonstrated that a more efficient removal of odourous compounds occurs with peat-filled biofilter.

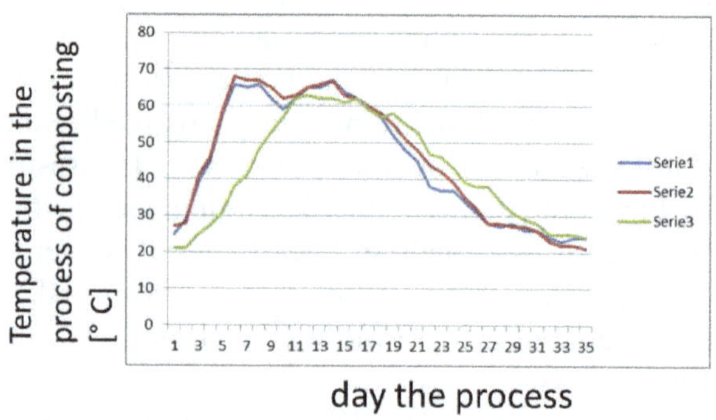

Figure 5. Figure temperature changes in the process of composting

A significant decrease in the content of gaseous pollutants produced in the composting process may be noted. Clearly better results were obtained using a peat deposit (Table 2).

This is due to increased porosity and thus greater space inhabited by microorganisms. The lower layer of the biofilter was filled with coarse filter material in the form of wooden frameworks. This was the first stage of purified biogas adaptation. The quality of the filter material also has an impact on the development of the biofilter. Seed analysis showed that the surface of peat developing biofilm faster than in the case of sawdust. The biodiversity of nutrients contained in the peat had a positive effect on the rate of settlement deposits. The method consists in the biological oxidation of H_2S to elemental sulfur by a specialized group of microorganisms from the family Thiobacillus, according to the reaction:

$$2H_2S\ 2S + O_2 + 2H_2O \quad (1)$$

They use a carbon dioxide as a carbon source. The product transformation is elemental sulfur and sulphates. The ammonia principle works by inhibiting the development of micro-organisms. The purpose of nitrifying bacteria is to provide a nitrogen source for the microorganisms.

Studies show that the use of biofiltration in technological processes is associated with a number of benefits. At the same time, this method is simple and does not require the application of excessive financial outlays. The odour is not only hydrogen sulfide and ammonia tests involve further analysis and selecting the bed for a process optimization. Time biogas flow through the fixed bed was forced air.

Table 2. Changes in the concentration of hydrogen sulfide and ammonia using a biofilter with peat and sawdust fillings

Contamination	Concentration before biofilter mg/m^3	Concentration after biofilter mg/m^3	Removal efficiency, %
The results obtained for deposits of peat*			
Hydrogen sulfide	3670	37	99
Amonnia	560	44	84
The results obtained for deposits of sawdust			
Hydrogen sulfide	3588	180	95
Amonnia	570	80	86

* The results of the tests present the arithmetic mean of three experiments, due to the long-term nature of the process and the scale of the sample.

CONCLUSION

Composting facilities are potential sources of odour nuisance but they can be effectively reduced through encapsulation devices and the use of deodourization methods. The biological methods, and in particular biofiltration, are increasingly popular due to their simplicity as well as low investment and operating costs, while being able to obtain high efficiency of treatment. However, proper selection of the device parameters (dimensions, layer height of the deposit, the appropriate distribution of treated gases) and proper operation, ensuring their smooth operation are important.

Biological methods, including biofiltration, are commonly used in the deodourization processes of gases, particularly in municipal buildings (sewage treatment plants, landfills). The conducted research indicates that biofiltration is an effective method of flue gas cleaning. The resulting gas purification efficiency, depending on the substances contained therein, ranges from 40 to 100%. An important issue related to the proper functioning of the biofilter, in addition to the use of appropriate design solutions, pre-treatment and conditioning of gases, is to provide optimal conditions for the growth and degradational activity of microorganisms. Providing certain regimes, concerning, among others temperature or humidity, affects the stable operation of the biofilter and increases the efficiency of gas purification related to odour-generating compounds.

Studies have shown that both fillings meet the criteria for the substrate to produce a biofilm. The effectiveness of odour removal in the case of peat and sawdust deposit is characterized by the appropriate parameters (pH, humidity, porosity) to reduce pollution by over 80%. According to the authors, the peat filling shows a reduction to a smaller extent but makes maintaining stable conditions for microorganisms easier. The main factor is that the peat maintains a constant humidity. During the experiments, it was also noticed that a significant increase in the biofilm appeared in the thermophilic phase of composting. It consisted of such factors as the higher content of nitrogen and sulfur compounds and the higher gas temperature at the entrance to the biofilter. A further part of the research will be conducted in the direction of determining the impact of the results obtained on other volatile pollutants.

Acknowledgement

Accomplished with BS/PB-401-301/11 request 6.

REFERENCES

1. Adamiak W., Kita U., Sówka I., Szklarczyk M., 2012. The role of microbial selection in the benzene biodegradation process, Chemical Industry, 91(5), 655–659.
2. Adler S.F. Use these guidelines to scale up and design biofiltration processes for the control of volatile organic compounds. CEP, 2001.
3. Burgess J.E., Parson S.A., Stuetz R.M., 2001. Developments in odour control and waste gas treatment biotechnology: a review, Biotechnology Advances, 19, 35–63.
4. Dicks R.M.M., Ottengraf S.P.P. 1991. Verification studies of a simplified model for the removal of dichloromethane from waste gases using a biological trickling filter. Bioprocess Eng., 93–99.
5. Gałwa-Widera M., Kwarciak-Kozłowska A., 2016. Ways of eliminating odors in the composting process. Rocznik Ochrony Środowiska, 18.
6. Jorio H., Bibeau L., Viel G., Heitz M. 2000. Effects of gas flow rate and inlet concentration on xylene vapors biofiltration performance. Chemical Engineering Journal,76(3), 209–221.
7. Kapusta K., 2007. Protection of fragranced air quality. World experience in light of the need of legal regulations in Poland, Scientific Work GIG, 4, 31–50.
8. Kośmider J., Mazur-Chrzanowska B., Wyszyński B. 2002. Odours. Wyd. Nauk. PWN, Warszawa.
9. Kwarciak-Kozłowska A. Bańka B. 2014. Biofiltration as a method of neutralization of odors generated during composting of biodegradable fraction of municipal and industrial waste, Engineering and Environmental Protection, 17(4), 631–645.
10. McNevin D., Barford J., 2000. Biofiltration as an odour abatement strategy, Biochemical Engineering Journal, 5(3), 231–242.
11. Pawlaczyk-Szpilowa M. Microbiology of water and sewage. PWN, Warszawa, 1980.
12. Ramirez A., Garcia-Aguilar B., Jones P., Heitz M., 2012. Improvement of methane biofiltration by the addition of non-ionic surfactants to biofilters packed with inert materials, Pro-cess Biochemistry, 47(1), 76–82.
13. Regulation of the Minister of the Environment of 24 August 2012 on the levels of certain substances in the air, OJ. from 18 September 2012, item 1031, Warsaw 2012.

14. Sówka I., 2011. Methods of identification of odorogenic gases emitted from industrial facilities, publishing house of the Wrocław University of Technology, monograph, Wrocław.
15. Szklarczyk M. 1991. Biological purification of waste gases. Ed. Wroclaw University of Technology, Monografia, Wrocław.
16. Szklarczyk M., Czemarmazowicz M. 1996. Materials of the 4th National Symposium on "Protection of Air in Industry". Lodz University of Technology, Łódź, 27–29 May.
17. Szynkowska M. I., Zwożdzak J., 2010.Contemporary problems of odors, Scientific and Technical Publishers, Warsaw.
18. Twardowski T., Michalska A. 2000. Dilemmas of modern biotechnology from the perspective of biotechnologist and lawyer. Ed. Organizer's House, Toruń.
19. Ulfik K., 2010. Biofiltration of volatile organic compounds by means of bacteria and microscopic fungi, Contemporary odor problems, ed. I.M. Szynkowska and J. Zwoździak, Scientific and Technical Publishers, Warsaw.
20. Warych J. 1998. Purification of gases. Processes and apparatus. WNT, Warsaw.
21. Wierzbińska M., 2010. Application of biofiltration methods for deodorization of waste gases using fibrous deposits. part. II – Factors determining the optimization of the biofiltration process. Air protection and waste problems.
22. Wierzbińska M, Modzelewski W.E. 2015. The use of biofilters for deodorising onerous gases. Inżynieria Ekologiczna, 41, 125–132.
23. Van Groenestijn J.W. Kraakman N.J.R., 2005. Recent developments in biological waste gas purification in Europe, Chemical Engineering Journal, 113, 85–91.

10

Stabilization/Solidification of Waste Containing Heavy Metals and Hydrocarbons using OPC and Land Trass Cement

Masrullita[1,3*], R.Y. Perry Burhan[2], dan Yulinah Trihadiningrum[1]

[1] Department of Environmental Engineering, Institut Teknologi Sepuluh Nopember, Surabaya 60111, Indonesia

[2] Departement of Chemistry, Institut Teknologi Sepuluh Nopember, Surabaya 60111, Indonesia

[3] Universitas Malikussaleh ACEH, Aceh 24351, Indonesia

* Corresponding author's e-mail: masrulita.nad@gmail.com

ABSTRACT

The stabilization/solidification process (S/S) is one of the alternative methods of treating B3 waste, especially heavy metal. The S/S uses cement as the solidification agent. The cement will bind heavy metal pollutants in a monolithic mass with a sturdy structure, thus inhibiting its movement. The presence of hydrocarbons affects the S/S strength. Therefore, it is necessary to add pozzolan material which can absorb hydrocarbon constituting the cement blocking component of pozzolan cement, i.e. Ordinary Portland Cement (OPC) combined with trass soil. This study aims to determine the maximum content of organic materials in the form of hydrocarbons can stabilize/solidify heavy metals contained in wastes containing hydrocarbons. This research is conducted in two steps. Stage I aims to obtain the optimum composition of the mixture. On the other hand, stage II is to determine the maximum content of hydrocarbons in percent weight that can stabilize/solidify organic wastes containing heavy metals – Cu, Cr, and Pb in artificial wastes. The composition of OPC and trass soil was varied at a ratio of 100: 0, 5:25, 50:50, 25:75 and 0: 100. The hydrocarbons used in step II were paraffin, added to the optimum composition of OPC and trass soil with a proportion of 2.5%, 5%, 5% and 10%. The S/S product quality test was performed, involving: compressive strength test, Toxicity Characteristic Leaching Procedure (TCLP) and paint filter test. Strength test was conducted using a compressive strength testing apparatus Toasters Universal Testing Machine Type RAT-200, MFG No. 20380 CAP 200 tf. TCLP test under US-EPA (method 1311). The method of analysis pertaining to heavy metal concentrations involved a colorimetric method for Cr (VI), neo-cuproine for Cu, and dithizone for Pb. The paint test refers to the US EPA 9095B method. The results showed that the optimum composition of OPC mixture: trass soil was 50:50, which is the composition used in stage II. The results of compressive strength test were 2770 tons/m^2. The TCLP test results for heavy metals Cu and Pb with hydrocarbon addition on Cu 10% and Pb 2.5% reached 0.076 and 0.076 mg/L, respectively. The result of the paint filter test indicates that there is no remaining free fluid.

Keywords: trass soil, heavy metal, ordinary Portland cement (OPC), stabilization/solidification

INTRODUCTION

Environmental pollution by hazardous toxic waste (B3) is a serious problem that has not been handled properly. The B3 wastes generally contain the heavy metals resulting from industrial activities. The industrial waste containing heavy metals may be inorganic waste that does not contain hydrocarbons, for example including electroplating, metallurgical, and smelting industrial wastes.. The waste generated from the oil and gas industry can be liquid, solid/mud or gas. Large amounts of oil sludge is generated during tank cleaning, crude oil storage, maintenance of related facilities and processing activities prior to crude oil sales to sea terminals. Oil sludge contains aromatic hydrocarbons (benzene, toluene, ethyl benzene and xylene), poly-aromatic hydrocarbons and heavy metals (Ayotamuno et al, 2007).

Oil-borne waste is a water-in-oil emulsion residue (W/O) containing a hydrocarbon and heavy metal compound. The most commonly found compounds are petroleum hydrocarbons (PHC) compounds, ranging from 5% to 86.2%, and other organic compounds such as Polycyclic Aromatic Hydrocarbons (PAH), alkanes, and phenols. On the other hand, the content of heavy metals such as lead is 0.001–0.12 mg/kg, copper 32–120 mg/kg, and chromium – 27–80 mg/kg (Hu et al., 2013). Phenol and PAH compounds are flammable, so they are classified into hazardous and toxic waste or B3 waste (Silva et al., 2012). Oily mud contains organic and inorganic contaminants. The organic contaminants are in the form of heavy metals, such as zinc, lead, copper, nickel, chromium, and mercury. The organic contaminants consist of Total Petroleum Hydrocarbon (Al Futaisi et al., 2007). Many processing techniques are used for the processing of waste oil industry, especially oily sludge. The methods with oil sludge incineration have been introduced to replace the hoarding and bioremediation methods. However, the method has limited concerns that the combustion process can release toxic gases which evaporate into the environment. Inorganic vaporous (VOC) vapors are formed and released. Toxic heavy metals also cannot be removed during the combustion process and will accumulate as solid particles of combustion areas (Dominguez et al., 2005).

Petroleum wastes generated by businesses or activities of oil, gas and geothermal or other activities producing petroleum waste are hazardous and toxic waste. The waste has the potential to cause pollution and environmental damage. Therefore, it is necessary to manage it well. Technological alternatives have been developed for treating the B3 waste through the physical, chemical, biological or combinational mechanisms. The B3 waste treatment technology must be in accordance with the characteristics of waste. One of the methods of managing petroleum and soil waste contaminated by petroleum methods can be carried out by the biological treatment. However, the biological treatment has the requirements of Total Petroleum Hydrocarbon / TPH contaminant contents. The requirements of the biologically treated petroleum waste are the maximum concentration of the initial TPH before the biological treatment process is lesser than 15% (Decree of the Head of Environmental Impact Management Agency no. 128 of 2003). The results of the stabilization/solidification process can be reused for building materials, construction materials, road materials, or materials for pile up (Desogus et al., 2013). Cement is one of the most widely used binder materials in the S/S method (La Grega et al, 2005). According to Paria and Yuet (2006), the S/S waste technique using cement is one of the B3 waste treatment alternatives with the aim of reducing the environmental pollution. The S/S process is very effectively used in the treatment of radioactive waste and heavy metals. The S/S product is an immobilized heavy metal trapped in the binder (Yang and Min, 2008). The results of the S/S process can be reused for building materials, construction materials, road materials, or materials for pile up (Desogus et al., 2013). The S/S process using cement has been applied to the organic wastes containing hydrocarbons. However, the presence of hydrocarbons can reduce the formation of crystal structures, thus affecting the strength of the S/S products (La Grega et al., 2005). The organic substances in the waste do not form strong chemical bonds with cement. Therefore, the resulting S/S product has a poor compressive strength. (Karamalidis and Voudrias, 2007).

Some studies show that natural pozzolan is widely used as a substitute for the portland cement because it is cheaper, reduces heat, and increases the chemical resistance (Ghricia et al., 2007). The addition of pozzolan can reduce the amount of cement used in the S/S process. (Segui et al., 2013). Ezziane et al. (2007) stated that the addition of special pozzolan to the cement can increase the compressive strength of mortar. Zain et l.. (2010) conducted the S/S research for immobilizing the petroleum sludge. The proportion of the mixture was investigated to find the ratio of cement, rice husk ash, and sludge. Chaff ash (RHA) is added to 5, 10 and 15% cement replacement. The performance of S/S products is measured by compressive strength and permeable porosity. The results show that the addition of rice husk ash (RHA) 5% shows the best performance associated with free pressure at 24.9 N / mm^2. Porosity was found to increase along with the RHA content. The morphology of the compacted cement surface was found to be in the range of 10 to 15 μm for 15% RHA. The addition of ground granulated blast furnace slag (GGBS), pulverized fuel ash (PFA), MgO and Zeolite is an efficient combination in immobilizing the waste containing heavy metals and organic contaminants (Wang et al., 2015).

In ASTM C618, pozzolan is mentioned as a material containing silica and alumina, where the pozzolan material itself does not have binding properties such as cement. In its delicate form and in the presence of water, the compound reacts chemically at a normal temperature to form a binding-like compound such as cement or having a cementic ability. In ASTM C 593–82, it is argued that pozzolan is divided into two kinds, namely natural pozzolan and artificial pozzolan. The natural pozzolan is a natural material that is a pile or sedimentation material from the volcanic ash lava that which contains active silica. Trass is an example of natural pozzolan. Pozzolan is made from stoves and waste utilization products processed into ashes containing reactive silica through the combustion process, such as fly ash (ash), rice huck ash, and micro silica (silica fume). According to Borowski and Hycnar (2016), fly ash in granulated pattern with phosphogypsum constitutes a promising material in the production of cement, for the hardening of cement mortar.

In this research, trass soil was used as OPC mixing. This is because land trass is one of natural pozzolan banya available in Indonesia. Trass is also defined as the rocks from eruptions and volcanoes that have been weathered, gray-white, bluish-gray, dark-gray, yellowish, and somewhat orange (BPPI, 1983; BBPI 1984). Trass is found as a sediment layer with a thickness of 2–8 m or more. According to Susilorini (2003), the Muria Kudus trass can be utilized as an alternative aggregate in concrete mixtures. Laboratory tests show that the SiO_2 content of the Muria Kudus trass is 42.02%, while the Al_2O_3 content is 28.08%. Overall, concrete with a fine aggregate mixture of the Muria Kudus trass shows better performance than normal concrete. The result of compressive strength test is equal to 29,802 Mpa, which exceeds from compressive strength requirement (19 MPa). The trass soil is added because most of its content consists of silica, alumina, or both elements. The results of the study indicate that trass soil potentially serves as a mortar mixture with an indication of a positive result on the strength and durability of concrete produced. The addition of trass soil can increase the compressive strength (Alina et al., 2007).

However, the problem is whether trass soils can stabilize/solidify heavy metals Pb, Cr (VI) and Cu contained in the wastes containing organic substances. Thus, further research is needed to analyze the ability of trass soil with OPC mixture to stabilize/solidify heavy metals in organic waste and to what proportion of hydrocarbon can S/S products of the same quality as the applicable standards be generated. Besides, it is expected to absorb hydrocarbons to improve the process effectiveness. It is also expected to exploit the potential of trash land that is widely spread throughout Indonesia. Indonesia has a lot of trass soil potential spread, but still little utilization, as a substitute for some or all of the cement.

METHODS

The stages of research conducted in this study include preparation of tools and materials, manufacture, testing of S/S specimen products, and data analysis. The materials used are portland ordinary cement mixture with trass soil on various composition variations, and artificial wastes containing Cu, Cr (VI) and Pb heavy metals. A set of tools was used for the manufacture and testing of S/S specimens. Artificial wastes are made up of two types of waste, i.e., heavy-duty heavy metal waste without containing hydrocarbons (inorganic waste) and artificial heavy-metal wastes containing hydrocarbons (organic waste). Heavy-duty artificial metals used in accordance with unreacted waste will be discharged to Category I landfill, wherein the heavy metal content of Cu is higher or equal to 1000 mg/kg wastewater (dry weight), Cr is higher or equal to 2500 mg/kg (dry weight) os waste, and Pb is higher or equal to 3000 mg/kg (dry waste). As for the artificial wastes of heavy metals containing hydrocarbons, preparation carried out by adding liquid paraffin p.a. Preparation of artificial waste of Cu heavy metals is performed using $CuSO_4.5H_2O$, Cr heavy metals using $K_2Cr_2O_7$, and heavy metals Pb using $Pb(NO_3)_2$.

The study was conducted in two stages. In the first step, the process of making S/S specimens on inorganic waste types with mixed cement mixture of OPC and trass is 100: 0, 75:25, 50:50, 25:75, 0: 100. The specimen preparation begins with a Normal Consistency test. The test aims to determine the amount of water required in the OPC and trass cement mixture. Normal consistency test refers to ASTM C187–11. The specimen is printed in a cube mold called a 5×5×5 cm specimen mold, and cured for 28 days. Specimen preparation refers to ASTM C 109. Furthermore, the S/S product test, which includes compressive strength, Toxicity Characteristic Leaching Procedure (TCLP) test of compressive strength tested by Toce Universal Testing Machine Type RAT-200, MFG No. 20380 CAP 200 tf. TCLP test based on US-EPA (method 1311).

The analysis heavy metal concentrations was performed according to Standard Method for examination of water and wastewater, by colorimetric method for Cr (VI), neocuproine for Cu, and dithizone for Pb. Paint test refers to the US EPA 9095B method The optimum composition obtained in the first phase of the study is used in the second stage. The second stage is done on the type of organic waste containing hydrocarbons. The waste used is also an artificial waste of heavy metals Cu, Cr, and Pb containing hydrocarbons. Paraffins are added with composition variations in percent mass, ie 2.5%, 5%, 7.5%, and 10%. The S/S result product is also tested with the same test as the first phase of implementation. In this step, the paint filter test is conducted, which aims to find out the free liquid content remaining in the sample. The paint test refers to the US EPA 9095B method. The data obtained were analyzed statistically using One Way ANOVA (Analysis of variance) method with Minitab 16.

RESULT AND DISCUSSION

Compressive Strength Test

On the basis of the research results it was observed that the type of waste affects the value of compressive strength. The value of compressive strength on inorganic waste type without hydrocarbon enhancers ranges from 56 tons / m^2 to 295 tons / m^2. The highest compressive strength is exhibited by the sponge with OPC composition: 100: 0 trass soil of 6830 tons/m^2. Conversely, the minimum compressive strength value of 40 tons/m^2 is shown by the specimen with 100% trass soil composition. The results of the study on the type of waste without containing hydrocarbons indicate that all compressive strength values in all specimens still meet the minimum limit set by the Head Decision BAPEDAL N0. 03 year 1995 that is equal to 10 tons/m^2. The value of compressive strength on non-hydrocarbon inorganic waste types can be seen in Figure 1. Determination of the effect of the organic matter in the form of hydrocarbon on the compressive strength value is also done by using a statistical test. The test was conducted using ANOVA One Way method with Minitab 16. The analysis result showed a significance at α <0,05 or Fcount> Ftable which means that there is a significant difference in the composition % ratio. If the value of Fcount> Ftable then H0 is rejected and H1 is interpreted significant, it means there is a difference from the comparison group. Conversely, the value of Fcount <Ftable then H0 is accepted and H1 is not interpreted significantly, meaning there is no difference from the comparison group (Agus, 2013). The results of the statistical analysis showed that the value of Fcount> Ftable with P value of 1.57425. Thus, it can be interpreted as significant, which means that the composition ratio indeed has an effect on the value of compressive strength.

The value of compressive strength in the study of organic waste types with the addition of hydrocarbons can be seen in Figure 2. It shows the effect of adding hydrocarbons on the value of compressive strength. The more hydrocarbons are added, the lower the compressive strength. The more hydrocarbons in the form of paraffin added, the lower the compressive strength. This is because the cement cannot coalesce with the added hydrocarbon to form cavities in the stabilized/solidified specimens. The presence of cavities will result in differences in density, thus

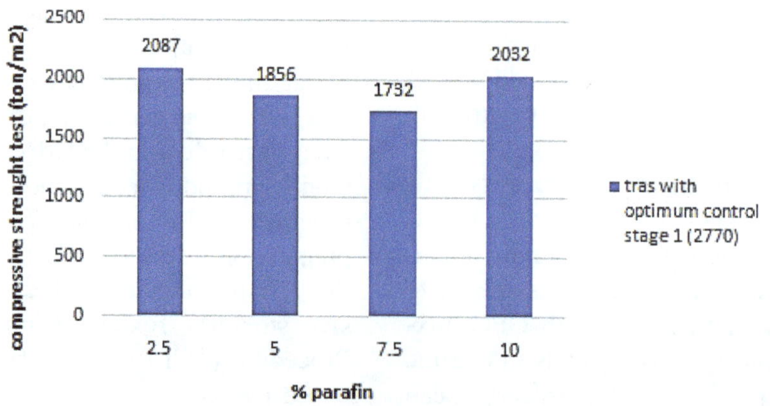

Figure 1 Strength Test Results of OPC and trass soil containing Heavy Metal (HV) without the addition of Hydrocarbons

Table. 1 ANOVA OPC Composition For Soil Trass Against Strong Press

Source of Variation	SS	Df	MS	F	P-value	F crit
Between Groups	4675008	1	4675008	0.23812	1.57425	4.964603
Within Groups	29696733	10	2969673			
Total	34371742	11				

affecting the density of the S/S specimen. The higher the density, the higher the value of the compressive strength. Leonard and Stegemann (2009) stated that the organic component does not match the cement matrix so that it can inhibit the hydration process. This hydration process is very important to determine the strength of cement. If the hydration of cement is disturbed, the compressive strength produced on cement becomes less than optimal.

TCLP Test Results

From the TCLP test results, leaching for heavy metals Pb, the best leaching was obtained on the OPC variation composition: trass soil 50:50 and 25:75, i.e. 0.114 and 0.105 mg/L. In the 75:25 variation of 0.12 mg/L, and on the 100% trass soil variations with the addition of heavy metals, the results still meet the quality standard (PP RI No 101 year 2014). This suggests that the S/S process with OPC variations and trass soils can immobilize Pb. Bajceta et l.. (2013) state that the S/S process can reach 99% Pb immobilization.

Leaching for Cu heavy metals with variations of OPC composition: best trass soil obtained at OPC composition: trass soil 25:75; 100% trass soil of 0.594 and 1 mg/L. In contrast, the lowest leaching was obtained in a variation of 100% composition of heavy metals, which amounted to 0.011 mg/L. The leaching results obtained between various composition variations indicate that the movement of Cu metal can be inhibited by the S/S process using OPC: trass soil. This is in accordance with research conducted by Bajceta et l.. (2013) that Pb, Cu can be immobilized up to 99% by using a modification of OPC and another pozzolan. The average value of heavy metal of S/S products with Tanah Trass can be seen in Figure 3.

In addition to the effects of heavy metal properties, curing time also affects the leaching from heavy metals. Pb and Cu result in lower leaching at curing for 28 days. However, Cr exhibits a higher leaching. The leaching of Cr started at 28 days of curing and the leaching effectiveness was achieved after 90 days of curing (Jain and Garg, 2008).

Determination of the influence of organic substances in the form of hydrocarbons on the value of compressive strength is also done by using statistical tests. The test was done by using ANOVA One Way method with Minitab 16. From the statistical analysis, it was found that in the table. 2 Fcount <Ftable with P value of 0.01021. so it can be interpreted as insignificant, which means that the composition ratio has no effect on leaching Pb. However, the results of leaching Pb still meet the quality standard. Table 3 showed from the results of statistical analysis that the value of Fcount <Ftable with P value of 0.03198. It can be interpreted as insignificant, which means that the composition ratio does not affect the leaching of Cu. However, Cu leaching results still meet the quality standards.

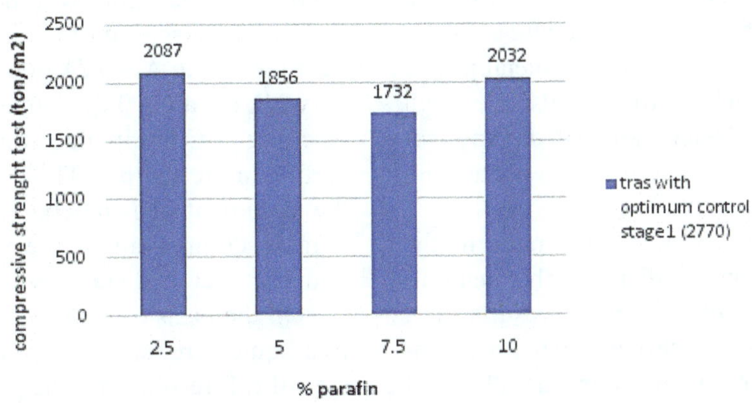

Figure 2 Strength test results of OPC and trass soil with the addition of hydrocarbons

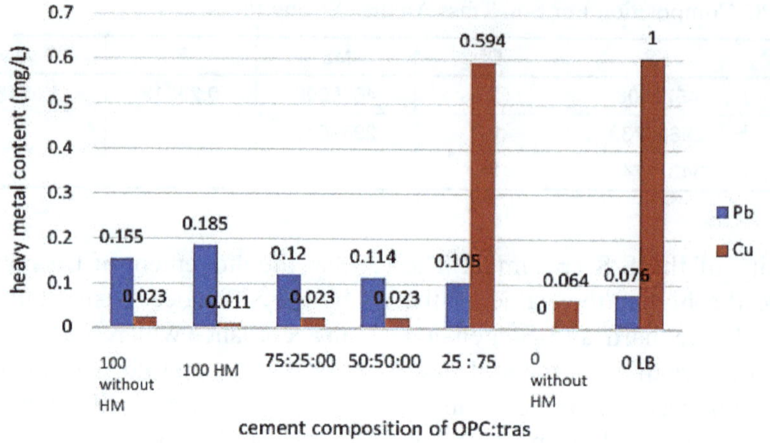

Figure. 3. Results of analysis Heavy Metal S / S Product OPC: Trass without the addition of Hydrocarbons

Table 2 ANOVA Composition of OPC For Land Trass Against Pb

Source of Variation	SS	Df	MS	F	P-value	F crit
Between Groups	8683.078	1	8683.078	9.261567	0.010211	4.747225
Within Groups	11250.47	12	937.5388			
Total	19933.54	13				

Table 3 ANOVA Composition of OPC For Land Trass Against Cu

Source of Variation	SS	Df	MS	F	P-value	F crit
Between Groups	5167.248	1	5167.248	6.200636	0.031984	4.964603
Within Groups	8333.417	10	833.3417			
Total	13500.66	11				

The TCLP test showed that the composition of OPC cement and 50:50 trass soil with paraffin addition, as shown in Figure 3, exhibits unstable heavy metal leaching result on each heavy metal type. This is due to the presence of organic components that interfere with the cement hydration process, as well as the value of compressive strength.

Due to the presence of organic components, the hydration process does not take place perfectly, so the ability of cement and its mixture with trass soil, is less effective in stabilizing/solidifying Pb and Cu. The results of heavy metal piging of Pb and Cu are unstable, probably also the result of reading the absorbance value of the spectrophotometer. The absorbance readings occur with unstable numbers.

The addition of organic substances in the form of hydrocarbons, indicating the instability of heavy metals of TCLP test results. The first phase of the study showed that Pb leaching was 0.01 mg/L. whereas in the Phase II study, the Pb leaching became unstable in the 0.076 mg/L range with the addition of 2.5% paraffin, 0.107 mg/L at 5%, 0.076 at 7.5% and 0.104 mg/L at 10%. From statistic test using ANOVA one way method (Table 4), the P value equal to 0.368 indicated that the addition of paraffin on the specimen did not affect the leaching of Pb. This is because Pb leaching on each sample is not significant.

In the study without the addition of Cu hydrocarbon, only 0.02 mg/L leached. On the other hand, in the research with the addition of hydrocarbon, the leaching of Cu increased. The statistical test using ANOVA one way method (Table 5) with P value 0.02 value indicated that the leaching of Cu influenced the addition of hydrocarbon in the sample. This is in accordance with the statement of Minocha et al. (2003) that the organic component of the waste may interfere with the chemical stability of the S/S specimen.

On the basis of table 6, it can be seen that no free liquids are left in the sample so that all samples of S/S results pass the paint test. This shows that the addition of paraffin mass up to 10% of the

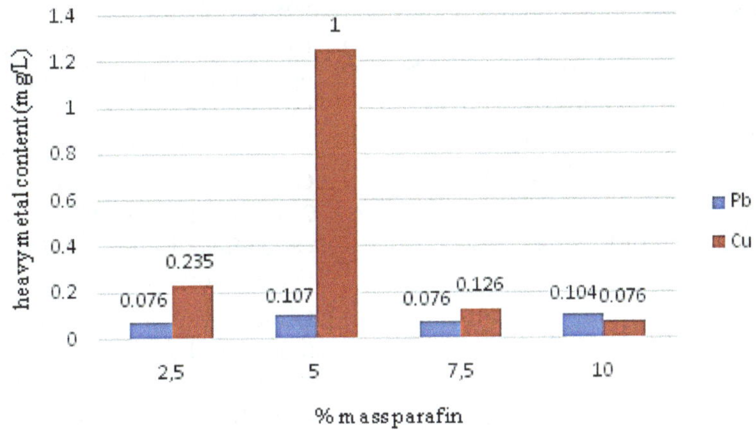

Figure 4. Results of heavy metal analysis of OPC 50: Trass 50 with the addition of hydrocarbons

Table 4 ANOVA% Paraffin For Trass Against Pb

Source of Variation	SS	Df	MS	F	P-value	F crit
Between Groups	82.23623	1	82.23623	26.31436	0.00684	7.708647
Within Groups	12.50058	4	3.125146			
Total	94.73681	5				

Table 5 ANOVA% Paraffin For Trass Against Cu

Source of Variation	SS	df	MS	F	P-value	F crit
Between Groups	73.80131	1	73.80131	22.04824	0.009339	7.708647
Within Groups	13.38906	4	3.347265			
Total	87.19037	5				

Table 6. Hasil Paint *Filter Test*

% mass parafin	Name object of tested		Result of paint filter test
2,5%	H	H1	No detection of free liquid
		H2	No detection of free liquid
5%	I	I1	No detection of free liquid
		I2	No detection of free liquid
7,5%	J	J1	No detection of free liquid
		J2	No detection of free liquid
10%	K	K1	No detection of free liquid
		K2	No detection of free liquid

weight of the cement mixture of portland: trass soil can still be bonded to cement, although it affects the value of compressive strength and heavy metal leaching results.

CONCLUSION

This type of waste will affect the quality of the stabilization/solidification process. The effect of organic substances in the form of hydrocarbons on the quality of SS products is seen based on the value of compressive strength test and heavy metal lattice content after the TCLP test. In the maximum proportion of 10% hydrocarbon content, the value of compressive strength still far exceeds the standard quality standard, which is 1700 tons/m^2. However, with the presence of hydrocarbons, it can affect the TCLP test, which results in an unstable outcomes of the metal leaching. The results of the paint test showed that no stains were left in the sample.

Acknowledgements

Thanks to Ristra Megawati Asri Pratiwi. ITS Civil Concrete Laboratory Technician, ITS Environmental Engineering Laboratory Technician and Technician of Water Quality Laboratory of ITS Chemical Engineering Industry who has assisted in the implementation of this Research.

REFERENCES

1. Al Futaisi, A. Jamrah, B. Yaghi, and R. Taha, 2007, Assessment of alternative Management Techniques of Tank bottom Petroleum Sludge in Oman. J. of Hazardous Materials, 141, 2007, 557–564.
2. Alina, B, Georgescu, M dan Zahanagiu, A. 2008, Properties of Blended Cements with Hazardous Waste Content. Journal of Applied Chemistry and Material Science, 53 (3), 229–237
3. APHA; AWWA; WEF 1998, Standard Methods for the Examination of Water and Wastewater. 20th Edition. L.S. Clesceri; A.E. Greenberg; A.D. (Eds.) Eaton, Washington, American Public Health Association.
4. ASTM C109. 2004. Tentang "Standard Test Method for Compressive Strength of Hydraulic Cement Mortars". Annual Books of ASTM Standards, USA.
5. Ayotamunu, M.J., Okparanma, R.N., Ogaji, S.O.T., dan Probert, S.D, 2007. Bioremediation of a sludge containing hydrocarbons. J. Appl. Energy. 84(9): 936–943.
6. Borowski G. & Hycnar J.J. 2016. The effect of granulated fly ashes with phosphogypsum on the hardening of cement mortar. Technical Transactions – Civil Engineering, Vol. 113, Iss. 2-B(7), 37–45.
7. Carpio, I.E.M., Machado-santelli, G., Sakata, S.K., Filho, S.S.F., dan Rodrigues, D.F. 2014. Copper Removal Using A Heavy-Metal Resistant Microbial Consortium in A Fixed-Bed Reactor. Water Research 62, 156–166.
8. Dominguez, A, Mendez, J.A. Inguanzo, M, 2005. Investigation Into the Characteristics of Oils Produced from Microwave Pyrolysis of Sewage Sludge. Fuel Process Technology.
9. Desogus, P., Manca, P.P., Orru, G., dan Zucca, A. 2013. Stabilization-solidification Treatment of Mine Tailing Using Portland Cement, Potasium Dihydrogen Phosphate, and Ferric Chloride Hexahydrate. Minerals Engineering 45, 47–54.
10. Ezziane, K., Bougara, A., Kadri, A., Khelafi, H., and Kadri, E. 2007. Compressive Strength of Mortar Containing Natural Pozzolan under Various Curing Temperature. Cement and Concrete Composites", 29, 587–593.
11. Hu, G., Li, J., dan Zeng, G. 2013. Recent Development in the Treatment of Oily Sludge from Petroleum Industry: A Review. Journal of Hazardous Material, Vol. 261, 470–490.
12. Gailius, A., Vacenovska, B., dan Drochytka, R. 2010. Hazardous Wastes Recycling by Solidification/Stabilization Method. Materials Science (Medziagotyra) 16(2).
13. Ghricia, M., Kenai, S., and Mansour, M.S, 2007. Mechanical Properties and Durability of Mortar and Concrete Containing Natural Pozzolana and Limestone Blended Cements. Cement and Concrete Composites 29, 542–549.
14. Government Regulation of the Republic of Indonesia No. 101Year 2014 on the Management of Hazardous and Toxic Waste.
15. Karamalidis, A.K dan Voundrias, E.A, 2007. Cement-based Stabilization/solidification of Oil Refinery Sludge Leaching Behavior of Alkanes and PAHs. Journal of Hazardous Material, 148, 122–135.
16. Decree of the Head of Environmental Impact Management Agency no. 128 of 2003 on the Procedures And Technical Requirements of Oil and Petroleum Wastewater Treatment Contaminated By Petroleum.
17. Decree of the Head of Environmental Impact Management Agency no. No. 03 of 1995 on Working Procedures of Stabilization / Solidification. Decision of the Head of the Controlling Agency.
18. Decree of the Head of Environmental Impact Management Agency No. 04 of 1995 on Procedures for Requirements for Stockpiling of Processing Requirements, Location Requirement for Used Processing and Location of Former Dumps of Hazardous and Toxic Waste Materials.
19. Decree of the Head of Environmental Impact Management Agency No. 09 of 1995 on Technical Requirements for Processing Hazardous and Toxic Waste Materials.
20. Leonard, Ş.A., dan Stegemann, J.A. 2009. Stabilization/solidification of Petroleum Drill Cuttings: Leaching Studies. Journal of Hazardous Materials 174, 484–491.
21. Minocha, A.K., Jain, N., dan Verma, C.L. 2003. Effect of Organic Materials on The Solidification of Heavy Metal Sludge. Journal of Construction and Building Materials 17, 77–81.
22. Research and Development Center Balai (BPPI), 1984. Research and Quality Testing of Tras in Central Java in order to Improve the Quality of Batak Products ountuk Development of Industrial Building Materials, Research Report of the Ministry of Industry Semarang.

23. Segui, P., Aubert, J.E., Husson, B., and Measson, M, 2013. Utilization of A Natural Pozzolan as The Main Component of Hydraulic Road Binder.Construction and Building Materials, 40, 217–223.

24. Senhadji, Y., Escadeillas, G., Mouli, M., Khelafi, H., dan Benosman. 2013. Influence of Natural Pozzolan, Silica Fume, and Limestone Fine on Strength, Acid Resistance and Microstructur of Mortar. Powder Technology 254, 314–323.

25. Shawabkeh, R.A. 2005. Solidification and Stabilization of Cadmium Ions in Sand-cement-clay Mixture. Journal of Hazardous Materials B125, 237–243.

26. Silva, L.J., Alves, F.C., dan Franca, F.P. 2012. A Review of the Technological Solutions for the Treatment of Oily Sludges from Petroleum Refineries. Waste Management and Research, 30(10), 1–15.

27. Susilorini, R, 2003. Trass Muria Kudus Explores the Potential of Nature as a Building Material of the People. Proceeding National Seminar on Environment, Unika Soegijapranata, Semarang, September 2003, P. A.1.1 – A.1.7

28. SNI 15–2049–2004 about "Portland Cement". National Standardization Agency.

29. U.S EPA. 1992. Toxicity Characteristic Leaching Procedure. U.S Environmental Protection Agencies.

30. U.S EPA. 2004. Paint Filter Liquid Test. U.S Environmental Protection Agencies.

31. Voglar, G.E., dan Lestan, D. 2011. Efficiency Modeling of Solidification/Stabilization of Multi-Metal Contaminated Industrial Soil Using Cement and Additives. Journal of Hazardous Materials 192, 753–762.

32. Zain, A.M., Shaban, G.Md., and Mahmud, H, 2010. Immobilization of Petroleum Sludge Incorporating Portland Cement and Rice Husk Ash. International Journal of Chemical Engineering and Applications, 1(3).

33. Zhang, J., Li, J., Thring, R.W., Hu, X., dan Song, X. 2012. Oil Recovery from Refinery Oily Sludge Via Ultrasound and Freez/Thaw. Journal of Hazardous Materials, 203–204, 195–203.

11

Status Assessment of Heavy Metals in Water of the Lepenci River Basin, Kosova

Pajtim Bytyçi[2,1], Osman Fetoshi[4*], Bujar H. Durmishi[3], Ferdije Zhushi Etemi[1], Hazir Çadraku[5], Murtezan Ismaili[2], Albona Shala Abazi[6]

[1] Department of Biology, Faculty of Mathematics and Natural Sciences, University of Prishtina "HasanPrishtina", George Bush, Prishtina 10000, Kosova

[2] Southeast European University (SEEU), Ilindenska no 335 Tetova, FYR Macedonia

[3] University of Tetova, Faculty of Mathematics and Natural Sciences, Department of Chemistry, Str. Ilindeni nn, 1200 Tetova, FYR Macedonia

[4] University of Applied Sciences, Faculty of Tourism and Environment, Ferizaj, 70000, Kosova

[5] University for Business and Technology, Faculty of Engineering Energy, Kalabira 10000, Pristina, Kosova

[6] University "HaxhiZeka", Faculty of Management in Tourism, Hotels and the Environment, Pejë, Kosova

[*] Corresponding author e-mail: osmanfetoshi@hotmail.com

ABSTRACT

Water represents an essential element for life and living things on earth. Aquatic ecosystems play a decisive role in the socio-economic development in urban and rural areas. In recent decades, there has been concern at the global level with regard to the deterioration of aquatic ecosystems due to the pollution, which is a product of mainly anthropogenic activity. Heavy metals pollution is worrisome for the ecological balance of the aquatic environment, affecting a variety of organisms. Therefore, the objective of this paper was to characterize the quality of water in the Lepenci River basin, to express heavy metals concentration, and to determine the surface water quality index in this basin. In order to achieve such an objective, water samples were collected at eight stations for analyzing the concentrations for heavy metals. Heavy metals were determined by means of atomic absorption spectrophotometry. A good correlation was found between Pb and Zn ($r = 0.84$), whereas the average negative correlation coefficient was shown between Mn and Cr ($r = -0.6513$). The heavy metals concentrations varied from 0.0092 to 0.1135 mg/L.The mean concentrations of heavy metals found in the river water were in the order of: Mn > Fe > Pb > Ni > Cd > Zn > Cr > Cu. The Water Quality Index varied from 57 to 81, with the average value of WQI = 68.1250, which ranks the surface water of this basin as fair. From the results we have concluded that the Lepenci River waters during the monitoring period have had low pollution from heavy metals.

Keywords: atomic absorption spectrophotometry, river water, heavy metals, WQI, quality, correlation.

INTRODUCTION

Water is an essential ingredient for the environment and its quality constitutes an issue of concern. Aquatic ecosystems, such as rivers, play an important role for rural and urban populations in many developing countries. However, in recent years, they have been subjected to various forms of degradation due to the pollution by internal waste, industrial leaks, agricultural waste, and bad fishing practices [Ndimele, 2008]. According to [Bakare et *al.*, 2003], industrialization and human activities have partially or fully turned the environment into a landfill site. Consequently, many rivers have been polluted and became harmful to the human life and the lives of other living things.

Resistant metals or metalloids with the density greater than 4.5 g/cm^3 are known as heavy metals and include lead, cadmium, mercury, iron, copper, zinc, nickel, and manganese [Anderson, 2003]. Some of them are essential elements without which the biochemical processes in living organisms would not be possible; however, when they exceed normal concentrations, they become harmful to organisms. Heavy metals pollution can cause disturbing effects on the ecological balance of the aquatic environment and affect a variety of organisms. This is of particular importance in ecotoxicology, as heavy metals are highly persistent and have the tendency to bioaccumulate and concentrate in the food chain; at higher concentrations, they become toxic to organisms. Heavy metal pollution in water can be monitored through measuring their concentrations [Senarathne and Pathiratne, 2007].

Heavy metals are of special importance to surface water pollution. Certain metals are useful, while others are harmful and toxic. Some metals, such as Ca and Mg are essential chemical elements, whereas other metals adversely affect water consumers. It was proven that Pb, Hg, and As are strong, toxic pollutants [BožoDalmacija, 2001]. The toxicity of heavy metals depends on the type of metal and compound, the amount that reaches the body and the duration of metal reaction. This group includes Hg, Pb, Cd, Cr, Cu, Ni, As and Zn.

Heavy metals are of particular importance for the environment, as they exhibit toxicity and persistence and are known for their bioaccumulation in food chains [WHO, 2000]. They cause adverse effects on plants and animals, and are dangerous to the human health. It is known that metal toxicity occurs mainly due to the presence of free metal ions.

Water quality directly affects the health of living things, reflecting the level of contamination of the aquatic ecosystem. Water quality monitoring is an important component of water management, for which the data analysis is needed to identify and characterize water quality issues. Assessment is the process through which water quality data is transformed into information. The information acquired from the monitoring is essential for assessing the water quality. Monitoring can also verify the water contamination based on which a corrective action can be undertaken. Surface waters should be monitored microbiologically, chemically, physically, and radiologicallyin order to determine the presence of pollutants. Therefore, effective monitoring and assessment of surface water quality are crucial to protecting the aquatic life and human health since wastewater consumption is one of the major causes of diseases. Thus, with a preventive approach, the quality of water can be managed.

Water quality assessment can be carried out in various ways. In 1965, Horton was the first to formulate a water quality index which was then used by many researchers for different types of water. A very powerful tool to this end is the Water Quality Index (WQI). WQI serves to sum up large amounts of water quality data in simple categories (good, bad) for management and public reporting [Durmishi et *al.*, 2012]. Researchers use different types of indices. The index aims at transforming numerous water quality data to simple, understandable, and usable information for the public. WQI represents a number from 0 to 100 where a higher value means better water quality and vice versa. The aim of the paper was to assess the level of heavy metal concentrations in water that will focus on the pollution status of the Lepenci river.

MATERIALS AND METHODS

Study area

The Lepenci river is located in the southeast of the Republic of Kosova. It originates in the Oshlak mountains at an altitude of 2212 m; its total length is 50 km and has a slope of 2.1%. Annual average flow is 8.4 m^3/s. The minimum flow of the river Lepenci 1.8 m^3/s, as shown at the Hydrometric Station in Hani i Elezit [MESP, 2010].

The Lepenci river represents the main catchment area in the southeastern region of Kosova. It has an area of 652 km^2 covering 5.98% of the territory of the Republic of Kosova. The area where the river Lepenci flows is characterized by the continental and mountain climate. The average annual rain is 861.4 mm, while the average annual temperature is 10.2°C [Kaçanik, 2011].

For realization of this work, the water samples were taken in spring, summer and autumn of 2017 in eight sampling points along the Lepenci river flow that included upper, middle and downstream. The sampling points were: SP1 (Prevallë), SP2 (Jezerc), SP3 (Brod), SP4 (Runjevë), SP5 (Nikë), SP6 (Gërlicë), SP7 (Kaçanik) and SP8 (Hani iElezit). The sample bottles were labeled with the date and source of sampling; they were kept in refrigerators at 4°C and transported under the appropriate procedure 21.

The analysis of water samples taken in river Lepenci was performed in the laboratory of the Kosova Hydro Meteorological Institute (KHMI). Heavy metals were determined by means of

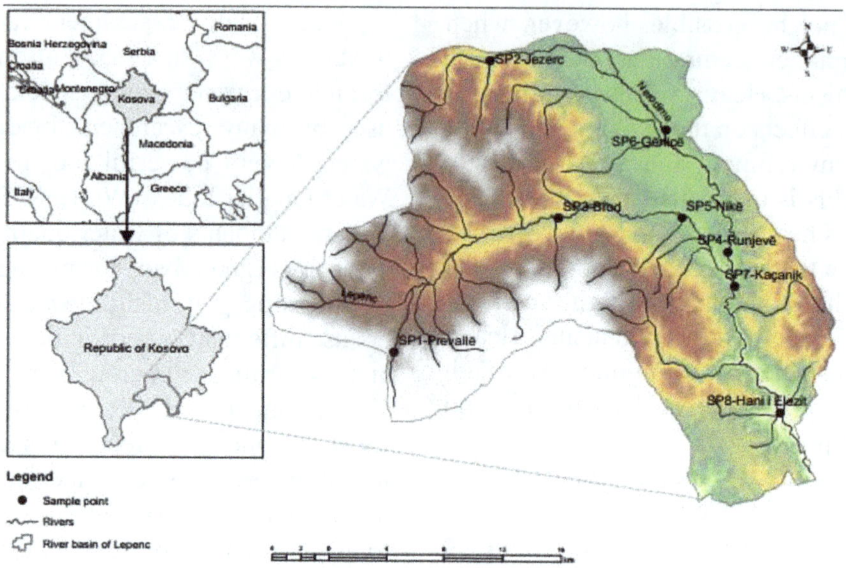

Figure 1. Map of monitoring stations

Atomic absorption Spectrophotometer of American label Perkin Elmer, type Analyst 400. The methods applied to these devices conform to standard methods such as: DIN, ISO and EN. Statistical calculations in this paper were conducted with Excel software, where the data is input and mathematical equations integrated in this program are extracted values: minimal, maximum, average, standard deviation and variation coefficient for the parameters measured for each sampling station that were shown graphically in the works. Data interpretations are performed based on the data obtained in the laboratory, their processing and comparison with the standards.

Canadian Water Quality Index

In order to assess the drinking water quality we have widely used the Water Quality Index (WQI) developed by the Canadian Council of Ministers of the Environment [CCME, 2001]. WQI consists of three measures of variance from selected drinking water quality objectives. These are: scope (F1), frequency (F2), and amplitude (F3).

The scope represents the water quality of the legal norm that does not meet the objectives during the period of interest and is expressed by the equation:

$$F_1 = \left(\frac{Number\ of\ failed\ variables}{Total\ number\ of\ variables}\right) \quad (1)$$

Frequency represents the percentage of individual tests that do not meet the objectives. It is expressed with the following equation:

$$F_2 = \left(\frac{Number\ of\ failed\ tests}{Total\ number\ of\ tests}\right) \cdot 100 \quad (2)$$

Amplitude represents the amount by which failed tests do not meet the objectives. It is expressed with the following equation:

$$F_3 = \left(\frac{nse}{0.01 \cdot nse + 0.01}\right) \quad (3)$$

where *nse* shows the normalized sum of excursions, that is, the summation by which individual tests do not meet the objectives.

WQI is then calculated according to the following equation:

$$WQI = 100 - \left(\frac{\sqrt{F_1^{\ 2} + F_2^{\ 2} + F_3^{\ 2}}}{1.732}\right) \quad (4)$$

The number 1.732 normalizes the resulting values to an extent between 0 and 100, where 0 represents the worst water quality and 100 represents the best water quality. Water quality is then ranked in one of the following six categories (Table 1).

RESULTS AND DISCUSSION

The results of heavy metals concentrations measured in the Lepenci river water at different stations are given in Table 2 and Figures 2–9.

Table 1. Categorization of water quality according to WQI values and description.

Class	WQI value	Description
Excellent	95–100	Water quality is protected with a virtual absence of impairment; conditions are very close to pristine levels. These index values can only be obtained if all measurements meet recommended guidelines virtually all the time.
Very good	89–94	Water quality is protected with a slight presence of impairment; conditions are close to pristine levels.
Good	80–88	Water quality is protected with only a minor degree of impairment; conditions rarely depart from desirable levels.
Fair	65–79	Water quality is usually protected but occasionally impaired; conditions sometimes depart from desirable levels.
Marginal	45–64	Water quality is frequentlyimpaired; conditions often depart from desirable levels.
Poor	0–44	Water quality is almost always impaired; conditions usually depart from desirable levels

Cromium (3+)

Cr is an essential micronutrient for animals and plants. It is considered as a relative biological and pollution significance element [Rajappa *et al.*, 2010]. Generally, the natural content of chromium in water is very low except for the regions with substantial chromium deposits. The concentration of Cr can result from industrial and mining processes. Fish are usually more resistant to Cr than other aquatic organisms, but they can be affected sub-lethally when the concentration increases [Krishna *et al.,* 2014].

The results from experimental measurements for the Cr^{3+} are shown in Table 2 and Figure 2. The recommended value for the Cr^{3+} according to the Romanian standards for assessing the ecological status of surface water (GD161) is 0.025 – >0.250 mg/L. During the three-season study, the Cr^{3+} concentration varied from 0.0110 to 0.0360 mg/L. The lowest value was measured at SP3 and SP4 stations in spring and fall, whereas the highest value was measured at SP5 station in spring. The average values in spring, summer, and fall were 0.023, 0.035, and 0.012 mg/L respectively, whereas the three-season average with

Table 2. Results of metal concentration and statistics

	Heavy metals, mg/L		Prevallë	Jezerc	Brod	Runjevë	Nikë	Gërlicë	Kaçanik	Hani i Elezit	Max	Min	Aver	SD	Cv
SP	Chromium	Cr^{3+}	< 0.003	< 0.003	0.011	0.016	0.036	< 0.003	< 0.003	0.029	0.036	0.011	0.023	0.011	0.49
SA	Chromium	Cr^{3+}	<0.003	0.035	<0.003	<0.003	<0.003	<0.003	<0.003	<0.003	0.035	0.035	0.035	0	0
AU	Chromium	Cr^{3+}	< 0.003	< 0.003	0.011	0.011	< 0.003	< 0.003	0.018	0.008	0.018	0.008	0.012	0.004	0.34
SP	Cadmium	Cd^{2+}	< 0.001	< 0.001	< 0.001	0.084	0.042	0.104	0.082	0.089	0.104	0.042	0.078	0.023	0.29
SA	Cadmium	Cd^{2+}	0.023	0.031	0.047	<0.001	0.075	0.053	<0.001	0.084	0.084	0.023	0.052	0.023	0.45
AU	Cadmium	Cd^{2+}	< 0.001	< 0.001	0.015	0.005	0.033	0.009	< 0.001	< 0.001	0.033	0.005	0.016	0.012	0.74
SP	Nickel	Ni^{2+}	0.014	0.015	0.068	0.053	0.018	0.155	0.119	0.047	0.155	0.014	0.065	0.051	0.78
SA	Nickel	Ni^{2+}	0.069	0.066	0.027	0.025	0.118	<0.006	<0.006	<0.006	0.118	0.025	0.064	0.038	0.59
AU	Nickel	Ni^{2+}	0.054	< 0.006	< 0.006	0.002	0.012	< 0.006	0.045	0.035	0.054	0.002	0.029	0.022	0.75
SP	Zinc	Zn^{2+}	0.005	0.007	0.01	0.055	<0.0003	<0.0003	0.075	0.008	0.075	0.005	0.03	0.036	1.01
SA	Zinc	Zn^{2+}	0.078	0.03	0.018	0.035	0.031	0.04	0.036	0.011	0.078	0.011	0.036	0.019	0.54
AU	Zinc	Zn^{2+}	0.014	0.050	0.026	0.003	0.002	0.038	0.031	0.168	0.168	0.002	0.05	0.053	1.07
SP	Manganese	Mn^{2+}	<0.002	<0.002	0.136	0.285	0.117	0.151	0.242	0.433	0.433	0.117	0.23	0.12	0.50
SA	Manganese	Mn^{2+}	0.088	0.106	0.094	0.095	0.058	0.460	0.053	0.115	0.46	0.053	0.15	0.13	0.84
AU	Manganese	Mn^{2+}	3.736	0.12	1.502	1.602	0.146	0.365	0.207	0.106	3.736	0.106	1.16	1.27	1.09
SP	Copper	Cu^{2+}	0.021	0.023	0.037	< 0.002	0.001	< 0.002	0.009	0.038	0.038	0.001	0.021	-0.015	-0.7
SA	Copper	Cu^{2+}	<0.002	<0.002	<0.002	0.001	<0.002	0.010	<0.002	<0.002	0.01	0.001	0.005	0.006	1.15
AU	Copper	Cu^{2+}	< 0.002	< 0.002	< 0.002	< 0.002	< 0.002	< 0.002	< 0.002	< 0.002	0	0	0	0	0
SP	Iron	Fe^{2+}	0.115	0.12	0.401	0.154	0.332	0.065	0.046	0.243	0.401	0.046	0.192	0.12	0.66
SA	Iron	Fe^{2+}	0.003	<0.002	<0.002	0.229	0.034	0.175	0.081	<0.002	0.229	0.003	0.108	0.09	0.88
AU	Iron	Fe^{2+}	< 0.002	0.03	0.008	0.112	0.075	0.128	0.199	0.165	0.199	0.008	0.103	0.06	0.67
SP	Lead	Pb^{2+}	0.047	0.053	< 0.002	0.181	< 0.002	0.295	0.341	0.385	0.0385	0.047	0.21	0.14	0.67
SA	Lead	Pb^{2+}	<0.002	<0.002	<0.002	<0.002	<0.002	<0.002	<0.002	<0.002	0	0	0	0	0
AU	Lead	Pb^{2+}	< 0.002	0.007	0.005	< 0.002	< 0.002	0.018	0.031	0.097	0.097	0.005	0.03	0.03	1.02

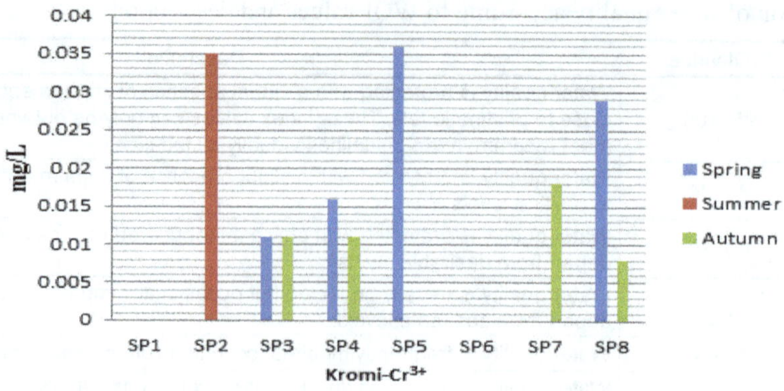

Figure 2. Variation of Cr values in water of Lepenci river

the standard deviation was 0.0233 ± 0.005 mg/L. The Cr^{3+} average concentrations at SP1-SP8 stations were 0.0030, 0.0137, 0.0083, 0.0100, 0.0140, 0.0030, 0.0080, and 0.0133 mg/L, respectively. In spring, the SP3 station showed the minimum value of 0.0110 mg/L, whereas the SP5 station showed the maximum value of 0.0360 mg/L. In summer, the SP2 station showed the minimum value of 0.0350 mg/L, whereas the SP2 station showed the maximum value of 0.0350 mg/L. In fall, SP7 and SP8 stations showed the minimum value of 0.0080 mg/L, whereas the SP7 station showed the maximum value of 0.0180 mg/L. The measurements for the Cr^{3+} in the water of Lepenc River basin were found to be within the minimum values which, if compared to the GD161 standards, were within the allowed limits, thus ranking the water of this river in the first class.

Cadmium (Cd^{2+})

Cd as pollutants in the water can reach from industrial discharges and waste ore. It is widely used in galvanization and production of batteries. Cd is chemically similar to Zn and found in water in the oxidation form +2. It occurs naturally as CdS and followed minerals of Pb and Zn. Food is the main source of Cd poisoning. The effects of acute Cd poisoning in humans are very serious, including hypertension, kidney damage, causing potential prostate cancer, etc. [Durmishi et *al*., 2016]. The physiological effects of Zn with Cd replacement in some enzymatic reactions hampered the normal functioning of enzymes. The toxic effect of Cd in water decreases with the increasing hardness of the water, due to rising water carbonates content. Heavy metal carbonates are less soluble in water and so a part of dissolved Cd is removed . It is a non-essential element, known to have a toxic potential. Cd is highly toxic and responsible for several cases of poisoning through food. Small quantities of Cd cause adverse changes in the arteries of human kidney. It replaces Zn biochemically and causes hypertension as well as kidney damage. It interferes with enzymes and causes a painful disease called Itai-itai [Rajappa*et al.*, 2010]. High concentration of Cd occurs at neutral and alkali pH.

The results from experimental measurements for the Cd^{2+} are shown in Table 2 and Figure 3. According to the Romanian standards for assessing the ecological status of surface water (GD161), the recommended value for the Cd^{2+} is 0.0005->0.005 mg/L. During the three-season study, the Cd^{2+} concentration varied from 0.0050 to 0.0840 mg/L. The lowest value was measured at SP4 station in fall, whereas the highest value was measured at SP8 station in summer. The average values in spring, summer, and fall were 0.0780, 0.0520, and 0.0160 mg/L respectively, whereas the three-season average with the standard deviation was 0.0487±0.0193 mg/L. The Cd^{2+} average concentrations at SP1-SP8 stations were 0.0083, 0.0110, 0.0210, 0.0300, 0.0500, 0.0553, 0.0280 and 0.0580 mg/L respectively. In spring, the SP5 station showed the minimum value of 0.0420 mg/L, whereas the SP6 station showed the maximum value of 0.1040 mg/L. In spring, the SP1, SP2, and SP3 stations showed the values below detection levels; water values at SP4, SP6, SP7, and SP8 stations, based on GD161 standards, resulted in classification to the fifth class, whereas the water at SP5 station belonged to the third class. In summer, the SP1 station showed the minimum value of 0.0230 mg/L, whereas the SP8 station showed the maximum value of 0.0840 mg/L. In summer, the analyzed values at SP4 and SP7 stations were below the detection levels; the water values at the SP1, SP2, and SP3 stations belonged to the third class, whereas Cd^{2+} values of water at SP5, SP6, and SP8 stations were categorized into the fifth category. In fall, the SP4 station showed the minimum value of 0.0050 mg/L, whereas the SP5 station showed the maximum value of 0.0330 mg/L. In fall, the Cd^{2+} values at SP1, SP2, SP7, and SP8 stations were under the detection levels; the water at the SP3, SP4, and SP6 stations belonged to the fifth category, whereas the water of the SP5 station belonged to the third category. According to the seasonal average at all stations, the water belonged to the fifth category.

Nickel (Ni^{2+})

Ni is an essential trace metal for several animal species, micro-organisms and plants; therefore, either deficiency or toxicity symptoms can occur when, too little or too much Ni is taken up, respectively. Although a number of cellular effects of Ni have been documented, a deficiency state in humans has not been described [Scoott-Fordsmand, 1997]. Ni and its compounds have many industrial and commercial uses, and the progress of industrialization has led to an increased emission of pollutants into ecosystems. Although Ni is omnipresent and vital for the function of many organisms, concentrations in some areas from both the anthropogenic release and naturally varying levels may be toxic to living organisms [Diagomanol et *al.*, 2004]. Ni cause toxic effects in the respiratory tract and immune system. The exposure of the general population to Ni mainly concernedthe oral intake, primarily through water and food. It is also known to affect non-occupationally exposed individuals, especially those handling stainless steel and nickel-plated articles of everyday use, because nickel is a common sensitizing agent with a high prevalence of allergic contact dermatitis [Kitaura et *al.*, 2003].

The results from the experimental measurements for the Ni^{2+} are shown in Table 2 and Figure 4. According to the Romanian standard for assessing the ecological status of surface water (GD161), the recommended value for the Ni^{2+} is 0.01 – >0.1 mg/L. During the three-season study, the Ni^{2+} concentration varied from 0.0020 to 0.01550 mg/L. The lowest value was measured at SP4 station in fall, whereas the highest value was measured at SP6 station in spring. Average values in spring, summer, and fall were 0.065, 0.064, and 0.029 mg/L respectively, whereas the three-season average with the standard deviation was 0.0527±0.0370 mg/L. The average concentrations values of the Ni^{2+} at SP1-SP8 stations were 0.0457, 0.0290, 0.0337, 0.0267, 0.0493, 0.0557, 0.0567, and 0.0293 mg/L, respectively. In spring, the SP1 station showed the minimum value of 0.0140 mg/L, whereas the SP6 station showed the maximum value of 0.1550 mg/L. In summer, the SP3 station showed the minimum value of 0.0250 mg/L, whereas the SP5 station showed the maximum value of 0.1180 mg/L. In fall the SP4 station showed the minimum value of 0.0020 mg/L, whereas the SP7 station showed the maximum value of 0.0540 mg/L. The measurements for the Ni^{2+} in the water of Lepenc River basin were found to be within the allowed recommended values of GD161 standards, thus ranking the water of Lepenci river in the first class.

Zinc (Zn^{2+})

Zn is found in natural waters in larger quantities compared with Cu, Pb, Cd and Hg, so the industrial discharge waters containing high concentrations of it. Zn is an essential element for the life of animal and human beings. It is found in virtually all food and potable water in the form

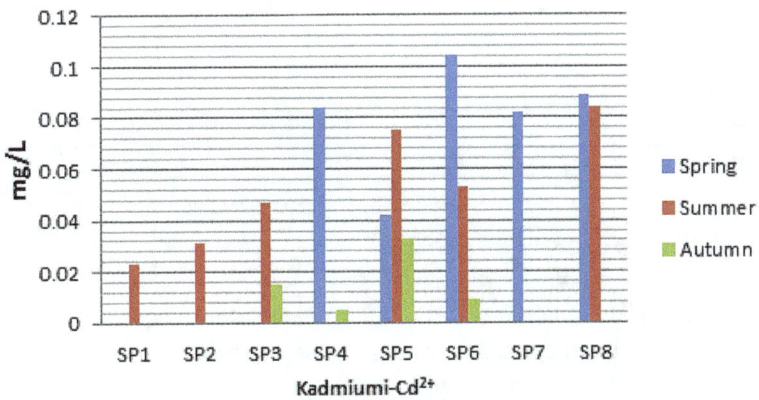

Figure 3. Variation of Cd values in water of Lepenci river

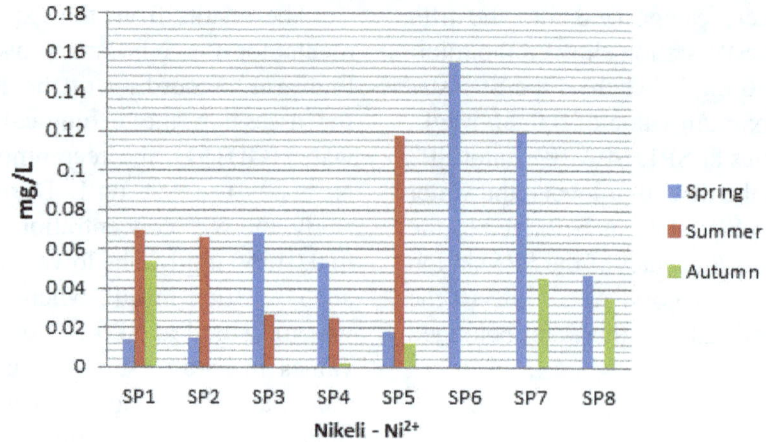

Figure 4. Variation of Ni values in water of Lepenci river

of salts or organic complexes [WHO, 2011]. The main sources of Zn pollution in the environment are zinc fertilizers, sewage sludge, and mining. Urban runoff, mine drainage, and municipal sewages are the more concentrated sources of zinc in water [Damodharan, 2013]. Zn is not accumulated in the body and it is activating enzymes. It affects bone growth, development and functioning of reproductive organs, etc. There are rare cases of Zn poisoning. Its action in the aquatic fauna depends on the hardness of water, saturated with oxygen and temperature. Salts of alkaline-earth elements reduce the toxicity of Zn, while increasing the temperature, whereas the lowering the concentration of dissolved oxygen increases the toxicity of Zn. It plays a vital role in the physiological and metabolic processes of many organisms. Other clinical signs of Zn toxicity have been reported as diarrhea, bloody urine, liver failure, kidney failure and anemia [Duruibe *et al.,* 2007].

The results from experimental measurements for the Zn^{2+} are shown in Table 2 and Figure 5. According to the Romanian standards for assessing the ecological status of surface water (GD161), the recommended value for Zn^{2+} is 0.1 – >1 mg/L. During the three-season study, the Zn^{2+} concentration varied from 0.0020 to 0.01680 mg/L. The lowest value was measured at SP5 station in fall, whereas the highest value was measured at SP8 station in fall. The average values in spring, summer, and fall were 0.030, 0.036, and 0.005 mg/L respectively, whereas the three-season average with the standard deviation was 0.0387±0.0360 mg/L. The average concentrations values of the Zn^{2+} at SP1-SP8 stations were 0.0323, 0.0290, 0.0180, 0.0310, 0.0111, 0.0261, 0.0473, and 0.0623 mg/L respectively. In spring, the SP1 station showed the minimum value of 0.0050 mg/L, whereas the SP7 station showed the maximum value of 0.0750 mg/L. In summer, the SP8 station showed a minimum value of 0.0110 mg/L, whereas the SP1 station showed the maximum value of 0.0780 mg/L. In fall, the SP5 station showed the minimum value of 0.0020 mg/L, whereas the SP8

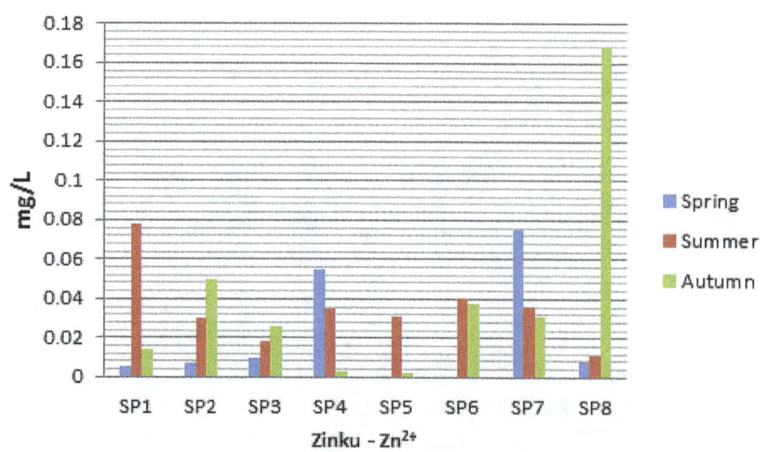

Figure 5. Variation of Zn values in water of Lepenci river

station showed the maximum value of 0.1680 mg/L. On the basis of the GD161 standards, the levels of Zn^{2+} in the water of Lepenc River basin during the three seasons were found to be within the recommended values, thus ranking the water of this river in the first class.

Manganese (Mn^{2+})

Mn is present in over 100 common salts and mineral complexes that are widely distributed in rocks, in soils and on the floors of lakes and oceans [Damodharan, 2013]. These Mn minerals include sulfides, oxides, carbonates, silicates, phosphates, arsenates, tungstates, and borates. However, the most important Mn mineral is the native black manganese oxide, pyrolusite (MnO_2). Mn is used for production of ferromanganese steels, electrolytic manganese dioxide for use in batteries, alloys, catalysts, antiknock agents, pigments, dryers, wood preservatives and coating welding rods [Bradi, 2005]. It is also used as an oxidant for cleaning, bleaching and disinfection (as potassium permanganate) and as an ingredient of various products [WHO, 2011]. Mn is an essential micronutrient present in all living organisms, because it functions as a co factor for many enzyme acticities [Suresh *et al.,* 1999]. Mn is a metal with low toxicity but has a considerable biological significance and seems to accumulate in fish [Kumar *et al.,* 2011]. According to [Krishna *et al.,* 2014], high Mn concentration interferes with the central nervous system of vertebrates; hence, the consumption of Mn contaminated fish potentially resulting in health risks to the consumers is a matter of concern. High concentration of Mn causes liver cirrhosis and also produces a poisoning called Manganese or Parkinson disease [Bradi, 2005].

The results from experimental measurements for the Mn^{2+} are shown in Table 2 and Fig. 6. According to the Romanian standards for assessing the ecological status of surface water (GD161), the recommended value for the Mn^{2+} is 0.05 – >1 mg/L. During the three-season study, the Mn^{2+} concentration varied from 0.0530 to 3.7360 mg/L. The lowest value was measured at SP7 station in summer, whereas the highest value was measured at SP1 station in fall. The average values in spring, summer, and fall were 0.2300, 0.1500, and 1.1600 mg/L respectively, whereas the three-season average with the standard deviation was 0.5133 ± 0.5067 mg/L. The average concentrations of Mn^{2+}at SP1-SP8 stations were 1.2753, 0.0760, 0.5773, 0.6607, 0.1070, 0.3253, 0.1673, and 0.2180 mg/L respectively. In spring, SP5 station showed the minimum value of 0.1170 mg/L, whereas the SP8 station showed the maximum value of 0.4330 mg/L. The Mn^{2+} concentration in the river waters at SP1 and SP2 stations was under the detection levels; water at SP3, SP5 and SP6 stations belonged to the second class, whereas the SP4, SP7, and SP8 stations belonged to the fifth class. In summer, the SP7 station showed the minimum value of 0.0530 mg/L, whereas the SP6 station showed the maximum value of 0.4600 mg/L. Compared to GD 161 standards, the water at SP1, SP2, SP3, SP4, SP5, SP7, and SP8 stations belonged to the first class, whereas the water at SP6 station belonged to the fifth class. In fall, the SP8 station showed the minimum value of 0.1060 mg/L, whereas the SP1 station showed the maximum value of 3.7360 mg/L. Compared to GD 161 standards, the water at SP1 and SP7 stations belonged to the fifth class, the water at SP2 and SP5 stations belonged to the first class, and the water at SP3 and SP4 stations belonged to the third class.

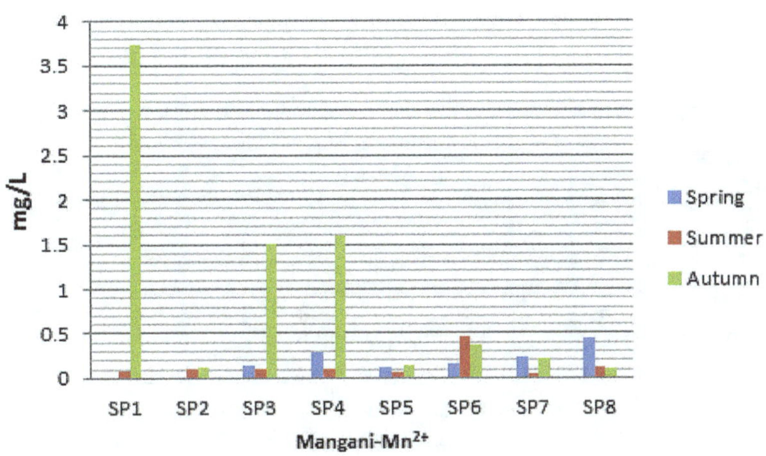

Figure 6. Variation of Mn values in water of Lepenci river

Copper (Cu^{2+})

Cu is an essential constituent of living systems and is widely distributed metal in nature. Cu can exist in aquatic environment in three forms, namely soluble, colloidal and particulate. Cu is a rare ingredient of natural waters. It is used in Cu pipes or dosing tanks with copper sulphate(II), which is used to prevent the growth of algae. It was proven that Cu is toxic to fish and other aquatic creatures, in those concentrations that do not pose a risk to man. It is known that an ion of Cu(II) is mainly poisonous chemical element. Copper is important for human life and other living things, so it plays an important role in metabolic processes, affecting a number of enzymes and hemoglobin synthesis. The toxicity of Cu in water depends on the alkalinity, pH value, the content of organic substances, etc. With the increase of these parameters, the concentration of ions of Cu(II) decreases in addition to the Cu toxicity to aquatic systems. High doses may also cause anaemia, liver and kidney damage, stomach and intestinal irritation [Tirkey *et al.*, 2012]. Copper ions (Cu^{2+}) are toxic to most life forms. Cu is highly toxic to invertebrates and moderately so to mammals in trace amounts.

The results from the experimental measurements for the Cu^{2+} during the three seasons are shown in Table 2 and Figure 7. According to the Romanian standards for assessing the ecological status of surface water (GD161), the recommended value for the Cu^{2+} is 0.02 – >0.1 mg/L. During the three-season study, the Cu^{2+} concentration varies from 0.0010 to 0.0380 mg/L. The lowest value was measured at SP4 and SP5 stations in summer, whereas the highest value was measured at SP8 station in spring. The average values in spring, summer, and fall were 0.0210, 0.0050, and 0.000 mg/L respectively, whereas the three-season average with the standard deviation was 0.0087±0.0070 mg/L. The average concentrations of Cu^{2+}at SP1-SP8 stations were 0.0083, 0.0090, 0.0137, 0.0017, 0.0017, 0.0047, 0.0043, and 0.0140 mg/L respectively. In spring, the SP1 station showed the minimum value of 0.0050 mg/L, whereas SP7 station showed the maximum value of 0.0750 mg/L. When compared to GD 161 standards, the obtained measured values were found to be under the detection levels at SP4 and SP6 stations. The water at SP1, SP2, SP5, and SP7 stations belonged to the first class, whereas the water at SP3 and SP8 stations belonged to the second class. In summer, the SP8 station showed the minimum value of 0.0110 mg/L, whereas the SP1 station showed the maximum value of 0.0780 mg/L. During this season we detected only two values at two stations (SP4 and SP6) where the water belonged to the first class, whereas at all other stations, the values were under the detection levels. In fall, the SP5 station showed the minimum value of 0.0020 mg/L, whereas the SP8 station showed the maximum value of 0.1680 mg/L. In fall, the Cu^{2+} concentration values at each station were under the detection levels.

Iron (Fe^{2+})

Fe is an essential metal for most living organisms and humans. It is a constituent of proteins and many enzymes, including hemoglobin and myoglobin. Fe is usually more abundant in freshwater environment than other metals, due to its high occurrence on Earth [Forstner et *al.*, 1979]. Fe deficiency can lead to anemia and fatigue, which are usually common among children under the age of five, pregnant women and

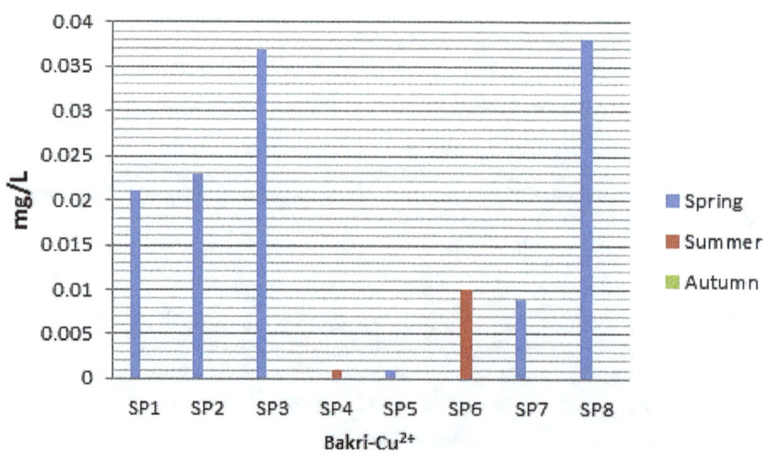

Figure 7. Variation of Cu values in water of Lepenci river

immuno-compromised individuals, thus making them vulnerable to numerous infections [Garvin, 2015]. Vuori [1995] reported that Fe has both direct and indirect effects on river ecosystems, as it affects lotic organisms by interfering with their normal metabolism and osmoregulation. He also noted that the combined effects of Fe contamination can reduce the occurrence and diversity of several aquatic organisms, including fish. High Fe concentration, together with its precipitate in aquatic ecosystems, do have negative effects on the behavior, reproduction and survival of aquatic animals [Gerhardt, 1992]. Fe is an important metal for the life of plants and animals. The soluble Fe shares two of its electrons (oxidation state +2), whereas when reacting with oxygen, it forms a rusty-brown precipitate of iron hydroxide that is insoluble. In waters rich with Fe, there is a bacterium known as "filamentous", which grows and multiplies. It has a highly negative impact on water quality, as it leads to the biomass collection in the distribution system. Fe is an essential metal of the hemoglobin structure and is used for the treatment of anemia, which is caused by the iron deficiency in the blood. It becomes toxic when used in large doses.

The results from experimental measurements for Fe^{2+}during the three seasons are presented in Table 2 and Figure 8. According to the Romanian standards for assessing the ecological status of surface waters (GD161), the recommended value for Fe^{2+} is 0.3 – >2 mg/L. During the three-season study, the Fe^{2+} concentration varied from 0.0030 to 0.4010 mg/L. The lowest value was measured at SP1 station in summer, whereas the highest value was measured at SP3 station in spring. The average values in spring, summer, and fall were 0.1920, 0.1080, and 0.1030 mg/L respectively, whereas the three-season average with the standard deviation was 0.1343±0.0900 mg/L. The average values of Fe^{2+} concentrations at SP1-SP8 stations were 0.0400, 0.0507, 0.1370, 0.1650, 0.1470, 0.1227, 0.1087, and 0.1367 mg/L respectively. In spring, the SP7 station showed the minimum value of 0.0460 mg/L, whereas the SP3 station showed the maximum value of 0.4010 mg/L. When compared to the GD 161 standards, the measured values resulted to be below detection levels at SP4 and SP6 stations. The water at SP1, SP2, SP5, and SP7 stations belonged to the first class, whereas the water at SP3 and SP8 stations belonged to the second class. In summer, the SP1 station showed the minimum value of 0.0030 mg/L, whereas the water at SP4 station showed the maximum value of 0.0030 mg/L. The values obtained at SP2, SP3, and SP8 stations were below detection levels. In fall, the SP3 station showed the minimum value of 0.0080 mg/L, whereas the SP7 station showed the maximum value of 0.1990 mg/L. In fall, the Fe^{2+} concentration values were below detection levels at all measuring stations. When compared to GD161 standard, the measured values at the three seasons resulted to be within the said standard, thus ranking the river water in the first class.

Lead (Pb^{2+})

Pb in the environment arises from both natural and anthropogenic sources. It is a natural constituent of air, water and biosphere. Pb is also a heavy metal that appears to be capable of +2 oxidation in aquatic environments and comes from various industrial sources and mines. Pb from gasoline (with tetraethyl lead), is the main source of atmospheric Pb and its large part passes in the water. Otherwise, it is rarely detected in natural waters. Pb is also a poisonous metal. In the case

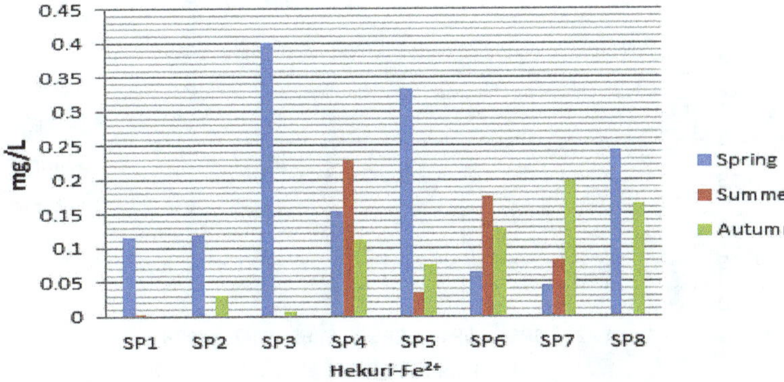

Figure 8. Variation of Fe values in water of Lepenci river

of humans, the poisoning with Pb causes disturbances in the functioning of the kidneys, reproductive system, liver and nervous system. Pb poisoning caused by the environmental exposure causes mental retardation to more children, while mild poisoning leads to anemia. The lack of Ca, Fe, Cu, Se and ascorbic acid in food conditions the accumulation of lead in the body. The largest amount of Pb accumulates is in the bones and it replaces Ca. Pb organic compounds easily reach the human body through the skin and central nervous system at risk. The toxic effect of Pb is also based on the great affinity to sulfur. Pb reduces the activity of enzymes, oxidation-reduction reactions in the cell and protein synthesis. It is known that Pb poisoning is greater in hard waters than in soft waters. Pb is a serious cumulative body poison. High levels of exposure may result in biochemical effects in humans which in turn cause problems in the synthesis of haemoglobin, effects on the kidneys, gastrointestinal tract, joints and reproductive system, and acute or chronic damage to the nervous system [Tirkey*et al.*, 2012]. The sources include burning of lead based petroleum fuels, organic and inorganic lead compounds now used in a variety of commercial products and industrial materials including plastics, storage batteries, bearing alloys, insecticides, ceramics, cable sheathings, sheeting, radiation shields and even some paints [Mutwiri, 2001].

The measurements from experimental results for the Pb^{2+} during the three seasons are presented in Table 2 and Figure 9. According to the Romanian standards for the assessment of ecological status of surface waters (GD161), the recommended value for Pb^{2+} is 0.005 – >0.05 mg/L. During the three-year study, the Pb^{2+} concentration varied from 0.0000 to 0.0970 mg/L. The lowest value was measured at SP1 – SP8 stations in summer, whereas the highest value was measured at SP8 station in fall. The average values in spring, summer, and fall were 0.2100, 0.0000, and 0.030 mg/L respectively, whereas the three-season average with the standard deviation was 0.0800 ± 0.0452 mg/L. The average values of Pb^{2+} concentrations at SP1-SP8 stations were 0.0170, 0.0207, 0.0030, 0.0617, 0.0020, 0.1050, 0.1247, and 0.1613 mg/L respectively. The Pb values in the water of Lepenc river basin fluctuate from a minimum value of 0.0470 mg/L (SP1) to a maximum value of 0.3850 mg/L (SP8) in the spring season. At two stations (SP3 and SP5) the values were below detection levels. In summer, the values of all monitoring stations were under the detection levels. In fall, the values ranged from a minimum of 0.005 mg/L (SP3) to a maximum of 0.097 mg/L (SP8). The values at SP1, SP4, and SP5 stations were below detection levels.

Coefficients of correlation related to the concentration of metals of the river Lepenci

The coefficients of correlation related to the concentration of waters in the Lepenci river were shown in Table 3. The results show that three values of coefficients of correlation appeared to be the most significant. Pb showed high a coefficient of correlation with Zn (r=0.8488), whereas Fe and Cd showed an average coefficient of correlation (r=0.6678). A average negative coefficient of correlation was found between Mn and Cr (r = -0.6513) as well as between Ni and Cr (r= -0.5277).

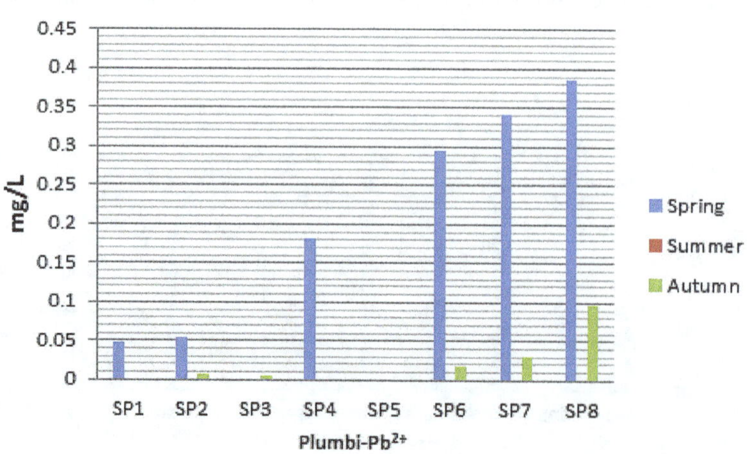

Figure 9. Variation of Pb values in water of Lepenci river

Table 3. Coefficients of correlation for metal concentrations in the Lepenci River

Correlation	Cr	Cd	Ni	Zn	Mn	Cu	Fe	Pb
Cr	1.0000							
Cd	0.1866	1.0000						
Ni	-0.5277	0.2196	1.0000					
Zn	0.0763	0.1914	-0.1559	1.0000				
Mn	-0.6513	-0.4941	-0.0573	-0.0885	1.0000			
Cu	0.0749	-0.1386	-0.4500	0.3621	0.0966	1.0000		
Fe	0.2811	0.6678	-0.1096	-0.0662	-0.3125	-0.2320	1.0000	
Pb	-0.0399	0.5742	0.1257	0.8488	-0.3152	0.0997	0.2511	1.0000

Water quality assessment of Lepenci River using WQI

The calculation of Lepenci River Water Quality Index was done using the Water Quality Index Desktop software developed by [Ramadani et *al.*, 2017]. The results for the factors F1, F2, and F3 and the WQI for the eight stations are given in Table 4, whereas the WQI values are graphically presented in Figure 10.

At SP1 station, the WQI value was found to be 80, since two parameters (Cd and Mn) did not meet the recommended value of the regulation for assessment of river water quality, and two tests failed: Cd with a value of 0.0230 mg/L in summer and Mn with a value of 3.736 mg/L in fall. At SP2 station, the WQI value resulted to be 81, since two parameters (Cd and Pb) did not meet the recommended value of the regulation for assessment of river water quality, and two tests failed: Cd with a value of 0.0310 mg/L in summer and Pb with a value of 0.0530 mg/L in spring. At SP3 station, the WQI value was 75, since two parameters (Cd and Mn) did not meet the recommended value of the regulation, and three tests failed: Cd with a value of 0.0470 mg/L in summer, Cd with a value of 0.0150 mg/L in fall, and Mn with a value of 1.5020 mg/L in fall. At SP4 station, the WQI value was found to be 65, since three parameters (Cd, Mn, and Pb) did not meet the recommended value of the regulation, and three tests failed: Cd with a value of 0.0840 mg/L in spring, Mn with a value of 1.6020 mg/L in fall, and Pb with a value of 0.1810 mg/L in spring. At SP5 station, the WQI value was 64, since two parameters (Cd and Ni) did not meet the recommended value of the regulation, and four tests failed: Cd with a value of 0.0420 mg/L in spring, Cd with a value of 0.0750 mg/L in summer, Cd with a value of 0.0330 mg/L in fall, and Ni with a value of 0.1180 mg/L in summer. At SP6 station, the WQI value amounted to 57, since three parameters (Cd, Ni and Pb) did not meet the recommended value of the regulation, and five tests failed: Cd with a value of 0.1040 mg/L in spring, Cd with a value of 0.0530 mg/L in summer, Cd with a value of 0.0090 mg/L in fall, and Pbwith a value of 0.2950 mg/L in spring. At SP7 station, the WQI value equalled 64, since three parameters (Cd, Ni and Pb) did not meet the recommended value of the regulation, and three tests failed: Cd with a value of 0.0820 mg/L in spring, Ni with a value of 0.1190 mg/L in spring, and Pb with a value of 0.3410 mg/L in spring. At SP8 station, the WQI value was 59, since two parameters (Cd and Pb) did not meet the recommended value of the regulation, and four tests failed: Cd with a value of 0.0890 mg/L in spring, Cd with a value of 0.0840 mg/L in summer; Pb

Table 4. Values obtained from the Water Quality Desktop and WQI calculation

Stations	F1	F2	F3	WQI
SP1- Prevallë	25	8.333334	20.88608	80
SP2- Jezerc	25	8.333334	17.97676	81
SP3-Brodë	25	12.5	31.23603	75
SP4-Runjevë	37.5	12.5	44.21459	65
SP5-Nikë	25	16.66667	53.10668	64
SP6-Gërlicë	37.5	20.83333	59.76530	57
SP7-Kaçanik	37.5	12.5	47.1482	64
SP8-Hani iElezit	25	16.66667	62.6401	59
			Average WQI:	68.1250

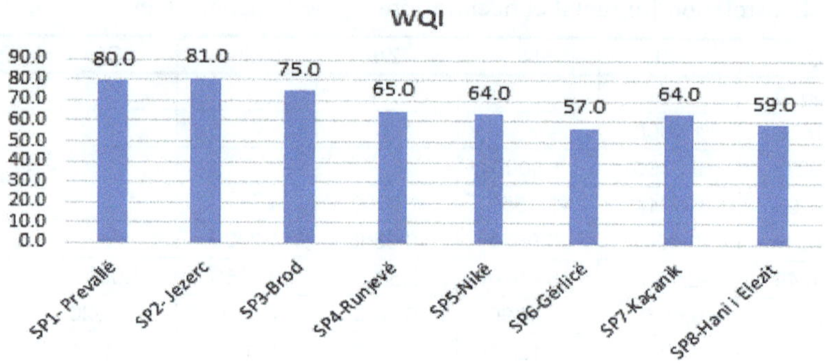

Fig. 10. WQI values at all measuring stations

with a value of 0.3850 mg/L in spring, and Pb with a value of 0.0970 mg/L in fall.

The results show that the river water at SP2 station had the best quality with a WQI value of 81 and belonged to the good category, whereas the river water at SP6 station had the worst quality with a WQI value of 57 and belonged to marginal category. Finally, based on these results, the average WQI value was calculated for the entire measurement period and it resulted to be 68.1250 which showed that the water of Lepenc River belonged to fair category.

CONCLUSIONS

Given the research conducted and the discussion of results we can draw the following conclusions:

1. Although Lepenci, being a mountain river, has a self-cleaning capability for its waters, slight heavy metals pollution was observed during the monitoring period;
2. Water pollution of Lepenci River with heavy metals assessed by the Romanian standards for the assessment of ecological status of surface waters (GD161) was low and the most severe heavy metals pollution was observed at the stations located at the bottom of the river;
3. During the monitoring period the water of Lepenci River was categorized in second-third class;
4. The sequence of average values with standard deviation of heavy metals concentrations in the water of Lepenc river expressed in mg/L was as follows: Mn (0.5133±0.5067) > Fe (0.1343±0.0900) > Pb (0.0800±0.0452) > Ni (0.0527±0.0370) > Cd (0.0487±0.0193) > Zn (0.0387±0.0360) > Cr (0.0233±0.005) > Cu (0.0087±0.0070);
5. Calculation of coefficients of correlation showed that it was the highest between Pb and Zn (r = 0.8488), average between Fe and Cd (r = 0.6678), and negatively average between Mn and Cr (r = -0.6513) and between Ni and Cr (r = -0.5277);
6. The Water Quality Index Desktop software was also used as a highly efficient tool for WQI calculation based on the guidelines of the Canadian Ministry of Environment;
7. On the basis of the WQI calculation using the Water Quality Index Desktop software, it was found that the river water at the SP2 station had the best water quality with a WQI value of 81 (the category: good), river water at SP6 station had the worst quality with a WQI value of 57 (category: marginal), whereas the average WQI for the entire measurement period was 68.1250 (category: fair);
8. Water Quality Index Desktop software has provided excellent and reasonable results;
9. We propose that the state authorities and institutions should support river water monitoring as an effective measure for examining their ecological status and protection against various types of pollution.

REFERENECES

1. Anderson D. 2003. Introduction to heavy metal monitoring centre for ecology and hydrology, Natural Environment Research Council, 13th October.
2. Bakare A.A., Lateef A., Amunda O.S., Afulabi R.O. 2003. The aquatic toxicity and characterization of chemical and microbiological constituents of water samples from Oba River, Odo-Oba, Nigeria. Asian J. Microbiol. Biotechnol. Environ. Science, 5, 11–17.
3. BožoDalmacija – Editor. 2001. Kontrola kvaliteta voda, Univerzitet u Novom Sadu, Insitut za

Hemiju, Kadetra za hemijsku tehnologiju i zaštitu životne sredine, 15–17, 253.

4. Bradi B.H. 2005. Heavy metals in the environment. In: Interface Science and Technology. Hubbard A. (Ed.), Vol. 6, Elsevier Academic Press, Neubrucke.
5. CCME. 2001. Canadian water quality guidelines for the protection of aquatic life: CCME Water Quality Index 1.0 User's Manual. Canadian Council of Ministers of the Environment.
6. City Kaçanik.2011. Municipality of Development Plan.
7. Damodharan U. 2013. Bioaccumulation of heavy metals in contaminated river water – Uppanar, Cuddalore South East Coast of India. htt//.dx.doi.org/10.5772/5334.
8. Diagomanol I.N.V., Farhang M., Ghazi-Khansari M., Jafarzadeh N. 2004. Heavy metals (Ni, Cr, Cu) in the Karoon waterway river, Iran. Toxicol. Lett., 151(1), 63.
9. Durmishi B.H., Abduli S., Reka A.A., Ismaili M., Shabani A., Durmishi A. 2016. Determination of the Content of Zn, Cu, Pb and Cd in the River Shkumbini (Pena) with Potentiometric Stripping Analysis. International Journal of Chemistry & Materials Sciences, 1(1), 17–32.
10. Durmishi B.H., Ismaili M., Shabani A., Abduli S. 2012. Drinking Water Quality Assessment in Tetova Region. American Journal of Environmental Sciences, 8(2), 162–169.
11. Duruibe J.O., Ogwuegbu M. C., Egwurugwu J.N. 2007. Heavy metal pollution and human biotoxic effects. International Journal of Physical Sciences, 2, 112–118.
12. Forstner U., Wittmann G.T.W. 1979. Metal Pollution in the Aquatic Environment. Springer-Verlag, Berlin, Germany.
13. Garvin K.S. 2019. Health Effects of Fe in Drinking Water. Available online: http://www.livestrong.com/article/155098-health-effects-of-iron-in-drinking-water (accessed on 24 November 2015).
14. Gerhardt A. 1992. Subacute effects of iron (Fe) on Leptophlebiamarginata (Insecta: Ephemeroptera). Freshwater Biol., 27, 79–84.
15. Kitaura H., Nakao N., Yoshida N., Yamada T. 2003. Induced sensitization to nickel in guinea pigs immunized with mycobacteria by injection of purified protein derivative with nickel. New Microbiol., 26(1), 101.
16. Krishna P.V., Rao K.M., Swaruparani V., Rao D.S. 2014. Heavy metal concentrations in Fish Mugilcephalus from Machilipatnam Coast and possiible health risks to fish consumers. British Biotechnology Journal, 4(2), 126–135.
17. Kumar B., Mukherjee D.P., Kumar S., Mishra M., Prakash D., Sigh S.K., Sharma C.S. 2011. Bioaccumulation of heavy metals in muscle tissue of fisheries from selected aquaculture ponds in east Kolkata etlands. Annals of Biological Research, 2(5). 125–134.
18. MESP 2010.The state of water report in Kosova. Prishtina.
19. Mutwiri N.M. 2001. Determination of cadmium, chromium, lead and mercury in honey samples from Mbeere, Meru and Kirinyaga districts unpublished, MSc Thesis, Chemistry Department, Egerton University, Nakuru, Kenya.
20. Ndimele P.E. 2008. Evaluation of phyto-remediative properties of water hyacinth (Eichhorniacrasspies) and biostimulants in restoration of oil-polluted wetland in the Niger Delta. PhD Thesis, University of Ibadan, Nigeria.
21. Rajappa B., Manjappa S., Puttaiah E.T. 2010. Monitoring of Heavy metal in groundwater of Hakinaka Taluk, India. Contemporary Engineering Sciences, 3(4), 183–190.
22. Ramadani E., Memeti A., Durmishi B.H. 2017. Water Quality Index: A New Automated Way of Measuring the Quality. International Journal on Information Technologies & Security, 3, 43–52.
23. Scott-Fordsmand J.J. 1997. Toxicity of nickel to soil organisms in Denmark. Rev. Environ. Contam. Toxicol., 148, 1.
24. Senarathne P. and Pathiratne K.A.S. 2007. Accumulation of heavy metals in a food fish, Mystusgulio in habiting Bolgoda Lake, Sri Lanka. Sri Lanka J. Aquat. Sci., 12, 61–75.
25. Suresh, B., Steiner W., Rydlo M., Taraschewski H. 1999. Concentrations of 17 elements in Zebra mussel (Dreissena polymorpha). Environmental Toxicology and Chemistry, 18, 2574–9.
26. Tirkey A., Shrivastava P., Saxena A. 2012. Bioaccumulation of heavy metals in different components of two Lakes ecosystem. Current World Environment, 7(2), 293–297.
27. Vuori K.M. 1995. Direct and indirect effects of iron on river ecosystems. Ann. Zool. Fennici., 32, 317–329.
28. WHO. 2000. Hazardous Chemicals in human and environmental health, WHO, Gineva, Switzerland.
29. WHO. 2011. Guidelines for drinking water quality, 4th edn. World Health Organization, Geneva, 2011, pp. 564.

12

The Content of Heavy Metals in Bottom Sediments of Selected City Rivers of the Podlasie Province

Mirosław Skorbiłowicz[1*], Elżbieta Skorbiłowicz[1], Małgorzata Górska[1]

[1] Bialystok University of Technology, Faculty of Civil and Environmental Engineering, Department of Technology in Engineering and Environmental Protection, ul. Wiejska 45A, 15-354 Bialystok, Poland

[*] Corresponding author's e-mail: m.skorbilowicz@pb.edu.pl

ABSTRACT

The purpose of the work was to determine the relationship between the state of the water environment quality of selected rivers (Zn, Cr, Pb, Cd and Cu tests in bottom sediments), and the sources of pollution resulting from the close proximity to the cities, through which they flow. The following rivers were selected for the study: the Biała river flowing through the city of Białystok, the Narew river flowing through the city of Tykocin and the village of Złotoria, the Supraśl river flowing through the village of Michałowo and Gródek as well as the Biała river flowing through the city of Bielsk Podlaski. The sediments were collected four times from the same points in 2016 in the period from July to October. The sediment samples were collected from the points located on rivers before and beyond towns and villages. The contents of the following metals were tested in the bottom sediment samples: Zn, Cr, Pb, Cd and Cu. The analyses were carried out applying the flame absorption spectrometry method. The statistical multivariate CA and FA analyzes were used. The highest contents of Zn, Pb and Cr were recorded in the bottom sediments from the following rivers: Biała (Białystok) and Biała inflow of Orlanka (Bielsk Podlaski), which resulted from the anthropogenic activity. The research also showed the impact of the agricultural activity due to the slightly elevated Cd level.

Keywords: heavy metals, river, bottom sediments

INTRODUCTION

The urban land use pattern and the urbanization degree have significant impacts on the heavy metal (HM) contamination, which have been extensively detected in soil, water, sediment, air, organisms, as well as city road dust [Giordano et al. 2005; Mohiuddin et al. 2010; Shi et al. 2008]. The deterioration of river water quality due to the indiscriminate discharge of untreated sewage, industrial waste and other anthropogenic activities like urbanization, population growths and land development along the river basin is one of the major concerns throughout the world [Kumar and Maiti 2015]. The rivers in the urban areas passing through the industrial zones are associated with the water quality problems because of the unregulated discharge of the domestic and industrial waste water into the water bodies that lead to the increased levels of heavy metal in the rivers [Lin et al. 2013; Silva et al. 2014]. The heavy metal contamination in the sediment is a significant environmental problem because of their toxicity, non-degradability and high bioaccumulation potential in biota and food chain [Sundaray et al. 2014)]. The sediments act as a sink of contaminants, receive and absorb the pollutants resulting from these sources [Liu et al. 2003], and have been recognized as an important indicator of water pollution. Heavy metals are among the most common environmental pollutants, and their occurrence in waters and biota indicate the presence of natural or anthropogenic sources. Their accumulation and distribution in soil, water and environment are increasing at an alarming rate causing deposition and sedimentation in water reservoirs and affecting the aquatic organisms as well [Cataldo et al. 2001; Hobbelen et al. 2004; Koukal et al. 2004; Okafor and Opuene 2007; Mohiuddin et al. 2010]. Heavy metals like Cr, Pb, Cd, etc. exhibit extreme toxicity [Miller et al. 2003; Harikumar et al. 2009]; heavy metals have become significant pollutants of many riverine systems [Dassenakis et al. 1998]. Heavy metals have a potential to contaminate soil and water, which can be dispersed and accumulated in plants and animals, and taken in by humans through consumption [Wcislo et al. 2002; Li et al. 2008]. The heavy metal contamination of sediments can critically degrade the aquatic systems [Charkhabi et al. 2005]. Their release from the sediment can make them enter the aquatic ecosystems and bring about severe problems [Mohammed and Markert 2006]. Despite the differences in the toxic effects of the metals with different metals as well as their concentrations and the time of exposure, their concentrations are reliable indicators of the ecosystem health [Singh et al. 2005]. The organic matter and fine grain content have often been considered two major apparent sediment properties associated with heavy metal enrichment in sediments [Laing et al. 2007]. Land use may also play crucial roles in the distribution of heavy metals in sediments. Land use is commonly classified as agriculture, forestry, industrial development, and urbanization. However, the types of land use resulting in heavy metal enrichment of sediments are not necessarily the same for different geographical locations. For example, Heikkila [1991] reported that high concentrations of organic matter and heavy metals in sediments were closely related to the increased intensity of agriculture, forestry, and peat harvesting in drainage basins. In contrast, Abrahim and Parker [2002] and Horowitz and Stephens [2008] demonstrated that the heavy metal concentrations in sediments tend to increase with the degree of industrial development and urbanization in the catchments. A successful evaluation for mixed accumulation of heavy metals from natural and anthropogenic sources typically requires normalizing methods for distinguishing the two different sources [Idris 2008]. The geochemical normalization approaches such as the enrichment factor (EF) and geo-accumulation index (Igeo) methods have been commonly used for the purpose.

The multivariate statistical techniques including the principal component analysis (PCA) and cluster analysis (CA) can be used as complementary tools for explaining the distribution of heavy metals in sediments. They statistically reduce the heterogeneity associated with the relationships among the sediment properties, metals, and other environmental factors. In particular, the PCA approach has been typically applied to identify the sources of heavy metals in sediments [Wang et al. 2011; Chabukdhara and Nema 2012]

The purpose of the work was to determine the relationship between the state of the water environment quality of selected rivers (Zn, Cr, Pb, Cd and Cu tests in bottom sediments), and the sources of pollution resulting from the close proximity to the cities through which they flow.

METHODS

As a research area, four rivers were selected in the Podlasie province: Biała (inflow of Supraśl), Narew, Supraśl and Biała (inflow of Orlanka), in which the dependence of the water quality (bottom sediment study) and close neighborhood of selected urbanized areas, through which they flow, were analyzed. In order to determine the extent of the negative impact on the river environment of anthropogenic activities resulting from the nature of development and infrastructure, agricultural and non-agricultural activities, the following towns were selected: Białystok for Biała river, Tykocin and Złotoria for Narew river, Michałowo and Gródek for Supraśl river as well as Bielsk Podlaski for Biała river (Orlanka tributary). The list includes both Białystok – aprovincial city with the largest number of inhabitants (296,000) and the most developed non-agricultural economic activity, as well as, to some extent, the direct opposite, i.e. Złotoria – a village without the sewage treatment plant, but with developed agricultural economy (Table 1).

The sediment samples were collected four times (from July to October 2016) from twelve measuring points located (in pairs) near each of the selected towns. The total number of samples was 48. It was assumed that the research material would be collected at a distance of 100 meters before and beyond the village, taking into account the course of the river. The bottom sediment collection points located on the rivers are marked as follows: before the village – A, while beyond the village – B (Figure 1).

The bottom sediments were collected in the littoral zone, in the top layer, where the deposition of the suspended material and accumulation of heavy metals occur [Bojakowska 2001]. Ten individual samples of bottom sediments from a depth of up to 5 cm below the surface of the water were collected at each designated measurement point.

Table 1. Basic characteristics of watercourses and urban areas

River/length [km]/average annual flowrate through urbanized area [$m^3 \cdot s^{-1}$]	Urbanized area	Number of residents	Industrial facilities	Sewage treatment plant [RLM]/average daily sewage amount [$m^3 \cdot d^{-1}$]
Biała/33/1,20	Białystok	296000	present	200000/100000
Górna Narew/199/17,56	Tykocin	2000	no	6000/500
Górna Narew/199/16,67	Złotoria	767	no	no
Supraśl/94 /1,22	Gródek	2900	no	3000/258
Supraśl/ 94/1,88	Michałowo	3107	no	2410/263
Biała/33 /0,14	Bielsk Podlaski	26000	present	45000/6000

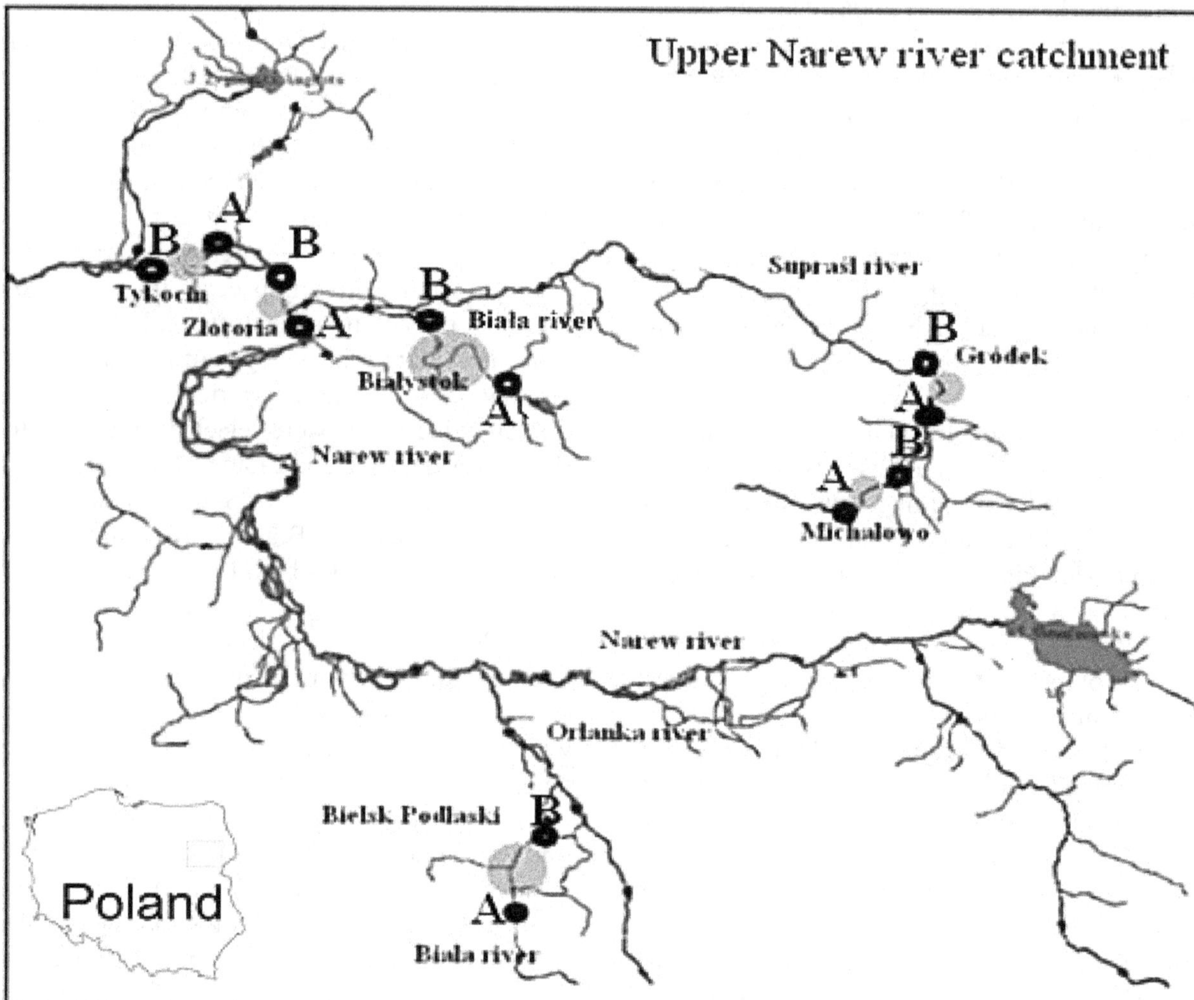

Figure 1. Location of measurement and control points on the examined rivers

After mixing the test material, a representative sample of approximately 1000 g was obtained. The samples were dried to an "air-dry" state and stored until assayed [Lis and Pasieczna 1995].

Before proceeding with the chemical analyses, the bottom sediment samples were dried at 40°C and sieved through a 0.2 mm sieve. The bottom sediments were mineralized in the MT1205-B2Magma Therm microwave system. Weighed samples (1 g) were poured in Teflon vessels with a mixture of concentrated HNO_3 (8 ml) and 30% H_2O_2 (2 ml) [Salminen et al. 2005]. The samples after filtration were quantitatively transferred to 50 ml graduated flasks. The following Zn, Cr, Pb, Cd and Cu metals were tested in mineralized solutions. The tests were carried out using flame absorption spectrometry method by means of the AAS ICE 3500 Thermo Scientific spectrometer.

The results of the sediment analyses were verified using a certified reference material for sediments NCS DC 73317a. The limit of quantification of the method [$mg \cdot dm^{-3}$] was: Cd – 0.01, Cr – 0.05, Cu – 0.035, Pb – 0.1 and Zn – 0.01. The obtained results of metals content were given in relation to air dry sediments and compared with the literature data. The content of organic substance was determined using the annealing method. The reaction of sediments in distilled water was examined with a potentiometric method.

In order to assess the degree of sediments contamination with heavy metals, the proposal to classify water sediments in Poland was used (Table 2).

In order to assess the quality of bottom sediments, the degree of sediment pollution was also employed, using a geochemical index (I_{geo}) and contamination factor (CF).

The geochemical index (I_{geo}) is defined using the formula [Müller 1979]:

$$I_{geo} = \log_2(\frac{C_m}{1,5GM}) \quad (1)$$

where: C_m – content of the analyzed metal ($mg \cdot kg^{-1}$),
GM – geochemical background ($mg \cdot kg^{-1}$).

As the geochemical background, the average contents in shales of the Earth's crust were assumed in $mg \cdot kg^{-1}$: Zn – 95, Cr – 90, Pb – 20, Cd – 0.3, Cu – 45 [Turekian and Wedepohl, 1961].

The bottom sediments were classified on the basis of the obtained geochemical index values (I_{geo}) according to the following ranges: $I_{geo} < 0$ – class 0 (non-contaminated sediments), $0 < I_{geo} < 1$ – class 1 (uncontaminated moderately (weakly) contaminated sediments), $1 < I_{geo} < 2$ – class 2 (moderately contaminated sediments), $2 < I_{geo} < 3$ – class 3 (moderately strongly contaminated sediments), $3 < I_{geo} < 4$ – class 4 (strongly contaminated sediments), $4 < I_{geo} < 5$ – class 5 (extremely strongly contaminated sediments), $5 < I_{geo} < 6$ – class 6 (extremely contaminated sediments).

Before performing the multidimensional analyses, the basic relationships in the prepared data set were checked. The calculations of Spearman's rank order correlation were made due to the number of some variables taken into analysis and the lack of normal distribution of numerical data. The statistic multivariate analysis (FA) and Ward cluster analysis (CA) were used for the analyses. In order to interpret the results of the factor analysis, it was assumed that the compounds of the variable (metal content) with the factor are strong when the absolute values of its charges are greater than 0.70. A similar value was used, among others, by Pucket, Bricker [1992] and Evans et al. [1996].

The cluster analysis was used to isolate the objects with similar contents of investigated metals in bottom sediments. This is a method that allows to isolate (classify) specific subsets of objects with mutually similar elements in a multidimensional space [Igras 2004]. Statistica software was used for the statistical calculations.

RESULTS AND DISCUSSION

The pH of bottom sediments ranged from 6.22 to 8.71 pH. The process of metals adsorption/desorption by various minerals and organic substance to a great extent depends on the reaction. Metals can be adsorbed under various pH conditions. For example, Cd and Zn have adsorption limits at higher pH than Cu, which makes these metals more mobile in the fluvial environment [Ciszewski and Aleksander-Kwaterczak 2015]. It should be emphasized that the solubility of micronutrients increases with decreasing the pH value [Small and Salomons, 1995]. The occurrence and rate of organic matter degrada-

Table 2. Classification of bottom sediments based on the geochemical criteria [Bojakowska and Sokołowska 1998, Bojakowska 2001].

Element	Geochemical background	Class I non-contaminated sediments	Class II moderately uncontaminated sediments	Class III contaminated sediments	Class IV strongly contaminated sediments
Zn	73	200	500	1000	>1000
Cr	6	50	100	400	>400
Pb	15	30	100	400	>400
Cd	<0,5	1,0	3,5	6,0	>6
Cu	7	40	100	200	>200

tion is also important for the binding of metals in sediments [Rubio et al. 2010]. In the tests carried out, the content of organic matter (OM) in the sediments of the examined rivers ranged from 1.14% to 16.70%. The largest variation in the average organic substance content (from four sampling dates) in the examined sediments occurred in the Biała river in Białystok city (A – 1.14%, B – 14.24%) and in Supraśl river in the city of Michałowo (A –5.87%, B – 11.52%). Figures 2 and 3 present the average contents of the investigated metals in the bottom sediments of the analyzed rivers. The obtained results of metal content analyses in particular samples showed the greatest variation in the case of the metals in the bottom sediments of the Biała river flowing through the city of Białystok, mainly for Zn, Cr, Pb (Zn A – 35.54 $mg \cdot kg^{-1}$, B – 184.82 $mg \cdot kg^{-1}$, Cr: A – 9.81 $mg \cdot kg^{-1}$, B – 27.78 $mg \cdot kg^{-1}$, Pb: A – 6.26 $mg \cdot kg^{-1}$, B – 153.26 $mg \cdot kg^{-1}$). Similar relations occurred for the Biała river (inflow of the Orlanka river), which was visible especially for Zn and Pb (Zn: A – 29.97 $mg \cdot kg^{-1}$, B – 88.54 $mg \cdot kg^{-1}$, Pb: A – 11.20 $mg \cdot kg^{-1}$, B – 16.70 $mg \cdot kg^{-1}$). However, in the bottom sediments from the Supraśl river near Michałowo, a considerable variation of Zn results was also obtained before (21.74 $mg \cdot kg^{-1}$) and beyond (48.44 $mg \cdot kg^{-1}$) the city.

First of all, the increased Zn and Pb contents in the bottom sediments of the Biała and Biała (tributary of Orlanka) rivers located in the research points beyond Białystok and Bielsk Podlaski are related to the city's infrastructure. Białystok and Bielsk Podlaski have the largest number of residents of about 296,000 and 26,000, respectively (Table 1), relative to other cities, which is also related to, among others, the largest number of motor vehicles. According to Dmochowski et al. [2011], toxic gases, dusts and aerosols containing heavy metals originate from communication routes. The greatest contamination of soils with automotive pollution is noticeable in the urban agglomerations where a large number of cars move in relatively small areas, and a high volume of traffic is conducive to high emissions. The research conducted by Plak et al. [2010] showed that the transport and the surface runoff from the roads with high traffic are the mainly source of

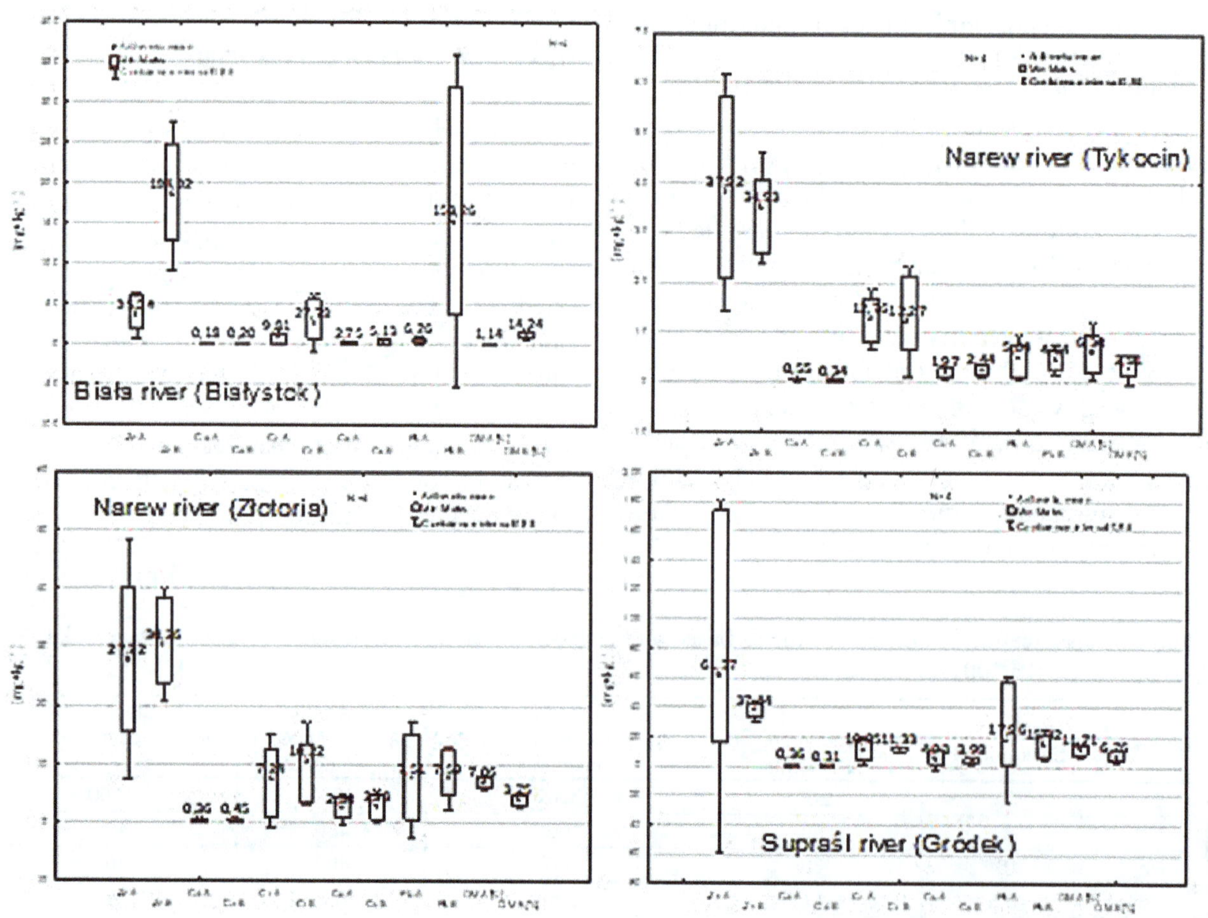

Figure 2. Contents of heavy metals and organic matter in the examined bottom sediments

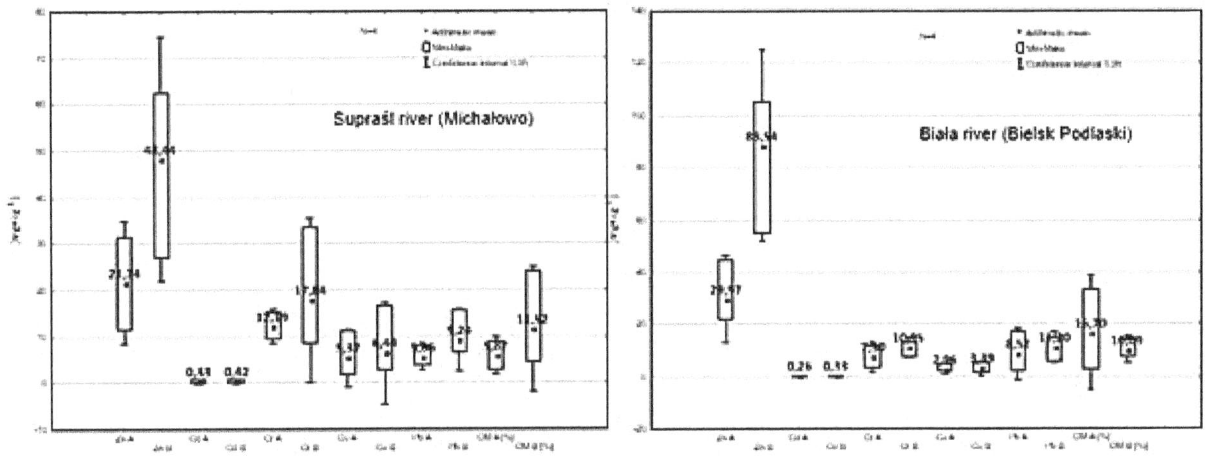

Figure 3. Contents of heavy metals and organic matter in the examined bottom sediments

lead, and – in smaller quantities – also zinc, copper, cadmium or nickel. The above-mentioned analyses are also confirmed by the Badyda's studies [2010], who states that the operation of the transport network causes pollution of soils in the immediate vicinity of busy roads. Another source of metals in urban agglomerations includes the pollution in the atmosphere. Rogula-Kozłowska et al. [2008], in their studies, proved that the highest concentrations of heavy metals are introduced into the atmosphere together with the so-called suspended dust, the largest emitter of which is the energy industry. In Białystok, there are also industrial facilities and a high capacity municipal sewage treatment plant. Bielsk Podlaski also has an industry and sewage treatment plant, but on a smaller scale. In other towns (Gródek, Tykocin, Złotoria and Michałowo), agriculture is the dominant branch of economy. A noticeable source of trace metals in the vicinity of Gródek and Michałowo may be the subsurface runoff from the fields fertilized with minerals and organics (slurry and manure). The possible reason for the presence of high Zn content in the sediment samples from rivers may be due to the huge discharge of the industrial waste and mixing of the runoff with fertilizers residue [Guo and He 2013; Mathivanan and Rajaram 2014; Benhaddya et al. 2014]. Michałowo does not have a developed industry and intense routes of car communication, which could explain the increase in the content of some metals in the Supraśl river sediments at point B, but it has a municipal sewage treatment plant, which can sometimes show certain irregularities in its operation. In the area of Michałowo, agricultural activity is conducted, which may be one of the reasons for the increased content of some metals in the Supraśl river sediments. For instance, the increased level of Cd in the Supraśl river sediments in the vicinity of Michałowo may indicate the transport of pollutants from the subsurface runoff and mineral fertilization from the agricultural land near almost all of the studied areas.

According to the Bojakowska and Sokołowska [1998] and Bojakowska [2001] classification, most of the sediments tested were considered uncontaminated. The enrichment of sediments in Cr, Pb and Zn occurred in the Biała river at the measurement point beyond the city of Białystok.

The accumulation indices for individual metals are presented in Table 3 and they are differentiated depending on a given heavy metal. The indices calculated for copper ranged from -6.76 to -2.01, for chromium from -7.34 to -1.27, which indicates the lack of contamination with these metals (class 0)in the examined rivers. The examined sediments in the analyzed rivers can also be regarded as uncontaminated with zinc due to the negative values of the indices, only four samples took positive values (0.29, 0.62, 0.80, 0.01), while the I_{geo} analysis for Pb showed 91.7% of samples in class 0, four samples of the bottom sediments from the Biała river beyond Białystok accepted the following values: 0.41 (class I), 1.06 (class II), 2.67 (class III), and 3.42 (class IV).

The material tested in the form of the numerical content of indices was subjected to a statistical analysis. No Spearman correlation was detected in most cases between the organic matter (OM) and metal contents in the sediments studied. Only in the samples taken beyond the villages, there was the correlation between organic substance (MO) and Zn content (r = 0.69) and Pb (r = 0.49). There

Table 3. Range of the accumulation index (I_{geo}) for metals in river sediments

River (City)	Stat.	I geo									
		Zn A	Zn B	Cd A	Cd B	Cr A	Cr B	Cu A	Cu B	Pb A	Pb A
Biała (Białystok)	min	-2.79	-0.15	-5.49	-3.14	-6.69	-4.03	-5.97	-4.74	-3.20	0.41
	max	-1.22	0.80	-0.01	-0.68	-3.15	-1.27	-3.88	-3.31	-1.72	3.42
Narew (Tykocin)	min	-2.77	-2.46	-0.06	-2.66	-4.06	-4.33	-5.95	-5.58	-4.98	-3.58
	max	-1.32	-1.81	0.71	0.33	-2.99	-2.66	-4.49	-4.23	-1.92	-2.17
Narew (Złotoria)	min	-3.21	-2.58	-3.35	-1.24	-7.34	-5.22	-6.52	-6.76	-5.84	-2.62
	max	-1.83	-1.89	0.40	0.39	-3.45	-3.35	-3.95	-3.82	-0.97	-1.25
Supraśl (Gródek)	min	-3.12	-2.09	-3.86	-2.35	-4.95	-3.71	-5.57	-4.85	-6.33	-2.26
	max	0.29	-1.73	0.23	0.31	-2.95	-3.34	-2.67	-3.53	0.96	-0.51
Supraśl (Michałowo)	min	-3.64	-2.39	-5.49	-5.35	-3.82	-3.97	-5.14	-4.62	-3.02	-2.21
	max	-2.17	-1.19	0.92	0.68	-3.15	-2.00	-2.61	-2.01	-1.96	-0.95
Biała dopływ Orlanki (Bielsk Podlaski)	min	-2.68	-1.37	-2.09	-3.91	-5.22	-4.17	-5.12	-5.41	-3.65	-2.33
	max	-1.68	-0.43	-0.30	0.20	-3.49	-3.37	-3.80	-3.64	-0.81	-1.02

was no association between the metals investigated and organic matter (OM), indicating that OM cannot be a major factor for the distribution of heavy metals in sediments. As a result of the statistical analysis of Ward aggregates (CA) and the content of heavy metals in the bottom sediments, the dendrogram shown in Figure 4 could be prepared. There are 5 clusters (groups of rivers) on the dendrogram. The first group contains only one object, i.e. the Biała river (tributary of Supraśl), which flows through Białystok. Sometimes, only one object in a cluster from Ward cluster analysis is referred to as an isolated object [Skorbiłowicz 2010]. The isolated object often has special properties in comparison with those forming other groups (aggregates). The other group included the remaining rivers. The city of Białystok is an urbanized area with a high degree of exposure of the Biała river to the anthropogenic impact, which increased the content of metals in its river sediments. The second group included other rivers, the bottom sediments of which are less susceptible to pollution from other urban areas. Nevertheless, two additional groups (III and IV) can be separated within group II. In group III, there is the Biała river (inflow of Orlanka), which is located in the zone of Bielsk Podlaski influence. Group III, like I, also includes an isolated object that has special properties. The bottom sediments of Biała (inflow of the Orlanka) are exposed to the pollution from Bielsk Podlaski, where industry and sewage treatment plant is present (Table 1). In group IV, there are two rivers: Narew and Supraśl, which flow through Złotoria, Tykocin, Michałowo and

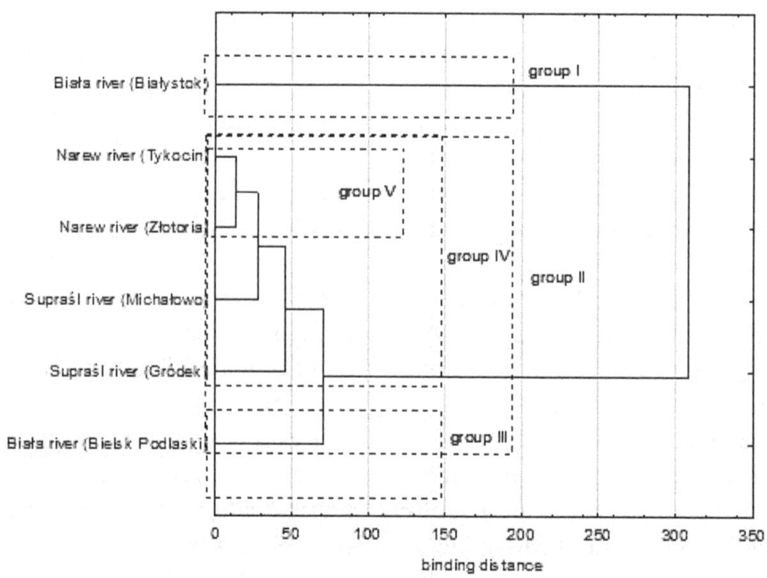

Figure 4. Ward agglomeration analysis of heavy metals in bottom sediments

Gródek, respectively. An interesting fact is group V, which includes the Narew river with two towns: Tykocin and Złotoria. The degree of impact of Złotoria and Tykocin on the Narew river sediments is smaller, as compared to the object from Group III and the objects from Group IV. The arrangement of objects in group V may be a consequence of a fairly close distance between the two towns along the Narew river. In the case of the Supraśl river, which flows through Gródek and Michałowo, the cluster analysis did not show similar results. Gródek and Michałowo located on the Supraśl river did not find themselves in one cluster (group) of cluster analysis. It has a supposed connection with the agricultural character of the areas adjacent to both towns, the intensity of impact of which is greater in the case of Michałowo. The impacts of agricultural activity may overlap the impact of urbanized areas, although the metal content in the Supraśl river sediments is not a threat to this ecosystem. Six-year studies performed by Skorbiłowicz [2012] upon the upper Narew river and its tributaries showed that the sediments of Supraśl and Narew river catchments are characterized by low contents of trace metals.

The work also looked for a compliance and confirmation of CA cluster analysis with the multifactor FA analysis. Two factors F1 and F2 were found on the basis of the FA analysis (Table 4). Factor F1 is the most dominant as it explains as much as 45% of the variability, as opposed to F2 explaining 20% of variability. The factor F1 correlates positively with Zn, Cr, Cu, Pb and OM in the bottom sediments located at points beyond the towns (designation B). The positive correlation of F1 with Zn, Cr, Cu, Pb and OM indicates the significant sources of these metals and organic matter. The FA analysis showed which metals leak from the analyzed and previously identified areas to the bottom sediments of rivers. It also indicated which metals migrate from other areas. The negative correlation between F1 and Cd in the sediments in both points before and beyond the cities is interesting, which excludes similar interpretation as in the case of other metals. Probably, there are other factors shaping the Cd content in the bottom sediments associated, for example, with the geochemical nature of sediments or unexplained processes connected with quite high mobility of this element. Large mobility of Cd can contribute to its easy mobilization from the bottom sediments and transport with river waters over further distances. Additionally, simultaneous processes of enriching and activating Cd from the bottom sediments may explain the negative correlation with the factor F1. The analyses showed a lower pH value in the river sediments at the points marked with the letter B, which may be one of the reasons for Cd mobilization. This issue certainly requires further research and analysis. The F2 factor has much less explanatory power (20%) and is associated with the sources of the bottom sediments enrichment in Zn, Cu and Pb of the studied rivers, but at the points marked with the letter A. This can be explained, for example, by the transport of metals in question from the upper sections of the examined rivers. Metals can migrate most often as bound to suspended matter, colloids or in a dissolved form, not only with the river water but also with the run-off from the surrounding areas.

Table 4. Results of multivariate analysis (FA)

Variable	Factor 1	Factor 2
Zn A		0.75
Zn B	0.91	
Cd A	-0.79	
Cd B	-0.74	
Cr A		
Cr B	0.94	
Cu A		0.73
Cu B	0.70	
Pb A		0.92
Pb B	0.91	
OM A [%]		
OM B [%]	0.92	
Explained variance [%]	45	20

The cluster analysis (CA) allowed to identify the urban areas, which contribute to the migration of metals to river ecosystems to the greatest degree. On the other hand, due to the multivariate analysis (FA), it was proven which metals migrate from the analyzed areas to the river ecosystems.

CONCLUSIONS

1. The highest contents of Zn, Pb and Cr were recorded in the bottom sediments from the rivers: Biała – inflow of Supraśl (Białystok) and Biała – inflow of Orlanka (Bielsk Podlaski). The influence of urbanization and anthropogenic activity such as: discharges of unpurified or insufficiently treated municipal and industrial sewage, surface runoff, emissions from communication routes, has been demonstrated. The changes pertaining to the content of metals in sediments were related to the nature of urbanized areas.

2. The research also showed the impact of agricultural activity (slightly elevated Cd level). The areas adjacent to the examined objects are in most cases the areas, on which agricultural activities related to mineral and organic fertilization are conducted.
3. Assessment of the bottom sediments contamination using the geo-accumulation index and classification of sediments in Poland according to Bojakowska and Sokołowska (1998) and Bojakowska (2001) showed that they are in most cases uncontaminated. The exceedances occurred in several samples of the bottom sediments from the Biała river beyond Białystok for Pb, Zn and Cr.
4. The statistic multivariate analyses allowed for the identification of the urban areas shaping the heavy metals content in the sediments of the examined rivers to varying degrees. It also turned out that they enabled to answer the question which metals migrate from the analyzed areas to the river ecosystems. The results of the conducted analyses complemented one another in explaining the origin of elements.

Acknowledgements

The research was carried out as part of the work No. S/WBiIŚ/3/2014 and financed from the resources for education by The Ministry of Science and Higher Education.

REFERENCES

1. Badyda A.J. 2010. Zagrożenia środowiskowe ze strony transportu. Nauka 4, Polska Akademia Nauk, Warszawa, 115–125.
2. Benhaddya M.L, Hadjel M. 2014. Spatial distribution and contamination assessment of heavy metals in surface soils of Hassi Messaoud, Algeria. Environ Earth Sci 71, 1473–1486.
3. Bojakowska I. 2001. Kryteria oceny zanieczyszczenia osadów wodnych. Przegl. Geolog. 49(3), 213–218.
4. Cataldo D. Colombo J.C., Boltovskoy D. Bilos C. Landoni P. 2001. Environmental toxicity assessment in the Parana river delta (Argentina): simultaneous evaluation of selected pollutants and mortality rates of Corbicula Fluminea (Bivalvia) early juveniles. Environ. Poll., 112 (3), 379–389.
5. Chabukdhara M., Nema A.K. 2012. Assessment of heavy metal contamination in Hindon River sediments: A chemometric and geochemical approach. Chemosphere, 87(8), 945–953.
6. Charkhabi A.H., Sakizadeh M., Rafiee G. 2005. Seasonal fluctuation in heavy metal pollution in Iran's Siahroud River. Environ Sci Pollut Res 12, 264–270.
7. Ciszewski D., Aleksander-Kwaterczak U. 2015. Zanieczyszczenie osadów metalami. Transport, akumulacja, remobilizacja, remdiacja, AGH, Wydawca AGH, Kraków, 2015, pp. 165.
8. Dassenakis M., Scoullos M., Foufa E., Krasakopoulou E., Pavlidou A., Kloukiniotou M. 1998. Effects of multiple source pollution on a small Mediterranean river. Appl. Geochem., 13 (2), 197–211.
9. Dmochowski D., Prędecka A., Mazurek M., Pawlak A. 2011. Ocena zagrożeń związanych z emisją metali ciężkich w aspekcie bezpieczeństwa ekologicznego na przykładzie ogródków działkowych w aglomeracji miejskiej, Polski przegląd medycyny i psychologii lotniczej, 3 (17), 257–265.
10. Evans C.D., Davies T.D., Wigington JRP.J., Tranter M., Kretschier W.A. 1996. Use od factor analysis to investigate processes controlling the chemical composition of four streams in Adirindack Mountaions. „J. Hydrol." 185, 297–316.
11. Giordano S., Adamo P., Sorbo S., Vingiani S. 2005. Atmospheric trace metal pollution in the Naples urban area based on results from moss and lichen bags. Environ. Pollut. 136, 431–442.
12. Guo R, He X. 2013. Spatial variations and ecological risk assessment of heavy metals in surface sediments on the upper reaches of Hun river, Northeast China. Environ Earth Sci 70, 1083–1090.
13. Harikumar P.S., Nasir U.P., Mujeebu Rahman M.P. 2009. Distribution of heavy metals in the core sediments of a tropical wetland system. Int. J. Environ. Sci. Tech., 6 (2), 225–232.
14. Hobbelen P.H.F., Koolhaas J.E., van Gestel C.A.M. 2004. Risk assessment of heavy metal pollution for detritivores in floodplain soils in the Biesbosch, The Netherlands, taking bioavailability into account. Environ. Poll., 129 (3), 409–419.
15. Idris A.M. 2008. Combining multivariate analysis and geochemical approaches for assessing heavy metal level in sediments from Sudanese harbors along the Red Sea coast. Microchemical Journal, 90(2), 159–163.
16. Igras J. 2004. Zawartość składnikow mineralnych w wodach drenarskich z użytków rolnych w Polsce. Monografie i Roprawy Naukowe. z. 13. Puławy.
17. Koukal B., Dominik J., Vignati D., Arpagaus P., Santiago S., Ouddane B., Benaabidate L. 2004. Assessment of water quality and toxicity of polluted rivers Fez and Sebou in the region of Fez (Morocco). Environ. Poll., 131 (1), 163–172.
18. Kumar A., Maiti SK. 2015. Assessment of potentially toxic heavy metal contamination in agricultural fields, sediment, and water from an abandoned chromite-asbestos mine waste of Roro hill, Chai-

basa, India. Environ Earth Sci 74(3), 2617–2633.

19. Li Y.L., Liu Y.G., Liu J.L., Zeng G.M., Li X. 2008. Effects of EDTA on lead uptake by Typha oreentalis Presl: a new lead-accumulating species in southern China. Bull Environ Contam Toxicol 81(1), 36–41.
20. Lin C., He M., Liu X., Guo W., Liu S. 2013. Distribution and contamination assessment of toxic trace elements in sediment of the Daliao river system, China. Environ Earth Sci 70(3), 163–3173.
21. Lis J., Pasieczna A. 1995. Atlas geochemiczny Polski 1:2 500 000. PIG, Warszawa.
22. Liu W.X., Li X.D., Shen Z.G., Wang D.C., Wai O.W.H., Li Y.S. 2003. Multivariate statistical study of heavy metal enrichment in sediments of the Pearl river Estuary. Environ Pollut 121, 377–388.
23. Mathivanan K., Rajaram R. 2014. Anthropogenic influences on toxic metals in water and sediment samples collected from industrially polluted Cuddalore coast, Southeast coast of India. Environ Earth Sci 72, 997–1010.
24. Miller C.V., Foster, G.D., Majedi B.F. 2003. Baseflow and stormflow metal fluxes from two small agricultural catchments in the coastal plain of Chesapeake Bay Basin, United States. Appl. Geochem., 18 (4), 483–501.
25. Mohammed MH., Markert B. 2006. Toxicity of heavy metals on Scenedesmus quadricauda (Turp.) de Brebisson in batch cultures. Environ Sci Pollut Res 13, 98–104.
26. Mohiuddin K.M., Zakir H.M., Otomo K., Sharmin S., Shikazono N. 2010. Geochemical distribution of trace metal pollutants in water and sediments of downstream of an urban river. Int. J. Environ. Sci. Tech., 7 (1), 17–28.
27. Müller G. 1981. Die Schwermettallbelastung der Sedimente des Neckars und seiner Nebenflusse: Eine Bestandsaufnahme. Chemiker Zeitung, "Chem., Techn. Chem., Chemiewirtschaft", 105, 6, 157–164.
28. Okafor E.C., Opuene K. 2007. Preliminary assessment of trace metals and polycyclic aromatic hydrocarbons in the sediments. Int. J. Environ. Sci. Tech., 4 (2) 233 – 240 pollutants and mortality rates of Corbicula Fluminea (Bivalvia) early juveniles. Environ. Poll., 112 (3), 379–389.
29. Plak A., Bartmiński P., Dębicki R. 2010. Wpływ transportu publicznego na zawartość wybranych metali ciężkich w glebach sąsiadujących z ulicami Lublina, Proceedings of ECOpole, Vol. 4, No. 1.
30. Puckett L.J., Bricker O.P. 1992. Factors controlling the major ion chemistry of streams in the Blue Rifge and Valley and Ridge physiographic provinces of Virginia and Maryland. Hydrol. Proc. 6, 79–98.
31. Rogula-Kozłowska W., Klejnowski K, Szopa S. 2008. Concentrations of 42 elements in atmospheric fine particles in Zabrze, Poland, Environment Protection Engineering, 34(4), 5–15.
32. Rubio B., Àlvarez-Iglasias P., Vilias F. 2010. Diagenesis and anthropogenesis of metals in the recent Holocene sedimetary record of the Ria de Vigo (NW Spania) Marine Pollution bulletin 60, 1122–1129.
33. Salminen R., Batista M.J., Bidovec M., Demetriades A., De Vivo B., Devos W., Duris M., Gilucis A., Gregorauskiene V., Halamic J., Heitzmann P., Lima A., Jordan G., Klaver G., Klein P., Lis J., Locutura J., Marsina K., Mazreku A., O'connor P.J., Olsson S.Å., Ottesen R.-T., Petersell V., Plant J.A., Reeder S., Salpeteur I., Sandström H., Siewers U., Steenfel T A., Tarvainen T. 2005. Geochemical atlas of Europe. Part 1 – Background Information, Methodology and Maps.
34. Shi G., Chen Z., Xu S., Zhang J., Wang L., Bi C., Teng J. 2008. Potentially toxic metal contamination of urban soils and roadside dust in Shanghai, China. Environ. Pollut. 156, 251.
35. Silva J.D., Srinivasalu S., Roy P.D., Jonathan M.P. 2014. Environmental conditions inferred from multi-element concentrations in sediments off Cauvery delta, Southeast India. Environ Earth Sci 71, 2043–2058.
36. Singh K.P., Malik A., Sinha S., Singh V.K., Murthy R. 2005. Estimation of source of heavy metal contamination in sediments of Gomti river (India) using principal component analysis. Water, Air and Soil Pollution, 166, 321–341.
37. Skorbiłowicz E. 2012. Studia nad rozmieszczeniem niektórych metali w środowisku wodnym zlewni górnej Narwi. Rozp nauk nr 222. Białystok: Oficyna Wydawnicza Politechniki Białostockiej, pp. 212.
38. Skorbiłowicz M. 2010. Czynniki i procesy kształtujące obieg składników mineralnych w wodach rzecznych zlewni górnej Narwi. Oficyna Wydawcza Politechniki Białostockiej, Białystok, pp. 150.
39. Smal H., Salomons W. 1995. Acidification and its long-term impact on metal mobility. In: Salomons W., Stigliani W. (Eds), Biogeodynamics of pollutants in soils and sediments, Springer-Verlag, Berlin, 193–212.
40. Sundaray S.K., Nayak B.B., Lee B.G., Bhatta D. 2014 Spatio-temporal dynamics of heavymetals in sediments of the river estuarine system: Mahanadi basin (India). Environ Earth Sci 71(4), 1893–1909.
41. Wang Y., Yang Z.F., Shen Z.Y., Tang Z.W., Niu J.F., & Gao F. 2011. Assessment of heavy metals in sediments from a typical catchment of the Yangtze River, China. Environmental Monitoring and Assessment, 172(1–4), 407–417.
42. Wcislo E, Ioven D, Kucharski R, Szdzuj J. 2002. Human health risk assessment case study: an abandoned metal smelter site in Poland. Chemosphere 47(5), 507–515.

13

The Effect of Organic Substrates Application on the Amylolytic Activity of Urban Soils

Urszula Wydro[1*], Elżbieta Wołejko[1], Agata Jabłońska-Trypuć[1], Andrzej Butarewicz[1], Tadeusz Łoboda[1]

[1] Bialystok University of Technology, Department of Chemistry, Biology and Biotechnology, Wiejska 45E, 15-351 Białystok. Poland

[*] Corresponding author's e-mail: u.wydro@pb.edu.pl

ABSTRACT

The aim of this work was to investigate the influence of fresh and thermally treated sewage sludge on the amylolytical activity of the urban soils. Two experimental areas on the lawns along the main roads of Bialystok were prepared in 2011 and 2015. The factors taken into account in the experiment involved: the type of sewage sludge (mechanically dewatered – SS and thermally dried "Granbial" – G), three doses of sewage sludge (0 – control; 14.5 and 29 t DM/ha) and two grass mixtures (Eko and Roadside). The number of amylolytic bacteria in grass mixtures and rhizosphere and amylase activity in soil were monitored twice during 2011 and 2012 vegetation season (in July and October). The main properties of soil (soil pH, granulometric composition, total organic carbon – TOC, available phosphorus and total nitrogen content) after the application of organic substrates were also analysed. Addition of the dewatered sewage sludge to soil resulted in an increase in the number of amylolytic bacteria (from 7.4 to 18.8 $\cdot 10^6$ cfu/g DM) in July, while in the soil treatment with thermally dried sludge, the increased number of the bacteria (from 11.03 to 44.68 $\cdot 10^6$ cfu/g DM) was observed in October. The amylases activity in the soil treated with SS exhibited the highest average value in July (2.11 mg Glc/g DM $\cdot$24 h), while in the soil treated with "Granbial", it was stable in studied period (from 1.65 to 2.18 mg Glc/g DM $\cdot$24 h).

Keywords: sewage sludge, rhizosphere, urban soil, amylases activity, amylolytic bacteria

INTRODUCTION

The urban soils in Poland are the least studied group of soils [Niedbała et al 2010]. According to Greinert [2000], urban soils are easily degraded because of a low content of humus and organic matter, low water capacity and a small amount of nutrients. The structure of these soils is not uniform and often is made of construction waste, such as rubble, brick and steel. Furthermore, soil has a compact structure as a result of using heavy equipment, which disturbs the air-water relations. In addition, the urban area located along communication routes is exposed to the pollution from transport. Wysocki and Stawicka [2005] have reported that the mechanical and chemical transformation of soil in the urban area causes the negative changes of its biological activity, which subsequently determines a proper biocenoses functioning. The transport contaminants can be harmful to soil microorganisms and interfere with their mutual relation [Bach and Frazik-Adamczyk 2005].

Lawns, as a basic element of green areas in cities, are the simplest and most natural coverage of soil surface. Grasses play an important role as a pioneer species in setting difficult areas. Lawns are essential for the protection and stabilization of soil, air purification, pollution absorption and microclimate improvement [Grejnert 2000, Wolski et al 2006]. Due to their function it is necessary to ensure proper quality of soil and growth of grasses.

The progressing urbanization has caused an increase in the amount of domestic and industrial wastes such as sewage sludge. The most common way for the management of this biosolid is its environmental recycling as a fertilizer or reclamation substrate. Sewage sludges are rich in organic matter, macro- such as nitrogen, phosphorus and microelements and can improve the physical, chemical and biological properties of soil [Jamali 2008, Jezierska-Tyś and Frąc 2008, Singh and Agrawal 2008, Sanches-Monedero 2004, Siuta 2003]. The above-mentioned properties support the use of sludge in the degraded soil reclamation in urban space, including the formation of lawns.

The amylolytic bacteria hydrolyze starch and glycogen by using amylases such as: α-amylase and β-amylase. The degradation products (i.e. glucose) are utilized by microorganisms as a source of carbon and energy and by plants as a source of nutrients [Kucharski et al 2000].

The aim of this work was to investigate the impact of unprocessed and thermally treated sewage sludge on the amylolytical soil activity, expressed as the number of amylolytic bacteria and soil amylases activity. Subsequently, the obtained results were compared with selected physicochemical properties of the soil.

MATERIAL AND METHODS

Two field experiments were prepared on the two lawns along the main roads of Białystok, (ul. Hetmańska and Popiełuszki)where two types of organic substrates were applied. The first experiment was established in 2010, whereas the second in 2014. In each study, one experimental area of 90 m^2 was divided into 18 plots, 5 m^2 each.

The stabilized mechanical dewatered municipal sewage sludge from the Municipal Wastewater Treatment Plant in Sokółka (SS) and organic fertilizer "Granbial" produced from the sewage sludge during thermal processing (at 200°C) in a Wastewater Treatment plant in Białystok (G) were used. In autumn 2010 and 2014, the studied areas were fertilized with municipal sludge and "Granbial" with 0 (C – control), 14.5 and 29 t DM/ha, respectively. The doses of sewage sludge were established according to the Regulation of Ministry of Environment [Directive of...2015], where the maximum sludge dose for two-year use for non-agricultural land reclamation does not exceed 30 t of dry matter/ha. The applied substrates were eligible for use in non-agricultural land reclamation in accordance with the Polish regulations [Directive of...2015]. The values of the studied parameters in soils and the substrates characteristics before starting the experiment are presented in Table 1 and 2.

In spring 2011 and 2015, two mixtures of lawn grasses were sown on the prepared plots, i.e. Eko from Nieznanice Plant Breeding Station (M1) which included 30% of *Lolium perenne* cv. Niga, 15% of *Poa pratensis* cv. Amason, 22.6% of *Festuca rubra* cv. Adio and 32.4%

Table 1. Selected physicochemical properties of the soils in the research plots at ul. Hetmańska and Popiełuszki, taken before the foundation of experiments I and II

Experiment	Location	pH	Soil category	P	Cd	Cr	Cu	Ni	Pb	Zn
		-	-	mgP$_2$O$_5$/100g	mg/kg DM					
I	Hetmańska	7.9	light	7.3	<0.5	10.9	9.5	4.5	12.4	40.9
	Popiełuszki	7.6	light	22.0	1.7	13.1	17.9	5.9	26.5	82.9
II	Hetmańska	8.1	light	5.7	0.5	12.1	16.7	6.3	7.5	50.8
	Popiełuszki	7.7	light	20.7	0.6	11.5	26.3	8.5	5.6	74.4
Limit value in the application of sludge to land reclamation for non-agricultural purposes					3.0	100	50	30	50	150

Table 2. Characteristics of the municipal sewage sludge from the sewage treatment plants in Sokółka and "Granbial" from Wastewater Treatment in Białystok used in experiments

Type of organic substrate	pH	Dry matter	Organic substance	N	P	N-NH$_3$	Ca	Mg	Cd	Cr	Cu	Ni	Pb	Zn
	-		% DM						mg/kg DM					
Sewage sludge (SS)	6.7	19.3	58.4	3.99	2.73	0.14	5.51	0.66	<0.5	58.0	194	22.0	23.5	1459
"Granbial" (G)	8.2	81.7	56.9	4.63	3.26	0.18	3.79	0.57	<1.25	76.6	198	30.1	26.0	1045
Limit value in the application of sludge to land reclamation for non-agricultural purposes									25	1000	1200	400	1000	3500

of *Festuca rubra* cv. Nimba, and grass mixture Roadside from Barenbrug (M2) which included 32% of *Lolium perenne* cv. Barmedia, 5% of *Poa pratensis* cv. Baron, 52% of *Festuca rubra rubra* cv. Barustic, 5% of *Festuca rubra commutata* cv. Bardiva (BE) and 6% of *Festuca rubra commutata* cv. Bardiva (NL).

General procedures regarding the main experiment are shown in Table 3.

Amylolytic activity analyses

The soil amylolytic activity after organic substrates application was assessed by determining the number of amylolytic bacteria in rhisosphere and soil amylase activity. Fresh samples of grass mixture roots with the soil for a number of amylolitic bacteria analyses and the soil from a depth of 0–20 cm for the analysis of the amylases activity were collected twice, as it was described in Table 3.

The number of amylolytic bacteria was determined according to the method described by Rodina [1968] on the medium with 2% of starch. The incubation of microorganisms was carried out at 27°C for 96 h. After the incubation time, the colonies grown in Petri dishes were covered by Lugol solution. For the estimation of amylolytic bacteria, the number colonies surrounded with yellow zone were counted. The mean number of colonies was expressed as cfu (colony-forming unit) per 1 g dry mater of soil and roots (cfu/g DM).

The soil amylases activity was determined with the method based on the analysis of reducing sugars amount produced after the enzymatic hydrolysis of starch. The soil samples were incubated with 1% starch solution in acetate buffer (pH 5.5) for 24 h at 50°C. The amount of reducing sugars produced, expressed in mg of released glucose (Glc) was determined with the colorimetric method described by Merci and Schinner [2005] for the invertase activity.

Physicochemical analysis

The granulometric composition of soil was analyzed by means of Analysette 22 NanoTec Plus (Fritsch) analyser. The values of pH in the soil samples taken from a depth of 0–20 cm were determined with the potentiometric method in a soil suspension and distilled water in the ratio of 1:2.5. The total nitrogen was measured by using Kjeldahl method based on wet oxidation of organic matter using H_2SO_4 and conversion of organic nitrogen to the ammonium form on the Vapodest 50s Analyzer (Gerhardt). The available phosphorus (expressed as P_2O_5) was determined by Egner-Riehm method in 0.04 solution of calcium lactate. The amount of available phosphorus was determined by UV-vis spectrophotometry Hach Lange DR 5000. The organic carbon (TOC) content was measured using Analytik Jena TOC multi NC analyzer.

Statistical analysis

The basic statistics (minimal and maximal values, median, means and standard deviation) for the studied physical and chemical parameters were calculated. In order to test the effects of the experimental factors (type and dose of organic substrates, sampling time, grass mixture) in the variables analyzed, the multi-way analysis of variance MANOVA with interaction term was performed. In the case of significant F-tests, the differences between group means were carried out by the Tukey post-hoc test at α=0.05. The correlation between the characteristics was calculated using Pearson correlation (with the level of significance established at $p < 0.05$). The computation and graphs were done by means of the Statistica 13.0 software.

Table 3. Scheme of experiment I and II

Year of study	Type and dose of sewage sludge	Mixture of grasses	Sampling time
2011	SS1 – mechanical dewatered, 14.5 t DM/ha	M1 – Eko	July and October
		M2 – Roadside	
	SS2 – mechanical dewatered, 29 t DM/ha	M1 – Eko	
		M2 – Roadside	
2014	G1 – thermally dried “Granbial”,14.5 t DM/ha	M1 – Eko	
		M2 – Roadside	
	G2 – thermally dried “Granbial”, 29 t DM/ha	M1 – Eko	
		M2 – Roadside	

RESULT AND DISCUSSION

The granulometric composition of the studied soils is shown in Table 4. Sand was the dominant fraction in the soils and its amount ranged from 53 to 70%, while the silt content was from 21 to 38% and clay – from 5 to14%. The pH values measured in water for the samples collected in July ranged from 7.1 (for soil G2 -treatment "Granbial" in dose 29 t DM/ha and M2-grass mixture Roadside) to 7.7 (soil with SS1 and M1) (Figure 1 A-B). Moreover, in October, the pH values in soil were similar to those in July and were from 7.1 to 7.8 for control with M2 and SS1 with M1, respectively.

Urban soils usually are neutral or alkaline. This is due to the presence of alkaline dusts, contaminants and limestone construction materials [Grejnert 2000]. An increase of soil pH has been reported also after sewage sludge fertilization, which was related to the use of calcium compounds for the sludge stabilization and hygenization [Singh and Agrawal 2008, Szymków and Świtecka 2013].

The average total organic carbon (TOC) contents in soil after treatment with sewage sludge and "Granbial" fertilizer are shown in Figures 1 C-D. The highest average TOC values in the soil collected in July were observed for the plots with "Granbial" in the dose 29 t DM/ha with Eko grass mixture (G2 with M2) and was 3.22% DM, while the lowest was found for control with M1 (1.86% DM). In the soil samples collected in October, the obtained results indicated that an average content of TOC decreased and ranged from 1.47 to 2.67% DM from the plots treated using SS1 with M2 and G2 with M1, respectively). The average total content of nitrogen in July ranged from 0.22% DM for the soil from plots supplemented using G1 with M2 grass mixture to 0.55% DM for the SS2 and M1 grass mixture, whereas the average N content in the soil collected in October was lower than in July and the highest content was recorded in SS2 with M2 (0.36% DM) whereas the lowest – in G1 with M1 (0.18% DM) (Figure 1 E-F).

The average content of available phosphorus in soil was presented in Figures 1 G-H. The values of this parameter obtained for the samples collected in July ranged from 20.06 to 67.93 mg P_2O_5/100 g DM for SS1 soil treatment with M2 grass mixture and G1 with M1, respectively. In the second time of sampling the highest average content of N in soil was recorded in G2 with M1 (60.57 mg P_2O_5/100 g DM), while the lowest in control with M2 (20.58 mg P_2O_5/100 g DM). Generally, higher average content of available P was recorded in the soil amended with "Granbial" fertilizer, compared to all other treatments.

Sewage sludge soil treatment modifies its chemical and biological properties, which was described in the literature data [Siuta 2003, Singh and Agrawal 2008, Napora and Grobelak 2014]. The changes were related to the sewage sludge origin, treatment processes and composition [Singh and Agrawal 2008, Mohamed et al. 2018]. Concerning the soil nutrients content, our results reveal that TOC and available P increased after dewatered sewage sludge and "Granbial" application, but in the case of SS such increase was observed only in July. The differences between the effects of studied substrates can be caused by varying the rate of the raw and thermally treatment sewage sludge mineralization. Wiater et al. [2003] reported that thermally dried granular sewage sludge mineralization occurs with lesser intensity than in the unprocessed sludge. According to Mohamed et al. [2018], forest soil fertilization with sewage sludge significantly influenced only the increase in the total phosphorus content. In the study of Kitczak and Czyż [2009], amendment of sandy soil with sew-

Table 4. Soil granulometric composition in control (C), soil treatments with 14.5 and 29 t DM/ha sewage sludge (SS1 and SS2, respectively) and 14.5 and 29 t DM/ha "Granbial" fertilizer (G1 and G2, respectively)

Treatment	Statistics	Sand [%]	Silt [%]	Clay [%]
C	mean	65	26	9
	range	53–74	21–35	5–13
SS1	mean	62	28	11
	range	57–68	23–32	9–11
SS2	mean	61	28	11
	range	56–65	26–31	9–14
G1	mean	68	27	6
	range	67–69	25–28	5–6
G2	mean	68	27	6
	range	61–70	25–32	5–6

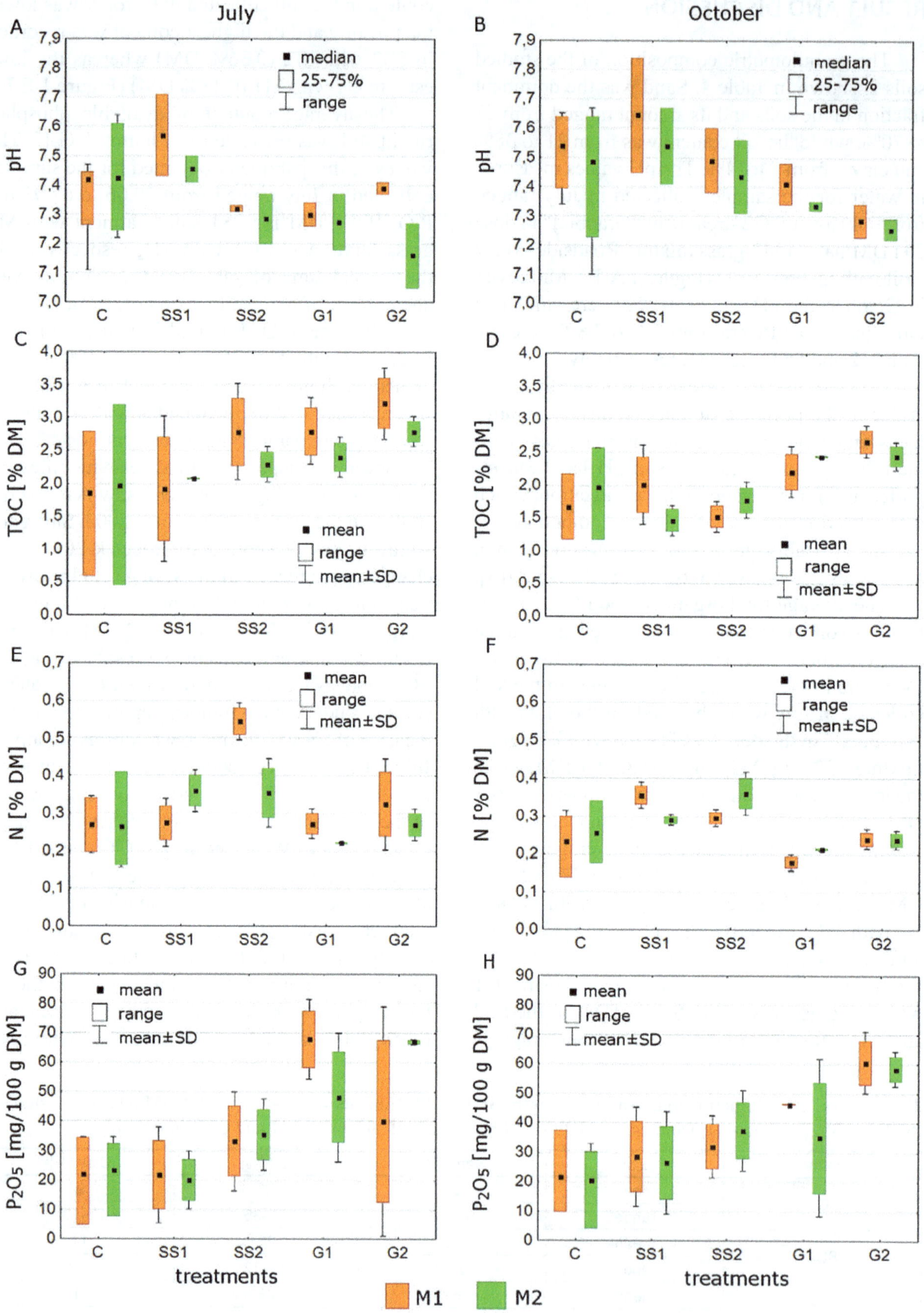

Figure 1. Selected physico-chemical parameters of soil in control plots (C), treated with 14.5 and 29 t DM/ha mechanical dewatered sewage sludge (SS1 and SS2, respectively) and 14.5 and 29 t DM/ha "Granbial" fertilizer (G1 and G2, respectively) and lawn grass mixtures of Eko (M1) and Roadside (M2) measured in July (2011 and 2015) and October (2011 and 2015) (SD - standard deviation)

age sludge in a plot experiment caused an increase of organic matter (by 40%), total nitrogen (from 0.1 to 0.4%) and phosphorus content (from 0.1 to 0.5%). Marando et al. [2011] conducted research in lysimetric experiment where they applied various types of sewage sludge (composed and thermally dried sludge at 80–90°C) for the reclamation of a soil from a working limestone quarry, situated in Spain. After 13 months of sludge addition, they observed a significant increase in C (from 0.5 to 1.7% and from 0.5 and 1.9% for composted and dried sludge, respectively) and in N content (from 0.07 to 0.14% and from 0.07 to 0.11% for composted and dried sludge, respectively). Moreover, the authors stated that application of composted sludge to soil resulted in much slower decomposition, compared to thermally dried sludge. Generally, in our study higher TOC, N and P contents for dewatered sewage sludge (SS) were obtained in the samples collected in July, while higher available P content for the samples with "Granbial" in October, as compared to the control and SS. In the case of TOC and N in soils treatment with G1 and G2, the average level of these parameters was stable during the period of study.

The impact of the experimental factors on the average number of amylolytic bacteria in the rhizosphere zone is presented in Figures 2 A-B. In this study, the grass mixture did not have significant effect on this parameter (Figure 2 A). The average total count of the studied bacteria was varied and the smallest value was recorded in the control plots with M1 (7.8×10^6 cfu /g DM of roots and soil), while the largest was observed in the samples from plots G2 with M2 (7.8×10^6 cfu/g DM of roots and soil).

There were significant differences in the average number of amylolytic bacteria connected with treatments and sampling time (Figure 2 B). The highest average number of these bacteria was found for the samples collected in October from G2 (44.6×10^6 cfu/g DM of roots and soil), whereas the lowest results were obtained for the samples taken in October from SS1 (2.5×10^6 cfu /g DM of roots and soil). The average number of amylolytic bacteria was significantly higher in the samples taken in October from plots treated with "Granbial" (G1 and G2) as compared to the control and soil with sewage sludge. It should be mentioned that the average number of the studied bacteria from SS1 and SS2 in July was higher (by about 500%) than in October, but these differences were not statistically significant. In the case of the soils treated with G1 and G2, this microbiological parameter was significantly higher in October as compared to July.

We found that the grass mixture with treatment did not significantly influence the amylases activity in soil (Figure 3A). The average amylases activity ranged from 1.3 to 2.2 mg Glc/(g DM · 24 h) for control with M1 and G2 with M1, respectively. The activity of the studied enzymes differed significantly depending on the treatment and sampling time (Figure 3B). The average activity of amylases in the soil samples collected in July was the lowest in the control (1.5 mg Glc/(g DM · 24 h)), whereas the highest in G2 (2.2 mg Glc/(g DM · 24 h)). Sig-

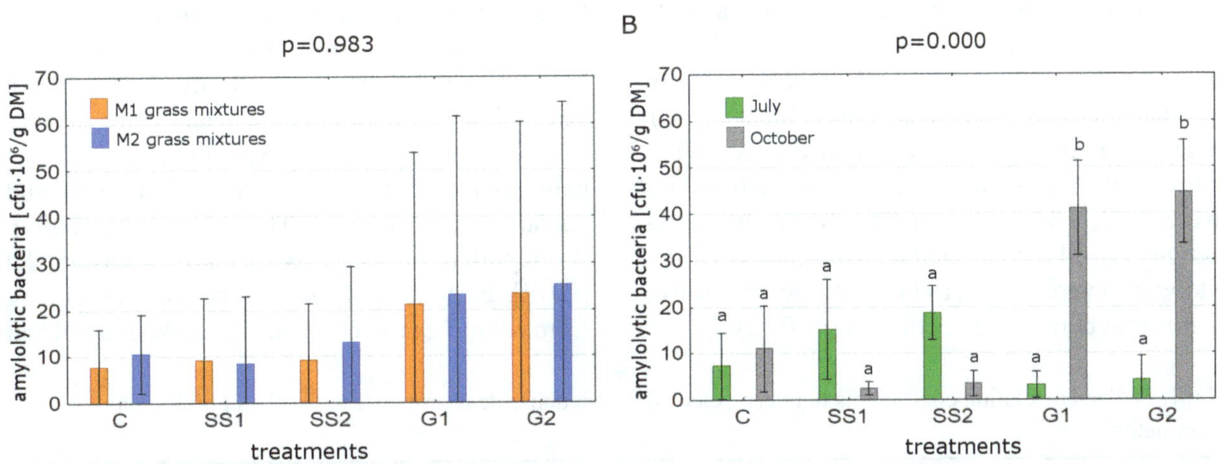

Figure 2. The average number of amylolytic bacteria in grass mixtures (M1 - Eko and M2 - Roadside) rhizosphere in samples collected in July and October from control plots (C), treated with sewage sludge in dose of 14.5 and 29 t DM/ha (SS1 and SS2, respectively) and "Granbial" fertilizer in dose of 14.5 and 29 t DM/ha (G1 and G2, respectively) (different letters for columns indicate significant differences between means for each treatment and grass mixture (A), treatment and sampling time (B), where $p<0.05$, assessment on Tukey's test for $\alpha=0,05$; the error bars represent 5% confidence interval)

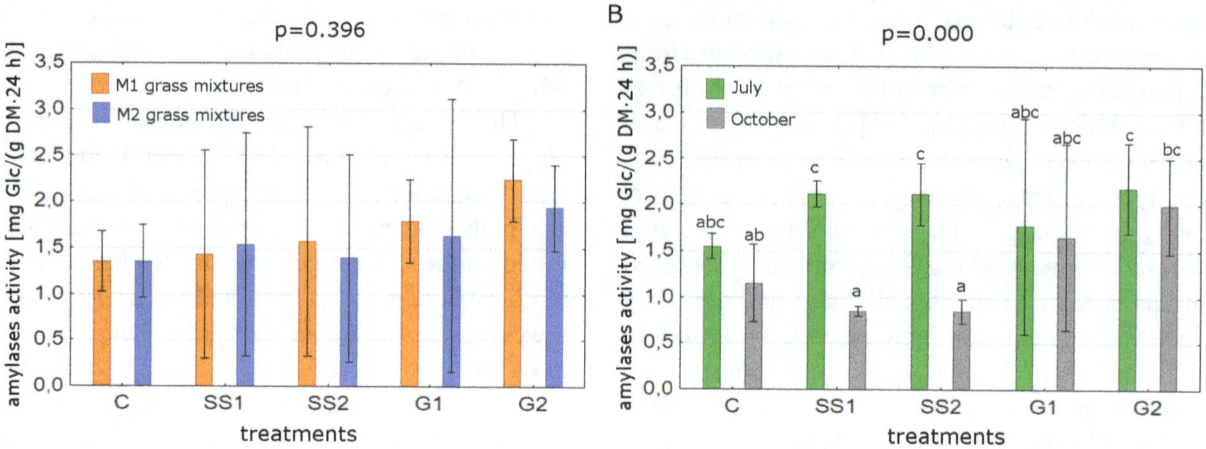

Figure 3. The average amylases soil activity in soil samples with grass mixtures (M1 - Eko and M2 - Roadside) collected in July and October from control plots (C), treated with sewage sludge in dose of 14.5 and 29 t DM/ha (SS1 and SS2, respectively) and "Granbial" fertilizer in dose of 14.5 and 29 t DM/ha (G1 and G2, respectively) (different letters for columns represent significant differences between means for each treatment and grass mixture (A), treatment and sampling time (B), where $p<0.05$, assessment on Tukey's test for $\alpha=0,05$; the error bars represent 5% confidence interval)

nificantly higher soil amylases activity in SS1, SS2 and G2 as compared to the control was observed. The results obtained in the second date of sampling (in October) ranged from 0.8 to 2.0 mg Glc/(g DM · 24 h) for SS and G2, respectively. It should be mentioned that the activity of these enzymes activity in SS1 and SS2 significantly decreased in October, approximately being halved as compared to July, while the soil amylases activity in G2 and G2 was at a similar level.

The correlation analysis presented in Table 5 showed significant negative relationships between the soil amylases activity and the pH measured in H_2O (r= -0.43). The content of clay also negatively correlated with the analyzed enzymes activity (r= -0.41). A positive correlation between the soil amylases activity and selected characteristics of soil such as total organic carbon content (TOC), available P and sand content, was observed. The correlation coefficients (r) for these relationships were: 0.53, 0.34, 0.30, respectively. The amylases activity of soil is connected with the amount of amylolytic bacteria in soil, as indicated by the positive significant correlation coefficient (r= 0.33).

The amylolytic activity in soil is an important part of carbon and energy circulation in the environment. In our study, the application of raw and thermally processed sewage sludge to soil intensified the activity of bacteria and starch decomposing enzymes. The literature data indicate that sewage sludge addition stimulated the growth of microorganisms and changed the enzymes activity. A direction and intensity of the modification depends on type of substrates, dose and studied enzymes [Napora and Grobelak 2014]. In addition, the microbial activity of soil is related to other factors i.e. soil pH, moisture, soil temperatures, content of major nutrient in soil, air temperatures and precipitation [Galus-Barchan and Paśmionka 2014].

In our study, a significant negative correlation between pH and amylases activity was obtained. The soil pH is important parameter, which decides about soil biological processes, ability of microorganisms for growth, availability of nutrients and contaminants [Su et al. 2004 and Rousk et al. 2009]. As reported by Galus-Barchan and Paśmionka [2014], bacteria best tolerate the soil pH ranging from 6.5 to 7.5; therefore, the high alkalinity of soil

Table 5. The correlation coefficients between selected physicochemical properties of soil and biological parameters

Variables	pH in H_2O	TOC	N	$P_{available}$	Sand	Silt	Clay	Amylolytic bacteria	Amylases activity
Amylolytic bacteria	-0.21	0.15	-0,28	0.24	0.25	-0,13	**-0.30***	-	**0.33***
Amylase activity	**-0.43***	**0.53***	0.03	**0.34***	**0.30***	-0.11	**-0.41***	**0.33***	-

* significant correlation coefficients for $p<0.05$

can inhibit the growth and activity of amylolytic bacteria. In the present study, a positive significant correlation between TOC and amylses activity was obtained, while no significant correlation between the number of amylolytic bacteria and TOC was found. According to Jorge-Mardomingo et al. [2013], microorganisms can behave differently in the soil amended with sewage sludge. The application of a very high dose of sludge may decrease the number of bacteria. It depends on the efficiency of using C from biosolids by microorganisms and the amount of labile and mineralizable fraction of organic matter in mix of soil-sludge. Moreover, the native microorganisms can be better adapted in the soil with lower amount of carbon, while a high dose of organic substrate may cause stress.

Jurkowski and Błaszczyk [2012] indicate that the highest growth of individual groups of microorganisms is at specific temperature and humidity where metabolic processes are the most efficient. In our study, the activity of amylolytic microorganisms in connection with sewage sludge treatment was varied, depending on the vegetation period. The air temperatures and rainfall as well as the type of sewage sludge can modify the number of bacteria and enzymes activity. Walczak et al. [2013] indicated that the air temperature regulates the biological activities of bacteria. However, too high temperature may inhibit the growth and cause cells damage. The water is also an important factor for the number and activity of bacteria and their growth is inhibited when the soil water content is less than 30%. The microorganisms are sensitive to the soil drought, even small amount of water below the optimal value results in an inhibition of their development.

The amylolitic activity may depend on the contaminants transport such as heavy metals and organic pollutants. Some metals e.g. As, Cd, Hg, have a strong bactericidal effect on the microbial growth, even at minimal concentration. However, the availability of micro- and macroelements from the amended soil is a function of the climatic conditions during the growing season, dose of sewage sludge and ratio of carbon and nitrogen [Gondek 2012].

In sum, the soil in urbanized areas is characterized by specific habitat conditions for microorganisms. As reported by Frączak [2010] and Tyszkiewicz [2014] any changes in soil properties influenced the biotic relation between microorganisms. Therefore, soil microbes and the enzymes activity are considered to be the most susceptible to the change of environmental parameters.

CONCLUSIONS

1. Supplementation of the studied soils with raw and thermally dried sewage sludge caused changes in the soil pH, TOC, available P and total N content. Higher values of TOC, total N and available P in soil with mechanical dewatered sewage sludge were observed in July, while the increase of available P in soil with thermally processed sludge was found in October.
2. The grass mixture did not have a significant effect on the number of amylolytic bacteria and amylases activity in the studied soils.
3. The time of sampling significantly influenced the number of amylolytic bacteria and amylases activity the in soil. In the case of the soil with dewatered sewage sludge, the significantly higher amylolytic activity was found in July, whereas in the soil with thermally dried sludge, these parameters were higher in October.
4. A significant negative correlation between the amylases activity and soil pH and positive correlation between TOC, P-available, sand content and amylases activity were found. The number of amylolytic bacteria was positively correlated with the sand content and amylases activity.

Acknowledgement

The research was financially supported by the Polish Ministry of Science and Higher Education, projects number MB/WBiIŚ/14/2017 and S/WBiIŚ/3/2015.

REFERENCES

1. Bach A., Frazik-Adamczyk M. 2006. Charakterystyka zagrożeń zieleni miejskiej ze szczególnym uwzględnieniem zieleni w ciągach komunikacyjnych. Opracowanie wykonane dla Urzędu Miasta Krakowa, Wydział Gospodarki Komunalnej i Ochrony Środowiska, Kraków.
2. Bourioug M., Girardclos O., Gillet F. Alaoui-Sehmer L., Pascale B.,Alaoui-Sossé B., Aleya L. 2018. Sewage sludge as a soil amendment in a Larix decidua plantation: Effects on tree growth and floristic diversity. Science of The Total Environment, 621, 291–301.
3. Frączek K. 2010. Skład mikrobiocenotyczny drobnoustrojów biorących udział w procesach przemian azotu w glebie w otoczeniu składowiska odpadów komunalnych. Woda-środowisko-Obszary Wiejskie, 10 (2), 61–71.
4. Galus-Barchan A, Paśmionka I. 2014.

Występowanie wybranych mikroorganizmów w glebie na obszarze Puszczy Niepołomickiej ze szczególnym uwzględnieniem grzybów pleśniowych. Polish Journal of Agronomy, 17, 11–17.

5. Gondek K. 2012. Wpływ nawożenia nawozami mineralnymi, obornikiem od trzody chlewnej i komunalnymi osadami ściekowymi na plon i niektóre wskaźniki jakości ziarna pszenicy jarej (Triticum aestivum l.). Acta Agrophysica, 19(2), 289–302.
6. Greinert A. 2000. Ochrona i rekultywacja terenów zurbanizowanych. Wyd. Politechniki Zielonogórskiej, Monografia, Zielona Góra.
7. Jezierska-Tyś S., Frąc M. 2008. Effect of sewage sludge on selected microbiological and biochemical indices of soil fertility in view of domestic and world wide studies: a review. Acta Agrophysica, 12(2), 393–407.
8. Joniec J., Furczak J. 2007. Liczebność wybranych grup drobnoustrojów w glebie bielicowej pod uprawą wierzby użyźnionej osadem ściekowym po drugim roku jego działania. Annales Universitatis Mariae Curie-Skłodowska Sec. E Agricultura, 62, 1, 93–104.
9. Jorge-Mardomingo I., Solver-Rovira P., Casermeiro M.A., de la Cruz M.T., Polo A. 2013. Seasonal changes in microbial activity in a semiarid soil after application of a high dose of different organic amendments. Geoderma, 206, 40–48.
10. Jurkowski M., Błaszczyk M. 2012. Charakterystyka fizjologiczno-biochemiczna bakterii fermentacji mlekowej. Kosmos. Problemy nauk biologicznych, 61(3), 493–504.
11. Kitczak T., Czyż H. 2009. Przydatność osadów komunalnych i mieszanek trawnikowych do zadarniania gruntów bezglebowych. Rocznik Ochrona Środowiska, 11, 465–472.
12. Kucharski J., Wyszkowska J., Nowak G., Harms H. 2000. Activity of enzymes in soil treated with sewage sludge. Polish Journal of Soil Science, 33 (1), 29–32.
13. Marando G., Jiménez P., Hereter A., Julià M., Ginovart M., Bonmati M. 2011. Effects of thermally dried and composted sewage sludges on the fertility of residual soils from limestone quarries. Applied Soil Ecology, 49, 234–241.
14. Napora A., Grobelak A. 2014. Wpływ osadów ściekowych na aktywność mikrobiologiczną i biochemiczną gleby. Inżynieria i Ochrona Środowiska, 4, 619–630.
15. Niedbała M., Smolińska B., Król K. 2010. Zanieczyszczenia gleb miejskich miasta Łodzi wybranymi pierwiastkami śladowymi. Zeszyty Naukowe Politechniki Łódzkiej. Chemia Spożywcza i Biotechnologia, 74, 29–38.
16. Rodina A. 1968. Mikrobiologiczne metody badania wód. Państwowe Wydawnictwo Rolnicze i Leśne, Warszawa.
17. Rousk J., Brookes P.C., Baath E. 2010. Investigating the mechanisms for the opposing pH relationships of fungal and bacterial growth in soil. Soil Biology & Biochemistry, 42, 926–934.
18. Rozporządzenie Ministra Środowiska z dnia 6 lutego 2015 r w sprawie komunalnych osadów ściekowych (Dz.U. 2015 poz. 257).
19. Sanchez-Monedero M.A., Mondini C., de Nobili M., Leita L., Roig A. 2004. Land application of biosolids. Soil response to different stabilization degree of the treated organic matter. Waste Management, 24, 325–332.
20. Schinner F., Mersi von W. 1990. Xylanase-, CM-cellulase- and invertase activity in soil, an improved method. Soil Biology and Biochemistry, 22, 511–515
21. Singh R.P., Agrawal M. 2008. Potential benefits and risks of land application of sewage sludge. Waste Management, 28, 347–358.
22. Siuta J. 2003. Uwarunkowania i sposoby przyrodniczego użytkowania osadów ściekowych. Inżynieria Ekologiczna, 9, 7–42.
23. Su D.C., Wong J.W.C., Jagadeesan H. 2004. Implication of rhizospheric heavy metals and nutrients for the growth of alfalfa in sludge amended soil. Chemosphere, 56, 957–965.
24. Szymków J., Świtecka M. 2013. Granulacja zhigienizowanych osadów ściekowych. Inżynieria i Aparatura Chemiczna, 3, 262–263.
25. Tyszkiewicz Z. 2014. Zróżnicowanie zbiorowisk grzybów wybranych gleb odwodnionych siedlisk bagiennych w dolinie rzeki Biebrzy. Inżynieria Ekologiczna, 40, 55–64.
26. Walczak M., Burkowska A., Brzezinska M., Kalwasińska A. 2013. Podstawy mikrobiologii w teorii i praktyce. Wydawnictwo Naukowe Uniwersytetu Mikołaja Kopernika w Toruniu, Toruń.
27. Wiater, J. Łukowski, A. Fitko, H., Stelmach, S., Sobolewski, A., Figa, J. 2003. Wstępne badania aplikacyjne granulowanych nawozów organiczno-mineralnych na bazie osadów ściekowych. Zeszyty Naukowe Politechniki Białostockiej. Inżynieria Środowiska, 16(2), 233–237.
28. Wolińska A. 2010. Aktywność dehydrogenazowa mikroorganizmów glebowych i dostępność tlenu w procesie reoksydacji wybranych mineralnych gleb Polski. Acta Agrophysica, 180(3), Rozprawy i Monografie.
29. Wolski K., Kotecki A., Spik Z., Chodak T., Bujak H. 2006. Ocena wstępna możliwości wykorzystania kilkunastu gatunków traw w stabilizacji skarp obwałowań składowiska „Żelazny Most" w Rudnej. Zeszyty Naukowe Uniwersytetu Przyrodniczego we Wrocławiu, Rolnictwo, 545, 293–299.
30. Wysocki C., Stawicka J. 2005. Trawy na terenach zurbanizowanych. Łąkarstwo w Polsce, 8, 227–236.

14

The Spatial Distribution of Heavy Metal Lead and Cadmium Pollution and Coliform Abundance of Waters and Surface Sediment in Demak

Agus Tjahjono[1*], Djoko Suwarno[2]

[1] Engineering Department, Semarang Merchant Marine Polytechnic (Politeknik Ilmu Pelayaran Semarang), Singosari 2 a Semarang, 50242, Central Java, Indonesia

[2] Civil Engineering Department, Unika Soegijapranata Chatolic University, Pawiyatan Luhur IV/1 Bendan Dhuwur, Semarang, 50234, Central Java, Indonesia

[*] Corresponding author's e-mail: a_agus_tjahjono70@yahoo.co.id

ABSTRACT

Various activities such as farming, industry, household waste, fish pond, and fishery can possibly affect the estuarine and coastal ecosystem. Demak waters, located to the east of Tanjung Emas Semarang, become an influential element of fishery industry that mostly produces anchovies and shrimps. This research was aimed at (1) analyzing the spatial distribution of heavy metal Pb and Cd in the surface water and sediment; and (2) investigating the abundance of coliform in Demak waters. Moreover, in this research, water pollution index was also analyzed and compared to other polluted water areas. It was found that the Pb content ranged from 0.6037 to 0.6647 $mg \cdot l^{-1}$ whereas the Cd content ranged from 0.064 to 0.1707 $mg \cdot l^{-1}$. Both lead and cadmium contents had surpassed the water quality standard. The Pb content in sediment ranged from 0.563 to 0.6823 $mg \cdot kg^{-1}$ whereas the Cd content reached 0.0047 to 0.0577 $mg \cdot kg^{-1}$. It showed that this sediment was not polluted by Pb and Cd. The analysis of water pollution index revealed that the pollution level found in Morosari, Rejo, Tuntang, and Lohbener estuary was low. On the other hand, Wedung and Serang estuary were heavily polluted. Compared to some previous studies, most of Pb and Cd contents in Demak waters were greater. However, lead and cadmium content in sediment was lower.

Keywords: Demak waters, spatial distribution of Pb and Cd in surface water and sediment, the abundance of Coliform, water pollution index

INTRODUCTION

Demak Regency, located in Central Java, geographically, lies in the coordinates of 6°43'26" South and 110°27'58"–110°48'47" East. This region extends over 89,743 ha. It borders with Jepara regency and the Java Sea to the north, Kudus and Grobogan regencies to the east, Grobogan and Semarang regencies to the south and Semarang to the west. Demak consists of 14 subdistricts. A few rivers flow through Demak, including Tuntang, Buyaran, and the biggest river in Demak, Serang [https://idwikipedia.org].

Demak waters are located to the east of Port of Tanjung Emas Semarang. It became an influential element of fishery industry, as it produces 308.556 tons of anchovies and 139.296 tons of shrimps [BPS, 2013]. However, high ecological pressure was also found in this area [Hastuti *et al.*, 2013].

On the basis of the previous studies carried out in Moro Demak waters, it was revealed that the Cd content in the surface water reached 0.116 mg/l whereas the Cd content in the sediment was 1.372 mg/l. It means that it had surpassed the water quality standard. When the concentration of heavy metal Pb in the water and sediment increases, the accumulation of heavy metal Pb in *Anadara granosa* will be elevated as well [Sitorus, 2004]. It leads to a bioaccumulation in the blood clam *Anadara granosa* clam [Wulandari *et al.*, 2009].

The coliform bacteria constitute the indicator of patogenic bacteria existence. The more coliform we found within the water, the worse the water quality would be. The coliform bacteria include various bar shaped bacteria, gram-negative bacteria, non-spore-forming bacteria, and lactose-fermenting bacteria which produce gas and acid when being incubated at 37°C in less than 48 hours [Widyaningsih *et al.*, 2016]. The existence of these bacteria is due to pollutants produced by human activities around the river.

Heavy metals are generally defined as the metals with relatively high densities, atomic weights, or atomic numbers. Heavy metals differ from other metals due to the effect they produce when being bound to the living organism. The heavy metal Pb is commonly used in the industry producing batteries and cables. However, the use of water-related contamination, the metal was added as an additive to the engine fuel to increase the octane value. The heavy metal Cd is used as a mixture in the chemical industry [Palar, 2012; Ridhowati, 2013].

This research was aimed at analyzing the spatial distribution of Pb and Cd in the surface water and sediment, as well as the abundance of Coliform in Demak waters. This research also analyzed the water pollution index and compared the results to the previous studies on water pollution.

MATERIALS AND METHODS

Research object

The research was held in May 2017 when the high tide occured at Demak waters, Central Java, Indonesia. The purposeful sampling was done at six stations including station 1, Morosari estuary (6°92'60" South, 110°46'07" East), Rejo estuary (6°85'20" South, 110°51'40" East), Tuntang estuary (6°83'24" South, 110°51'76" East), Lohbener estuary (6°77.09' South, 110°56'85" East), Wedung estuary (6°74.54' South, 110°53'24" East), and Serang estuary (6°71'99" South, 110°54'37" East).

In each station, the sampling was done three times. The Morosari estuary was taken as the sampling location because firstly, it was mangrove area which is the fishery habitat and secondly, absorption of heavy metal possibly occurs in this area. Other estuaries were also used as the sampling locations since they were close to the fishermen settlement and fish farming.

Data collection methods

The materials of this research included the sample of sea water and sediments which contained Pb, Cd, and Coliform taken from Demak waters. In this research, primary and secondary research data were applied. The primary data included the data which were taken directly when the research took place including seawater, pH, DO, TDS, water turbidity, water depth, salinity, current, and the water transparency.

The determination of heavy metal concentration (Pb and Cd) in the seawater.

Firstly, 500 mililiters of sea water was put into teflon separatory funnel. The pH scale was then set at pH 4 by adding liquid HCl. APDC (Amonium Pirolidin Ditio Carbamat) and NaDDC (Natrium Dietil Ditio Carbamat) were also added before it was shaken for one minute. Afterwards, 25 ml of MIBK (Metil Iso Butil Keton) was also added and the solution was shaken again for 30 seconds. The aqueous phase was then separated. Subsequently, 10 ml of distilled and deionized water was poured into the upper phase. It was shaken and the aqueous phase was then separated. After that, 1 ml of HNO_3 was added, shaken, and left standing for about an hour. Then, 19 ml of distilled and deionized water was added, shaken, and aqueous phase was collected to be analyzed by using furnace AAS (Atomic Absorption Spectroscopy) with argon gas and Pb and Cd lamps [Sitorus, 2004; APHA, 2012].

The sample of sediment was dried at 105°C for 24 hours. Then 10–20 grams of the dried sediment was put into a centrifuge tube (polyethylene), 500 ml of distilled and deionized water was added and shaken. It was afterwards centrifuged for 30 minutes in 2000 RPM. The aqueous phase was then discarded. It was dried at 105°C (oven-dry weight) for 24 hours. 1 gram of sediment was put into a sealed teflon beaker. Subsequently, 5 ml of aqua regia and 6 ml of HF (Hygrogen Fluoride) were added. It was then heated at 130 °C until it was dried up. Next, 9 ml of distilled and deionized water was added. It was then analyzed by using furnace AAS with argon gas and Pb and Cd lamps based on the types of the metal which were being analyzed [Sitorus, 2004; APHA, 2012].

The coliform bacteria determination used MPN (Most Probable Number) method [SNI-01–2332–1991; Ijong, 2015]. The method was as follows:

- Presumtive test of coliform

The first step of the presumtive test was to prepare trisalt solution with 10^{-1} to 10^{-3} dilution and to shake until it was homogenous, and to prepare nine tubes of LTB (Lauryl Tryptose Broth) which contained Durham tube for one sample. Then, 1 ml of sample water was added into tube 10^{-1} and shaken until it was homogenous by using vortex. Next, 1 ml of the solution was taken from tube 10^{-1} and put into tube 10^{-2}, and the process was continued until tube 10^{-3}.

By using a sterile pipette, 1 ml of solution of each dilution was moved into each of the three LTB tubes. The process was continued until the sixth sample. The tubes were then incubated at 35°C for about 24–48 hours. Positive tubes would produce bubbles in the Durham tube.

- Confirmative test of Coliform

The method used in this test included moving the cultures from the positive LTB tubes to the tubes with BGLB (Briliant Green Lactose Bile) 2%broth which contained the Durham tube by using inoculation needle. The tubes were incubated for 24–48 hours at 35°C. The positive tubes were those which produced gas in Durham tube. The numbers of the positive tubes were compared based on the MPN value in order to find out the numbers of coliform within the sample.

- Completed test

By using a loopful, the typical colony above was moved to the lactose broth. Then, the formation of acid and gas was re-observed. When the result was positive, aseptically, one eye of loopful was taken and scratched to the NA, and subsequently incubated at 37°C overnight. The growing colony was then used in Gram-Coloring test. The test was regarded as positive if Gram-Negative was obtained and stem cells were formed.

Data analysis

The data analysis method used to examine the density of Coliform bacteria [SNI 2897–2008, Bambang *et al.*, 2014] was as follows:

$$\textit{Coliform Density} = \\ = \textit{value} \text{ of MPN table} \times \frac{1}{\textit{Median dillution}} \quad (1)$$

The data analysis of (1) the heavy metal Pb and Cd content in water and sediment, and (2) the coliform abundance was done by using a comparative analysis of temperature, pH, DO, TDS, salinity, and turbidity. In this analysis, the Spearman's rank-order correlation was applied. This correlation analysis would show the correlation between two variables. The analysis employed SPSS software for Windows, version 22.

The status of water pollution was determined by using the water pollution index based on Ministry of Environment Decree No. 115 year 2003 about the guidelines for water quality assessment. [Nemerow & Sumitomo, 1970].

$$\mathrm{PI_j} = \sqrt{\frac{\left(\mathrm{C_i}/\mathrm{L_{ij}}\right)_M^2 + \left(\mathrm{C_i}/\mathrm{L_{ij}}\right)_R^2}{2}} \quad (2)$$

where: L_{ij} – standard water quality parameter for each parameter at specified water quality purpose (j)
C_i – measured water quality parameters
PI_j – the pollution index for a specified water quality purpose (j)
$(C_i/L_{ij})_M$ – maximum value of C_i/L_{ij}
$(C_i/L_{ij})_R$ – average value of C_i/L_{ij}

The pollution level based on the water pollution index of water quality standard as written in Ministry of Environment Decree no. 115 year 2003 is $0 \le PI_j \le 1$ (meet quality standard/good), $1 \le PI_j \le 5$ (lightly polluted), $5 \le PI_j \le 10$ (moderately polluted) dan $PIj > 10$ (heavily polluted).

RESULTS AND DISCUSSION

On the basis of the findings in the six stations at Demak waters, it was revealed that the Pb content at riverbed was higher than what was found at the surface of the water. The Pb content found at surface of the water ranged from 0.6037 to 0.6647 $mg \cdot l^{-1}$. The highest concentration of Pb was found at the Serang estuary. The concentration was thirteen times higher than the water quality index 0.05 $mg.l^{-1}$. Serang estuary is the biggest river on which the fishermen often go back and forth to the sea. The condition matched with the high concentration of TSS, which proved that a high concentration possibly decreased the photosyntesis activity of both macro and micro marine plants. It led to the decrease of the oxygen which killed fish [Jiyah *et al.*, 2017]. On the other hand, the concentration of Pb in the riverbed was about 0.6207 and 0.7163 $mg \cdot l^{-1}$; the highest concentration occurred at the estuary of Tuntang river. The concentration surpassed the minimum requirement of the water quality index; it was fourteen Times higher than what was required (Figure 2).

The Pb within seawater was able to accumulate to the benthic. The largest accumulation was found in hard branching coral which was followed by soft coral *Alcyonaceae*. For example, the accumulation of *Acropora aspera* and *Pocillopora damicornis,* within ten days, turned into 0.895 mg·kg^{-1} [Samawi *et al.*, 2014; Panuntun *et al*, 2012]. The pollution of Pb led to the highest accumulation in the sediment. The accumulation was followed by biota and seawater [Usman *et al.*, 2013]. The metal accumulation additionally showed the toxic effect of Pb to *Cromileptus altivelis*. The effect of Pb caused the damage of hepar cell (concentration 0.15 ppm) in the form of fatty degeneration, hydropic degeneration, hemorrhage, congestion, and necrosis hepatica (liver cell death) [Triadayani *et al.*, 2010]. The accumulation of Pb in mullet fish *Mugin cephalus* (Linnaeus, 1758) in Cilacap waters was three times higher than those which were taken from 0-500 m and >1000 m from the pollution source point [Heriyanto & Subiandono, 2011].

The content of Pb in the sediment possibly affected the organisms living at the riverbed, such as the *Strombus canarium* snail. The greater content of the heavy metal in the sediment, the higher content of Pb found in the organism [Nasution & Siska, 2011]. Blood clam *Anadara granosa* and mangrove shell *Polymesoda bengalensis* were also found in the sediment. When the content of Pb within the sediment reached 0.823 mg·kg^{-1}, an accumulation of both organisms was twice higher (1.750 mg.kg^{-1}) [Amriani *et al.*, 2011]. Moreover, *Anadara granosa* was highly prone to chromium accumulation[Suprapti, 2008].

The content of Cd found in the surface water was about 0.064 and 0.1707 mg·l^{-1}; the highest concentration was found in the estuary of Rejo river. The concentration of Cd had surpassed the water quality index beingseventeen times higher. However, the concentration decreased as it went to the east. The condition did not occur in the estuary of Lohbener river. The concentration in the riverbed was about 0.055 and 0.12 mg·l^{-1}; Cd was also found at the estuary of Tuntang river. The concentration surpassed the water quality index; it was 12 times greater than the index (Figure 3).

The Cd affected the fish such as Gulama *Sciaena russelli* living in Dumai waters. The accumulation of Cd was mostly found in the fins and gills. In terms of Cd, Cu, Pb, and Zn, the highest accumulation was found in the gills, since it constitutes the respiratory organ where gas exchange occurred [Nurrachmi *et al.*, 2011].

Cd also correlated to green shell *Perna viridis* in Jakarta bay. In this area, the numbers of spermatozoa (r = 0.64) which influenced the development of the spermatozoon were affected to the greatest

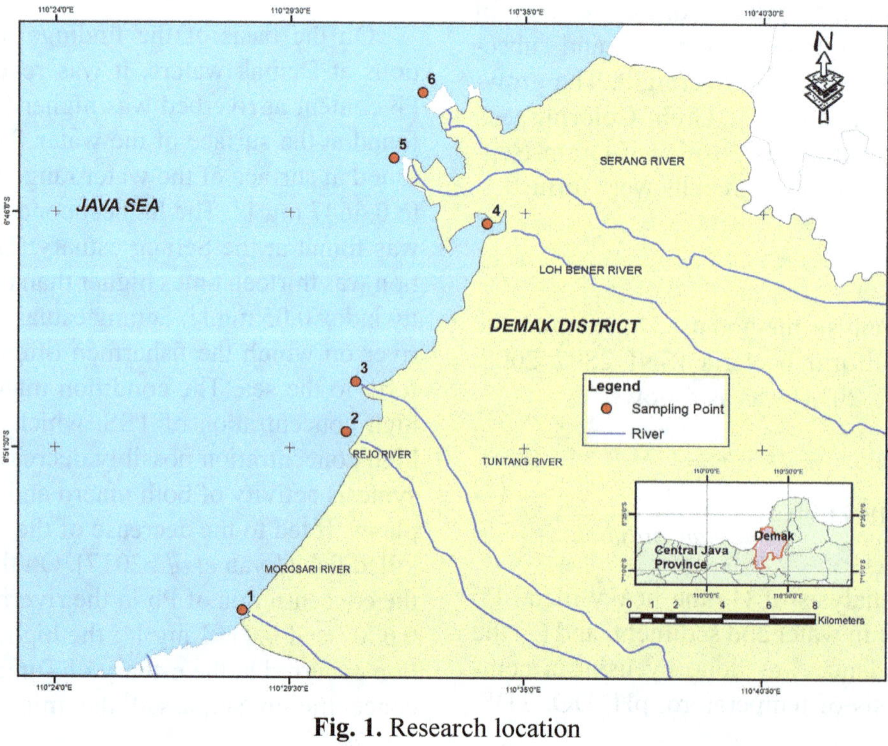

Fig. 1. Research location

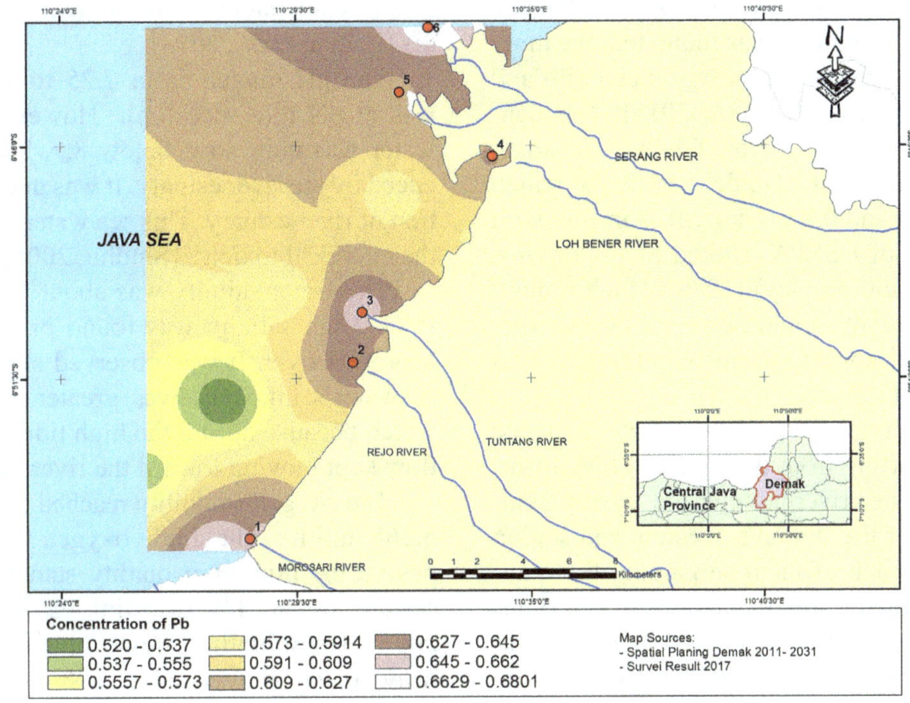

Fig. 2. The distribution of Pb ($mg \cdot l^{-1}$) in Demak waters

degree. It was followed by the sex cells (r = 0.60) [Jalius *et al.*, 2008]. Bioaccumulation also occurred in green shell *Amusim pleuronectes* which was found in Wedung waters, Demak the content in sediment reached 0.4694 $mg \cdot kg^{-1}$ whereas the content on soft tissue became 12.61 times higher (5.9212 $mg \cdot kg^{-1}$) [Azhar *et al.,* 2012].

The Pb content in the surface sediment ranged from 0.2563 to 0.6823 $mg \cdot kg^{-1}$ whereas the Cd content ranged from 0.0047 to 0.0577 $mg \cdot kg^{-1}$; the maximum concentration occurred in Lohbener estuary (Figure 4 and 5). The correlation between the sediment particle (sand, mud and clay) and the heavy metal (Pb and Cd) showed that the

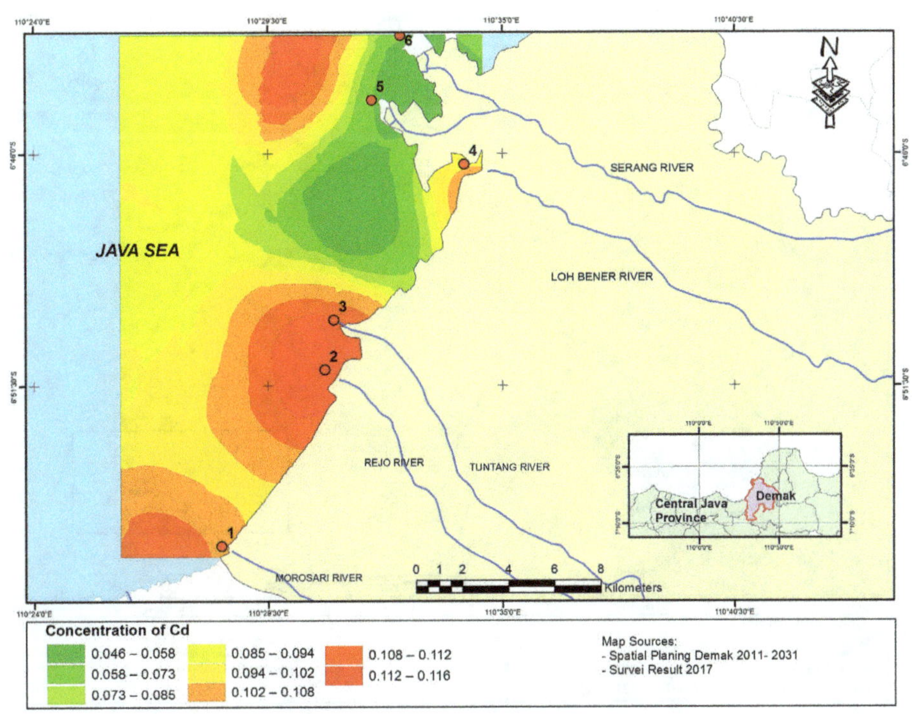

Fig. 3. The distribution of Cadmium ($mg \cdot l^{-1}$) in Demak waters

heavy metal was positively correlated with sand, mud, and clay fraction. It was found that the more mud there was, they more heavy metal (Pb and Cd) we found [Kinasih *et al.*, 2015]. The content of Pb and Cd was below CEQG (Canadian Environmental Quality Guidelines) 2002 which are 30 mg·kg^{-1} and 0.7 mg·kg^{-1}. It is in line with the guidelines of USEPA (United States Environmental Protection Agency) (1989) which require that the Pb content should be below 40 mg·kg^{-1}. It means that according to this condition, it was regarded as not polluted.

The distribution of coliform bacteria in the six stations starting from the estuary of Morosari river to Lohbener riverhad met the quality standard, except for the station 5 Wedung and station 6 Serang, as each of them surpassing the quality standard of Coliform abundance by 400 times (Figure 6). It was influenced by the polluted ecosystem. Moreover, the household waste produced by human activities also affected the coliform abundance [Kunarso, 2011].

The analysis of the quality and condition of Demak waters included temperature, pH, DO, TDS, turbidity, salinity, current, water transparency, and depth. The water temperature of the six estuaries was about 29.08°C and 30.15°C, which is common tropical water temperature. The solubility of oxygen decreased as the water temperature increased. The increase of water temperature also possibly elevated the content of some materials [Usman *et al.*, 2013].

The pH ranged from 7.25 to 7.7475 which was at pH tolerance limit. However, the pH of water was quite low, below 8, which was influenced by the river estuary. It was due to the condition of river estuary. This sea water pH influenced the chemical toxicity [Sindhu, 2005].

The water salinity was about 18.5 and 35.45 PSU. High salinity was found in the estuary of Morosari river. It was observed showed that the seawater's influence was greater than the fresh water, because during the high tide, the sea water flow kept moving toward the river estuary.

The oxygen solubility reached 5.06 mg·l^{-1} and 6.268 mg·l^{-1}. The high oxygen solubility surpassed the minimum quality standard for aquaculture (3 mg·l^{-1}) [Murtini & Peranginangin, 2006]. The solubility showed that the presence of heavy metals influenced the respiration of aquatic organism. The shortage of oxygen could possibly occur due to the increase of the nutrient supply and organic material such as fish farm waste or fertilizer run-off [Rangkuti *et al.*, 2017].

Total Dissolved Solids (TDS) is defined as any dissolved and colloid materials which are in the form of chemical compound that are not filtrated by the filter with the diameter of 0.45 [Rao, 1992]. The content of TDS was 3.861 mg·l^{-1} and 6.786 mg·l^{-1}. The maximum TDS was found in the estu-

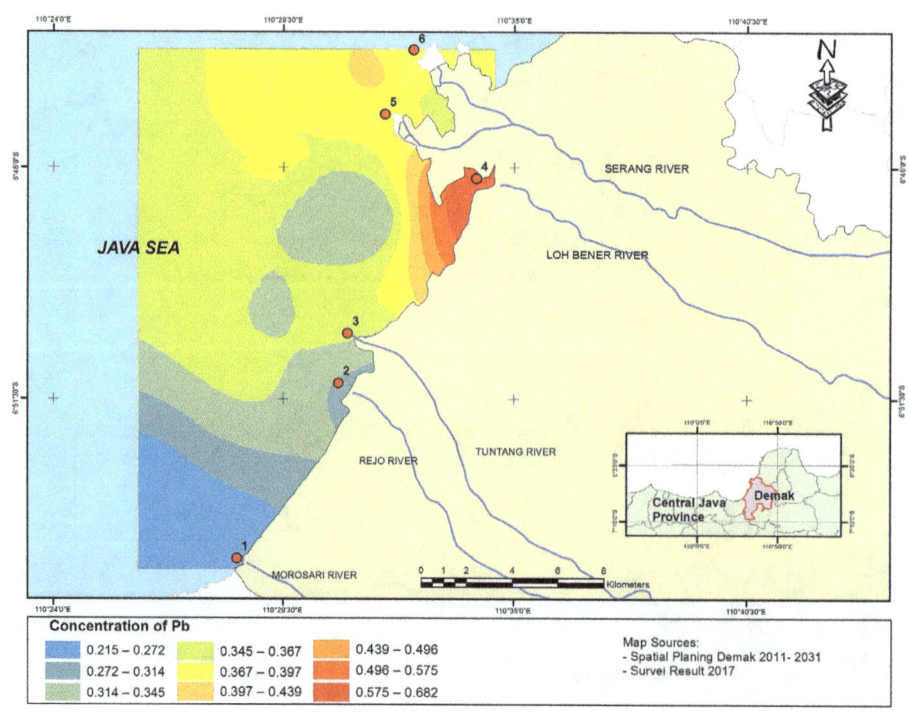

Fig. 4. Pb content in surface sediment (mg·l^{-1}) in Demak waters

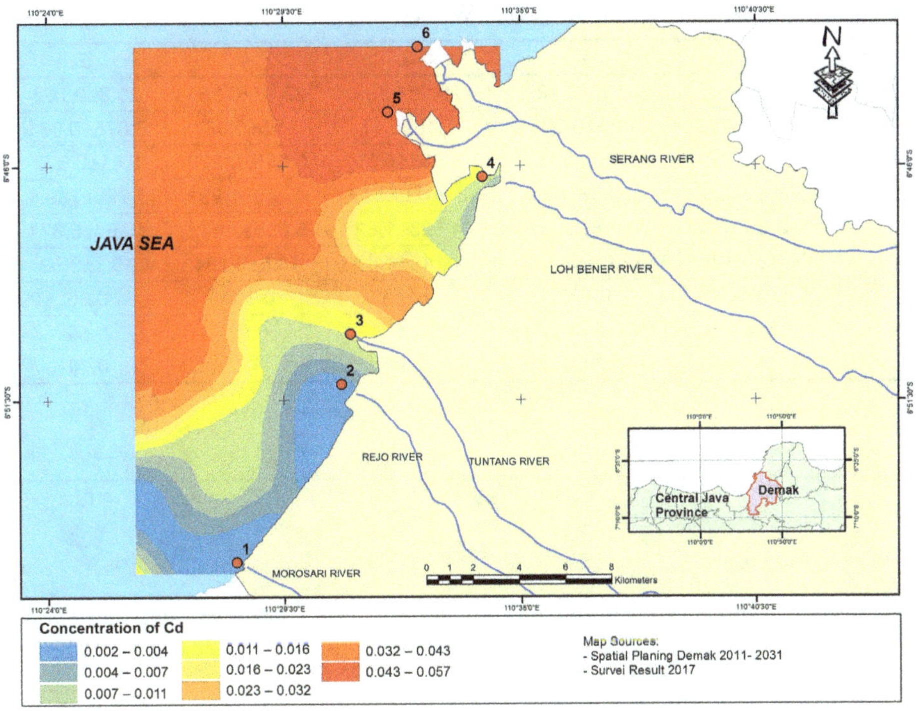

Fig. 5. Cd content in surface sediment (mg·l-1) in Demak waters

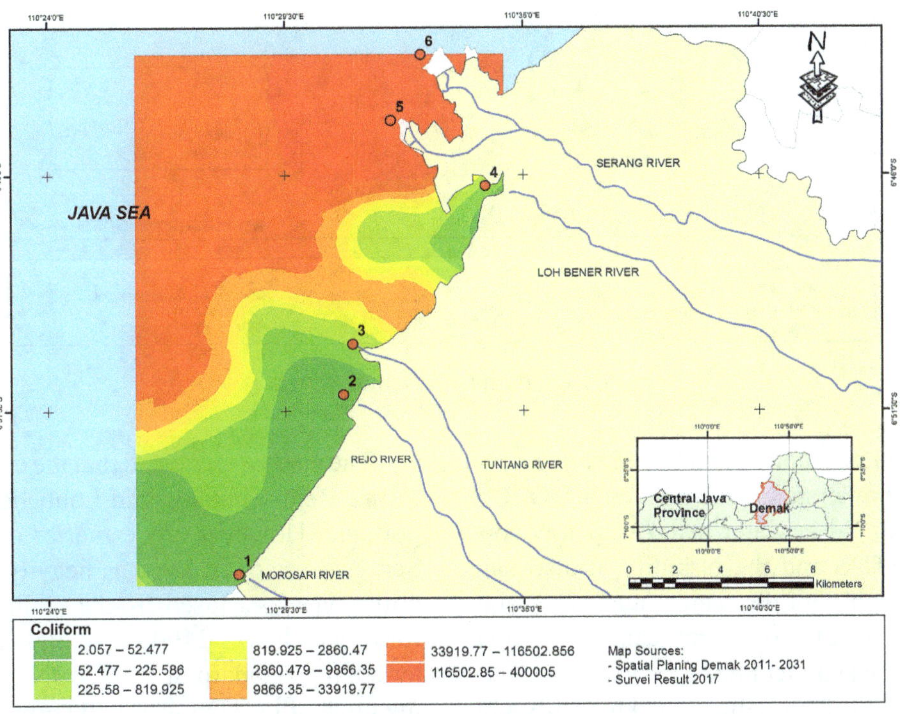

Fig. 6. Coliform abundance (MPN.-1000 ml) in Demak waters

ary of Morodemak river. The river depth ranged from 0.74 to 3.84 m and the water transparency was also low (Table 1).

The flow pattern during a high tide showed that the water moved to the land and the south east. When the observation took place, there was a high tide and a low tide. The highest tide occurred at 09:00 and the diurnal tide occurred at 19:00. This flow pattern caused the creation of the sediment and solution, despite spreading out to the sea, they gathered downstream and at the estuary (Figure 7).

On the basis of the analysis, it was revealed that Pb in water showed a low correlation with the

Table 1. The quality and physical condition of Demak Waters

Parameter/ Station	1	2	3	4	5	6
Temperature (^{0}C)	29.08±0.492	29.35±0.129	29.98±0.39	29.95±0.351	29.8±0.762	30.15±0.47
pH	7.458±0.069	7.38±0.008	7.43±0.17	7.25±0.087	7.375±0.081	7.7475±0.05
DO (mg.l^{-1})	5.06±0.312	6.268±0.102	6.033±0.35	5.635±0.232	5.51±0.872	6.2475±0.59
TDS (mg.l^{-1})	6,786±286.3	4,596±398.9	3,861±2,553	4,240±1,991	5,796±739.1	5,310±308
Turbidity (NTU)	3.623±1.539	1.918±0.319	4.503±1.82	5.035±2.413	10.64±6.371	32.5±4.79
Salinity (ppt)	35.45±0.129	17.65±0.129	19.15±0.47	18.5±0.294	19.1±0.52	18.75±0.13
Current (m.s^{-1})	0.083	0.125	0.143	0.333	0.25	0.125
Transparency (m)	1.05	2.45	1.8	0.91	0.59	0.47
Depth (m)	2.7	3.84	2.9	1.7	0.74	1.2

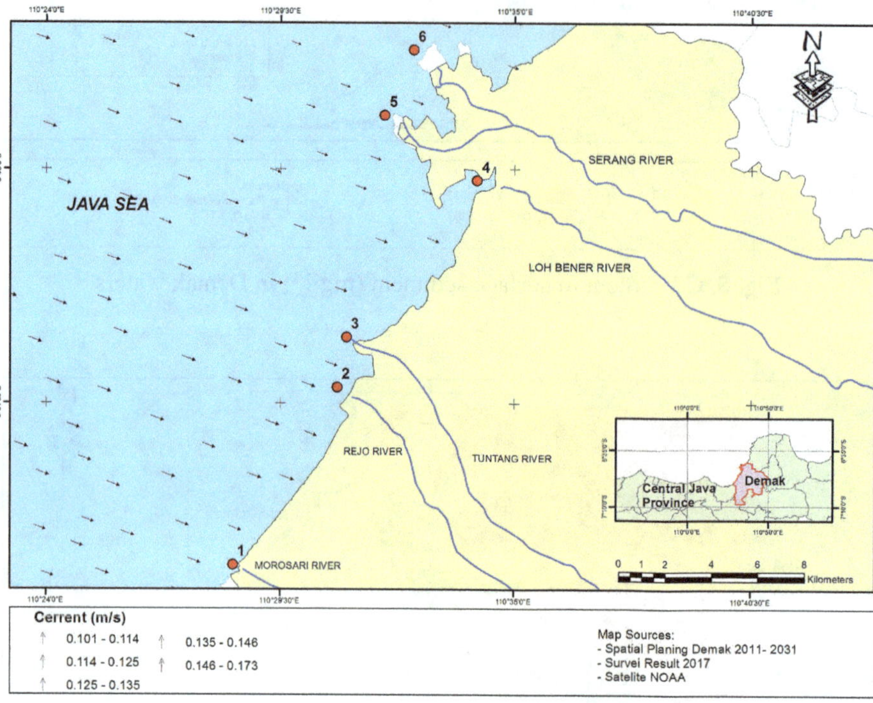

Fig. 7. Currents in Demak waters

temperature, TDS, salinity, and turbidity. However, Pb had a moderate correlation with DO. On the other hand, Cd in water had a very low correlation with TDS and the turbidity. It also had a low correlation with the temperature, DO, and salinity. Moreover, it has a significant correlation to pH. The Pb in sediment had a very low correlation with pH, DO, TDS, salinity, and turbidity. On the other hand, it had a moderate correlation with the temperature. It was also found that the Cd in sediment had a very low correlation with pH, DO, TDS, salinity, and turbidity. However, the Cd in sediment was highly correlated with the temperature. Coliform was also highly correlated with the temperature but had a low correlation with pH, TDS, and salinity. Moreover, it had a very low correlation with DO (Table 2).

The analysis revealed that the estuaries of Morosari, Rejo, Tuntang, and Lohbener were lightly polluted. However, the estuaries of Wedung and Serang were considered as heavily polluted. The pollution was caused by the abundance of coliform, which was 2400 times higher than the water quality standard of 1000 MPN·ml^{-1}. Moreover, the content of TDS also had surpassed the Class II water quality standard, based on Government Regulation No. 82 year 2001 about managing the quality of water and controlling water pollution. This regulation defines the requirement of water usable for fisheries. The highest content of TDS was found in the Morodemak estuary, whereas the lowest was found in the Rejo estuary (Table 3).

Comparing to other studies, the Pb content in Demak rivers was higher than what was found in

Table 2. Spearman correlation between heavy metal and Coliform toward physicochemical variables

	Temperature	pH	DO	TDS	Salinity	Turbidity
Pb in water	-0.314	0.086	0.429	-0.429	-0.143	-0.429
Cd in water	0.257	0.943**	0.2	0.143	0.314	0.143
Pb in sediment	0.543	-0.6	-0.029	-0.314	-0.371	-0.314
Cd in sediment	0.714	-0.371	0.143	-0.371	-0.429	-0.371
Coliform	0.771	0.257	-0.029	0.029	0.2	0.029
			** significant at 0.01 (2 tailed)			

Table 3. Pollution Index Test

No.	Parameters	1	2	3	4	5	6	L_{ij}	1	2	3	4	5	6
		C_{imax}							PI_j					
1	Turbidity	4.96	2.25	6.09	7.14	16.14	36.6	4	3.14	2.95	2.95	2.85	145.95	146.45
2	pH	7.56	7.39	7.58	7.33	7.45	7.8	7						
3	TDS	7050	4962	6193	6058	6472	5592	1000						
4	DO	5.34	6.36	6.32	5.82	6.27	6.81	6						
5	Salinity	35.5	17.9	19.5	18.9	20	18.9	34						
6	Cd	0.15	0.171	0.12	0.13	0.064	0.065	0.01						
7	Pb	0.7	0.648	0.72	0.64	0.64	0.695	0.05						
8	Coliform	4	3	1,100	1,100	2,400,000	2,400,000	1,000						

some other water areas in Indonesia, such as Babon waters and estuary, port of Parepare, Katingan & Kahayan waters, Kendari and Kelabat bay. Moreover, the Pb content in Demak waters was 1640 higher than the Pb content in Kelantan, but was equal to the one in Jinzhou bay. The content of Cd in Demak water was also greater than in those areas, except for Kelabat and Jinzhou bay. However, the content of Pb in the sediment found in Demak was lesser than in the areas above, except for Tanjung Tias waters. The Cd content in the sediment investigated in this research was also lower than the contents which revealed in previous studies (Table 4).

Table 4. Comparison of the current research to the previous studies

Waters	Concentration in Sea Water ($mg \cdot l^{-1}$)		Concentration in Surface Sediment ($mg \cdot kg^{-1}$)		References
	Pb	Cd	Pb	Cd	
Demak estuary	0.6506±0.0229	0.0938±0.0264	0.3918±0.1493	0.0247±0.0202	Recent study
Babon estuary	0.02901±0.0114	NA	NA	NA	Kartikasari *et al.*, 2002
Tangerang coastal	NA	NA	7.822±6.2786	0.1±1.4629	Hariyadi *et al.*, 2017
Ambon island	NA	0.01–0.03	NA	0.17–0.32	Rumahlatu, 2011
Morodemak	NA	0.011	NA	1.372	Wulandari *et al.*, 2009
Parepare port	0.3468±0.2591	NA	46.5199±12.4573	NA	Usman *et al.*, 2013
Katingan & Kahayan	1.2942±0.1869	0.0053±0.0003	6.1861±1.3302	0.06±0.0037	Harteman *et al.*, 2008
Kendari bay	0.0125±0.005	NA	0.7605±0.0587	NA	Amriani *et al.*, 2011
Betahlawang	NA	NA	4.78–6.26	2–3.1	Kinasih *et al.*, 2015
Sayung estuary	NA	NA	19.899±2.4741	NA	Amalia *et al.*, 2014
Tanjung Tias	BDL	0.067	0.126	0.094	Achyani & Salim, 2014
Kelabat Bay	3.925±1.1471	0.825±0.9743	NA	NA	Arifin, 2011
Tanjung Emas port	NA	NA	10.88±3.09	1.05±2.29	Tjahjono *et al.*, 2017
Yalujiang estuary	0.0004–0.0018	0.0008–0.0013	12.3–29.9	0.05–0.41	Li *et al.*, 2017
Kelantan estuary	NA	NA	52.0133±12.1	0.0733±0.0306	Wang *et al.*, 2017
Jinzhou Bay	0.61	0.92	NA	NA	Wang *et al.*, 2012

Remarks: NA (Not Avaliable), BDL (Below Detection Limit)

CONCLUSION

The lead content in Demak waters was higher in the riverbed than the content found on the river surface. The content of Pb in the river surface ranged from 0.6037 to 0.6647 $mg \cdot l^{-1}$ whereas the content of Cd in the river surface ranged from 0.064 to 0.1707 $mg \cdot l^{-1}$. Both the content of Pb and Cd surpassed the water quality standard. The Pb content in the sediment ranged from 0.2563 to 0.6823 $mg \cdot kg^{-1}$ whereas the Cd content ranged from 0.0047 to 0.0577 $mg \cdot kg^{-1}$. Both contents found in the sediment revealed that the sediments were not polluted. The pollution index test showed that Morosari, Rejo, Tuntang and Lohbener estuaries were considered as lightly polluted, whereas Wedung and Serang estuaries were heavily polluted. Most of the Pb and Cd contents in Demak waters were higher than in the previous studies. However, the Pb and Cd content in the sediment considered in the current research was lower than what had been found in the previous studies.

Acknowledgment

First of all, the writer's deepest gratitude goes to Funding of Budget Project Instructions (DIPA) PIP Semarang year of 2017 without whom this research would never be conducted. The writer also would like to express gratitude to the team of Faculty of Fisheries and Marine Diponegoro University, especially to Vian and Andreas who had willingly helped in collecting the samples. Last but not least, to Health Laboratory Semarang which had done the measurement of heavy metal content in both seawater and sediment and coliform abundance, thank you. This research could not have been completed without your help.

REFERENCES

1. Achyani R., Gazali S. 2014. Evaluation and accumulation of heavy metals on Anadara Granosa and Anadara inaequivalvis in Tarakan waters (In Indonesian). Jurnal Harpodon Borneo, 7(1), 19–26.
2. Amalia R., Widada S., Hariyadi. 2014. Analysis of heavy metal Lead in riverbed sediment of Sayung estuary, Demak regency (In Indonesian). Jurnal Oseanografi, 3(2), 167–172.
3. Amriani, Hendrarto B., Hadiyarto A. 2011. Bioaccumulation of heavy metals Lead (Pb) and Zinc (Zn) in blood cockles (Anadara granosa L.) and mangrove cockles (Polymesoda bengalensis L.) in Kendari Bay waters (In Indonesian). Jurnal Ilmu Lingkungan, 9(2), 45–50.
4. APHA (American Public Health Association). 2012. Standard methods for examination of water and waste water, 22nd edition. Street NW (US) : Water Environment Federation, pp. 1496.
5. Arifin Z. 2011. The concentration of heavy metals in water, sediment, and biota in Kelabat Bay, Bangka Island (In Indonesian). Jurnal Ilmu dan Teknologi Kelautan Tropis, 3(1), 104–114.
6. Azhar H., Widowati I., Suprijanto J. 2012. Study of the contents of heavy metal Pb, Cu, Cd, Cr in simping scallop (Amusim pleuronectes), water, and sediments found in Wedung waters Demak and maximum tolerable intake analysis in humans (In Indonesian). Journal of Marine Research, 1(2), 35–44.
7. Bambang A.G., Fatimawali, Kojong N. 2014. Analysis of Coliform contamination and Escherichia coli identification in water refills of Manado city depots (In Indonesian). Universitas Sam Ratulangi, Manado. Jurnal Ilmiah Farmasi, 3(3).
8. BPS (Biro Pusat Statistik). 2013. Area of Demak regency in 2013 (In Indonesian). https://demakkab.bps.go.id, diakses 14 January 2018.
9. BPS (Biro Pusat Statistik). 2015. Fish production based on type of fresh fish sold in 2 fish auction in Demak regency 2015 (In Indonesian). https://demakkab.bps.go.id acessed January 14th 2018.
10. Effendi H. 2003. Assessment of water quality for water source and environmental management (In Indonesian). Penerbit Kanisius, Yogyakarta.
11. Hariyadi S., Muflih A., Mashar A., Widigdo B., Wardiatno Y. 2017. Spatial distribution of some heavy metals in the sediments of Tangerang coastal waters, Banten Province, Indonesia. AES Bioflux, 9(2).
12. Hartenan E., Soedharma D., Winarto A., Sanusi H.S. 2008. Detection of heavy metals in waters, sediments, and fish fins of Badukang (Arius caelatus HAN A. Maculatus) at the esturies of the Kahayan and Katingan rivers, Central Kalimantan (In Indonesian). Berita Biologi, 9(3), 275–283.
13. Hasan S. 2004. Analysis of characteristics of aquatic environment affecting on the accumulation of heavy metals Lead in the body of blood clam in the coastal waters of eastern North Sumatra (In Indonesian). Jurnal Ilmu-Ilmu Perairan dan Perikanan Indonesia, 1(1), 53–60.
14. Hastuti E.D., Anggoro S., Pribadi R. 2013. The influence of vegetation type and density on the content of Cd and Cr sediment in coastal areas of Semarang and Demak (In Indonesian). Prosiding Seminar nasional Pengelolaan Sumber daya alam dan lingkungan.
15. Heriyanto N.M., Subiandono E. 2011. Absorption of heavy metal pollutants (Hg, Pb and Cu) by mangrove species (In Indonesian). Jurnal Penelitian

Hutan dan Konservasi Alam, 8(2), 177–188.

16. Ijong F.G. 2015. Fisheries & marine microbiology (In Indonesian). Rineka Cipta, Jakarta.

17. Jalius, Setiyanto D.D., Sumantadinata K., Riani E., Ernawati Y. 2008. The heavy metal of accumulation and its effects to spermatogenesis on the green mussel (Perna viridis). Jurnal Ilmu-Ilmu Perairan dan Perikanan Indonesia, 15(1), 77–83.

18. Jiyah, Sudarsono B., Sukmono A. 2017. Distribution analysis of Total Suspended Solid (TSS) in coastal waters of Demak regency using LANDSAT imagery (In Indonesian). Jurnal Geodesi Undip, 6(1).

19. Kartikasari V., Tandjung S.D., Sunarto. 2002. The accumulation of heavy metals Cr and Pb on mangrove Avicennia Marina at Babon estuary the border of Semarang and Demak regency of Central Java (In Indonesian). Manusia dan Lingkungan, IX, Nov, 137–147.

20. Keputusan Menteri Negara Lingkungan Hidup Nomor 115 Tahun 2003 tentang [Guidelines on Water Quality Status Determination] [In Indonesian].

21. Keputusan Menteri Negara Lingkungan Hidup Nomor 51 Tahun 2004 tentang Water Quality Standard for Coastal water (In Indonesian).

22. Kinasih A.R.N., Purnomo P.W., Ruswahyuni. 2015. Analysis of the relationship texture sediment with organic materials, heavy metal (Pb and Cd) and macrozoobenthic at Betahlawang river, Demak. Diponegoro Journal of Maquares Management of Aquatic Resources, 4(3), 99–107.

23. Kunarso D.H. 2011. The quality of waters in Makassar strait viewed from bacteriological aspect (In Indonesian). Biosfera, 28(1), 32–34.

24. Li H., Li L., Ye S., Li H., Fan J.. 2017. Assessment of nutrient and heavy metal contamination in the seawater and sediment of Yalujiang Estuary. Marine Pollution Bulletin, 117, 499–506.

25. Murtini J.T., Peranginangin R. 2006. Heavy metal contents in Asiatic Hard Clams (Meritrix meritric) and sea water in Banjarmasin (In Indonesian). Jurnal Perikanan, VIII (2), 177–184.

26. Nasution S., Siska M. 2011. The content of heavy metal lead (Pb) in sediments and snails (Strombus canarium) in the coastal waters of Bintan island (In Indonesian). Jurnal Ilmu Lingkungan, 5(2), 82–93.

27. Nemerow N.L. & Sumitomo H. 1970. Benefits of water quality enhancement. Report no. 16110 DAJ, prepared for the U.S. Environmental Protection Agency.

28. Nurrachmi I., Amin B., Habibi M.N. 2011. Bioacumulation of Cd, Cu, Pb and Zn in some parts of Gulama (Sciaena russelli) body from Dumai waters, Riau (In Indonesian). Maspari Journal, 02(01–10).

29. Palar H. 2012. Pollution and toxicology of heavy metals (In Indonesian). Rineka Cipta, Jakarta.

30. Panuntun P., Yulianto B., Ambariyanto. 2012. Accumulation of heavy metal Pb on coral Acropora aspera: a preliminary study (In Indonesian). Journal of Marine Research, 1(1), 153–158.

31. PP (Peraturan Pemerintah) Nomor 82 Nomor 2001 about Water quality management and pollution control (In Indonesian).

32. Priyatno D. 2008. Independently learning SPSS (Statistical Product and Service Solution) for data analysis & statistical test (In Indonesian). Mediakom, Yogyakarta.

33. Priyatno D. 2010. Easy and fast technique to perform data analysis with SPSS and question and answer on exams (In Indonesian). Penerbit Gava Media, Yogyakarta.

34. Rangkuti A.M., Cordova M.R., Rahmawati A., Yulma, Adimu H.E. 2017. Coastal and marine ecosystems of Indonesia (In Indonesian). Bumi Aksara, Jakarta.

35. Rao C.S. 1992. Environmental pollution control engineering. Wiley Eastern Limited, New Delhi, pp. 431.

36. Ridhowati S. 2013. Identifying Metal Pollution (In Indonesian). Graha Ilmu, Yogyakarta.

37. Rumahlatu D. 2011. Concentrations of heavy metals Cadmium in water, sediment and Deadema setosum (Echinodermata, Echinoidea) in the waters of Ambon island (In Indonesian). Ilmu Kelautan, Juni, 16(2), 87–85.

38. Samawi M.F., Tambaru R., Husain A.A., Burhanuddin A.A. 2014. Distribution of heavy metal Lead (Pb) in benthic organisms and sediments of Bonebatang waters (In Indonesian). Prosiding Simposium Nasional Kelautan dan Perikanan 2014, Universitas Hasanuddin Makassar.

39. Sindhu S.F. 2005. The content of heavy metals Cu, Zn and Pb in water, tilapia (Oreochromis niloticus) and common carp (Cyprinus carpio) in floating net cages, Saguling Reservoir (In Indonesian). Skripsi, diterbitkan Jurusan Manajemen Sumberdaya Perairan FKIP IPB, Bogor.

40. SNI (Standar Nasional Indonesia) 2897–2008. Method of microbial contamination assessment in meat, eggs, milk and dairy products (In Indonesian).

41. SNI (Standar Nasional Indonesia)-01–2332–1991. Determination of Coliform and Escherichia coli (In Indonesian).

42. Suprapti N.H. 2008. Chromium content in waters, sediments and blood clams (Anadara granosa) in coastal area around the estuary of Sayung River Morosari Village Demak Regency, Central Java (In Indonesian). BIOMA, 10(2), 36–40.

43. Tjahjono A., Bambang A.N., Anggoro S. 2017. Pollution assessment in surface sediments of trace metal in port of Tanjung Emas Semarang. Advanced Science Letters, 23, 2215–2219.

44. Triadayani A.E., Aryawati R., Diansyah G. 2010.

The influence of metal Lead (Pb) in humpback grouper's liver tissue (Cromileptes altivelis) (In Indonesian). Maspari Journal, 1, 42–47.

45. USEPA. 1989. Sediments classification methods compendium. Draft final report. United States Environmental Protection Agency, Watershed Protection Division, USA.

46. Usman S., Nafie N.L., Ramang M. 2013. Quantitative Distribution of heavy metals Pb in water, sediment and Crimson Snapper (Lutjanus erythropterus) around Parepare waters (In Indonesian). Marina Chimica Acta, 14(2).

47. Wang A-j, Bong C.W., Xu Y-h., Hassan M.H.A., Ye X., Bakar A.F.A., Li Y-h., Lai Z-k., Xu J., Loh K.H. 2017. Assessement of heavy metal pollution in surficial sediments from a tropical river-estuary-shelf system: a case study of Kelantan River, Malaysia. Marine Pollution Bulletin, http://dx.doi.org/10.1016/j.marpolbul.2017.08.010.

48. Wang J., Liu R.H., Yu P., Tang A.K., Xu L.Q., Wang J.Y. 2012. Study on the pollution characteristics of heavy metals in seawater of Jinzhou Bay. Procedia Environmental Sciences, 13, 1507–1516, doi 10.1016/j demak, accesed Januari 14th 2018.

49. Wulandari S.Y., Yualianto B., Santoso G.W., Suwartimah K. 2009. The content of heavy metals Hg and Cd in water, sediment and blood clam (Anadara granossa) using Neutron Activation Analysis Method (APN) (In Indonesian). Ilmu Kelautan, 14(3), 170–175.

15

Water Quality Assessment of the Morava e Binçës River based on the Physicochemical Parameters and Water Quality Index

Milaim Musliu[1], Astrit Bilalli[2*], Bujar Durmishi[3], Murtezan Ismaili[1], Halil Ibrahimi[4]

[1] South East European University, Faculty of Contemporary Sciences and Technologies, Tetovë, Republic of Macedonia

[2] University of Peja "Haxhi Zeka", Faculty of Agribusiness, Pejë, Kosovo

[3] Faculty of Natural Sciences and Mathematics, State University of Tetova, Ilinden str n.n., Tetovo, Republic of Macedonia

[4] University of Prishtina "Hasan Prishtina", Faculty of Mathematics and Natural Sciences, Department of Biology, Prishtina, Kosovo

* Corresponding author's e-mail: astritbilalli@yahoo.com

ABSTRACT

The quality of surface waters is being impacted by the anthropogenic and natural pollution, thus limiting the usage of this water for drinking, industry, agriculture, recreation and other purposes. The water quality indices are intended to provide a single value for the water quality of a source or a stream that reduces the large amount of parameters in a simpler expression and enables an easy interpretation of the monitoring data. During 2017, seventeen physicochemical parameters were measured in spring, summer, autumn and winter, in five locations along the Morava e Binçës River in Kosovo. For the assessment we employed the Water Quality Index (WQI) which uses the physicochemical parameters for the evaluation of the water quality. The findings of this study ascertain that MB1 station had the best quality with a value of WQI 88 and is classified in the Good Category, whereas the lowest quality of water was found on in MB4 station with a value of WQI 65 and it is thus classified in the Fair Category. Finally, the average WQI was calculated for the entire measurement period and it resulted in a value of 77.60 indicating that the Morava e Binçës River waters belong to the Fair Category.

Keywords: Physicochemical parametres, River waters, Water Quality, WQI, Morava e Binçës

INTRODUCTION

One of the major world problems is a drastic reduction of the available water resources [Durmishi, 2005; Ismaili and Durmishi, 2006]. Nowadays, many people worldwide suffer from the lack of safe and quality water, which is essential for popular needs. In numerous countries, water resources are depleted faster than they can be replenished and not enough to meet the demands of modern human life. Environmental pollution and especially the contamination of water sources in this regard becomes an issue to be addressed immediately. The increased urbanization, industrialization, the modernization of agriculture and the increase in traffic all contribute to the global pollution, which in turn requires accurate monitoring and information about the quality of water resources. The concerns for the quality of water come from the global social trends, population growth and development activities, which have been the cause of pollution. Moreover, inadequate management of water systems can cause serious problems in the water availability and quality of water [Krishnan *et al.*, 2007]. Hence, it is necessary to evaluate the quality of the river water.

Water quality can best be described by the physicochemical parameters and biological parameters. The physicochemical parameters play an important role in the system restoration maintenance and self-regulation of water quality [Barakat *et al.*, 2012]. Some correlations between these parameters can be done and useful conclusions can be drawn to show the quality of water. The physicochemical parameters of water and the dependence of all life process of these factors make it desirable to take as an environment [Anbarasu and Anbuselvan, 2017]. A very powerful tool for water quality assessment is the Water Quality Index (WQI). WQI serves to sum up large quantities of water quality data under simple conditions (e.g. good, bad) for the management and public consistent reporting [Durmishi *et al.*, 2012]. WQI represents a mere number of 0–100, where a higher value indicates a better water quality and vice versa.

The purpose of this article was to evaluate the water quality of the Morava e Binçës river by measuring some physicochemical parameters and WQI.

MATERIALS AND METHODS

Study area

The basin of the Morava e Binçës River is located in the south and southeast of the Republic of Kosova. This river is created by two other smaller rivers which flow to the northern slopes of Karadaku of Skopje. The length of the Morava e Binçës River in the territory of Kosova is around 60 km [MESP, 2015] with a surface of the basin 1,175 km^2. In the upper part, the river is fast with a great tilt, deep and narrow bed where vertical erosion is emphasized. In the further course, this river passes through the Gjilan area with a length of 24 km [MESP, 2010]. In this part it has a wider bed and a smaller tilt.

Physicochemical parameters

The physicochemical parameters were measured during all four seasons of 2017. The parameters were measured in five stations (locations) in the mainstream of Morava e Binçës river. The locations were selected starting from the source area – the first locations in Korbliq (MB1), the second locations in Viti (MB2) the third locations in Uglarë village (MB3), the fourth locations in Pogragjë (MB4) and the fifth locations in Dheu i Bardhë village (MB5). The first sampling stacion is located at the spring area of this river, while other sampling stations are located in the mid-

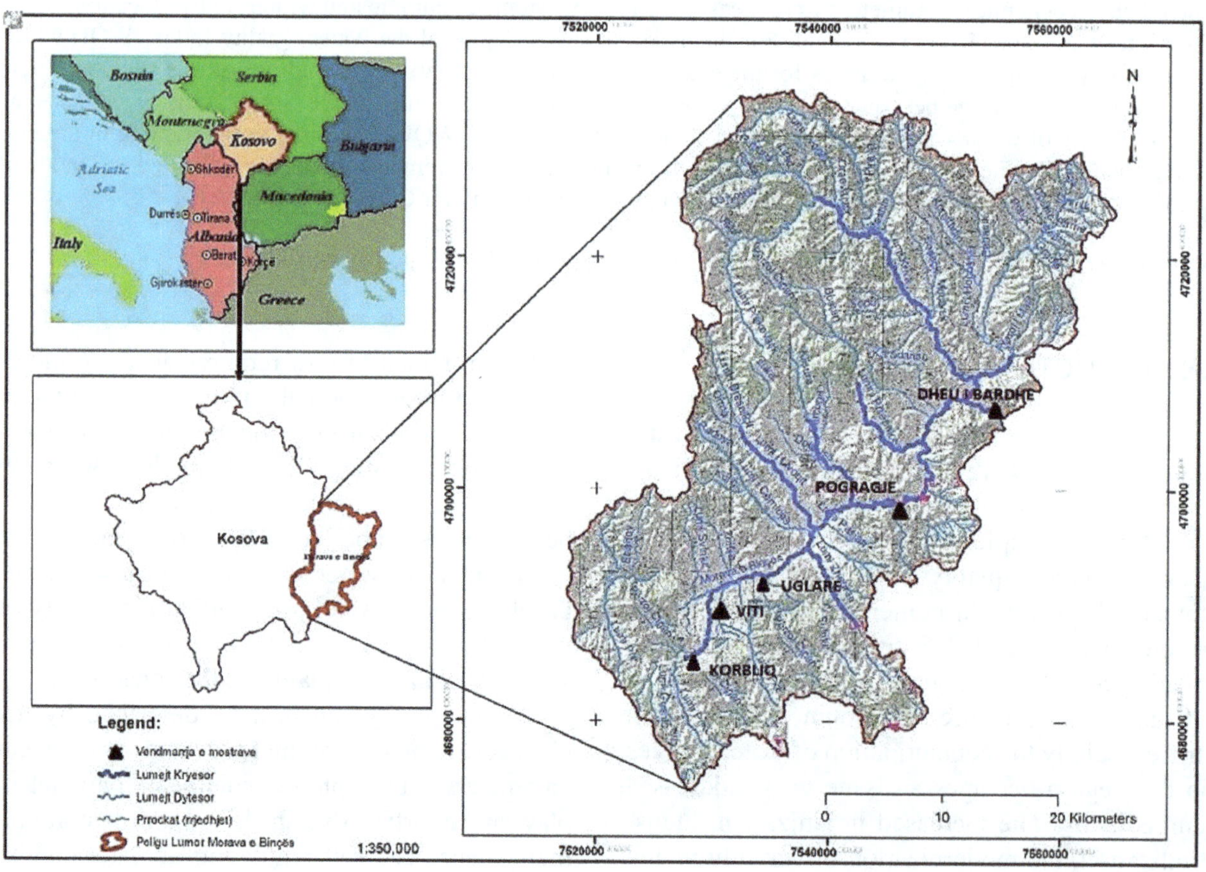

Figure 1. The Map of the Morava e Binçës River with indicated sampling stations

stream and downstream area of the river where there is a serious discharge of different pollutants into the river.

Water sample collection and analysis

The water sampling for analysis is carried out by using equipment and containers based on the ISO 5667–6 standard, which sets the principles that will be applied in designing the sampling programs of sampling techniques and treatment of water samples taken from rivers as well as physical and chemical assessment and process. The samples were taken in four time periods: spring, summer, autumn and winter during 2017, in five stations (locations) in the mainstream of Morava e Binçës. At each station were measured seventeen physicochemical parameters.

The analyses of the samples taken from the water of Morava e Binçës were carried out in the laboratory of the Hydrometeorological Institute of Kosova in Prishtina. The water quality parameters were defined by using the following measuring equipment: WTW 350i for electrical conduct, AQUALITIC/PC COMPACT for turbidity, the measures for pH value were carried out with the HI 98130 pH-meter, the dissolved oxygen and oxygen saturation was performed with HI 9146, spectrophotometer detergents with the type of SECOMAN Pastel model UV, phosphates and ammonia with SECOMAN PRIM LIGHT, etc.

RESULTS AND DISCUSSION

The results of this research are presented in the Tables 1–2 and in Figures 2–5.

Water Temperature (WT)

The variation of WT concentration has been in the range of 7.50–25.30 °C. The lowest value was measured in MB1 station (winter), while the highest one in MB2 station (summer). The average values in winter, spring, summer and autumn were 8.82, 16.14, 21.6 and 14.00°C respectively, whereas the annual average with a standard deviation of 15.140±5.227°C. The average values of WT in MB1-MB5 stations were 13.500, 15.775, 16.225, 15.525, and 14.825°C, respectively. The WT values of Morava e Binçës basin resulted within the recommended values of GD161 regulation and the water of this river was classified as first class (Fig. 2).

Turbidity (TUR)

The concentration variation of TURB was in the range of 0.100–51.200 NTU unit. The lowest value was measured in MB1 station (winter), while the highest one in MB4 station (autumn). The average values in winter, spring, summer and autumn were 0.240, 17.072, 21.302 and 21.386 NTU units, respectively, while the annual average with a standard deviation were 15.000±17.138 NTU units. The average values of TUR in MB1-MB5 stations were 3.140, 4.205, 22.55, 29.65, and 15.328 NTU units. The TUR values that exceeded the GD161 Regulation were 47.500 (MB3, summer), 32.100 (MB3, autumn), 28.800 (MB4, sprind), 38.800 (MB4, summer), 51.200 (MB4, autumn) and 34.300 (MB5, spring) NTU units. The TUR values in Morava e Binçës water were within the recommended values of GD161 regulation and the water of this river was classified as first class (Fig. 2).

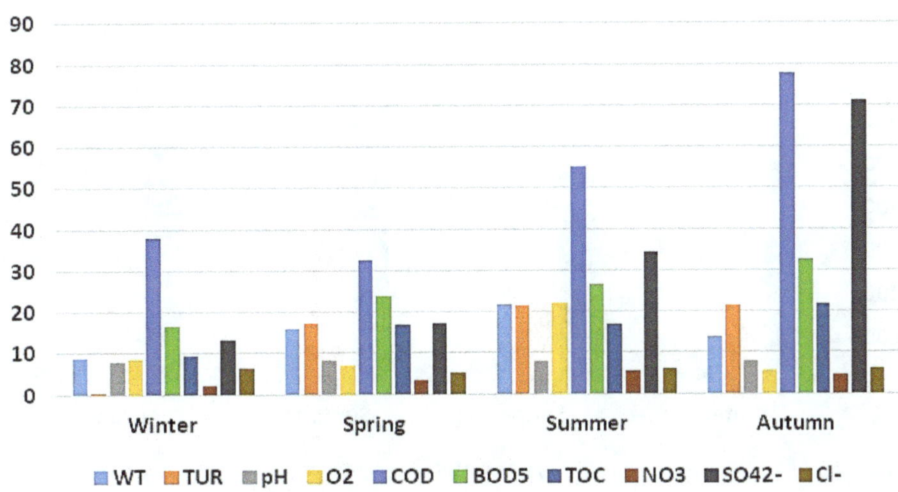

Figure 2. Anual average values of some parameters

Electrical Conductivity (EC)

EC variation was in the range of 220.00–955.00 µS/cm in all four seasons. The lowest value was measured in MB1 station (winter), while the highest one in MB5 station (summer). The average values in winter, spring, summer and autumn were 322.00, 371.40, 638.60 and 627.00 µS/cm respectively, while the annual average with standard deviation was 489.75±229.91 µS/cm. The average value of EC in MB1-MB5 stations were 268.75, 357, 617, 590.5, and 615.5 µS/cm respectively. There were no values of EC that exceeded the GD161 Regulation. The EC values in the water of Morava e Binçës river were within the recommended values of GD161 Regulation and the water of this river was classified as first class (Fig. 3).

Water Soluble Materials (WSM)

The WSM variation was in the range of 107.00–550.00 mg/L in all four seasons. The lowest value was measured in MB1 station (winter), while the highest one in MB5 station (summer). The average values of WSM in winter, spring, summer and autumn were 167.40, 185.60, 322.40 and 345.40 mg/L respectively, while the annual average with standard deviation was 255.20±131.18 mg/L. The average value of EC in MB1-MB5 stations were 38.50, 168.50, 314.75, 304.00, and 350.25 mg/L respectively. There were no values of WSM that exceeded the GD161 Regulation. The WSM values in the water of Morava e Binçës river were within the recommended values of GD161 Regulation and the water of this river was classified as first class (Fig. 3).

Hydrogen ion concentration (pH)

The pH variation was in the range of 7.680–8.310. The lowest value was measured in MB3 station (summer), while the highest one in MB1 station (autumn). The average values of pH in winter, spring, summer and autumn were 7.774, 8.170, 7.856 and 8.012 respectively, while the annual average with standard deviation was 7.953±0.202. The average value of pH in MB1-MB5 stations were 8.050, 7.973, 7.853, 7.960 and 7.923 respectively. There were no pH values that exceeded the GD161 Regulation. The pH values in the water of Morava e Binçës river were within the recommended values of GD161 Regulation and the water of this river was classified as first class (Fig. 2).

Dissolved Oxygen (DO)

The DO variation was in the range of 3.600 – 86.200 mg/L. The lowest value was measured in MB3 station (autumn), while the highest one in MB5 station (summer). The average values of DO in winter, spring, summer and autumn were 8.466, 7.106, 22.078 and 5.598 mg/L respectively, while the annual average with standard deviation was 10.812±17.870 mg/L. The average value of DO in MB1-MB5 stations were 10.0500, 6.8175, 5.0825, 6.195, and 26.1825 mg/L respectively. Two values of DO were found to be under the recommended norm of GD161 Regulation: 3.600 mg/L (MB3 in autumn) and 3.940 mg/L (MB4 in summer). The other values of DO in the water of Morava e Binçës river resulted within the recommended values of the GD161 Regulation and the water of this river was classified as first class (Fig. 2).

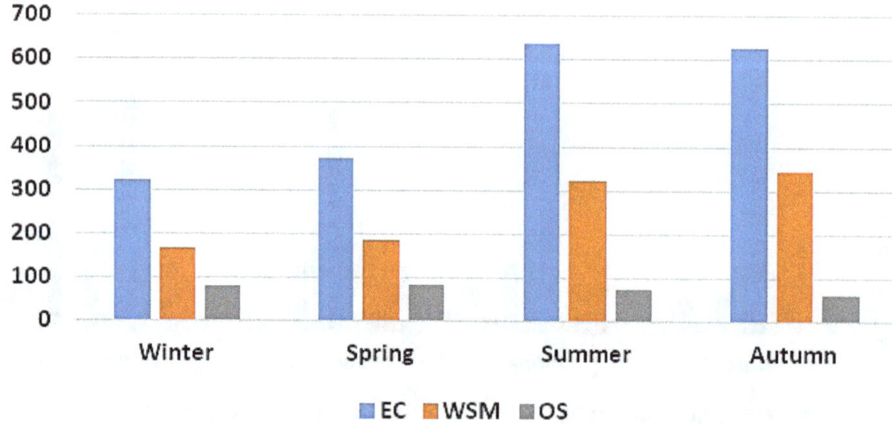

Figure 3. Average annual values of EC, WSM and OS

Oxygen Saturation (OS)

The OS variation was in the range of 39.90–92.40 %. The lowest value was measured in MB3 station (autumn), while the highest one in MB5 station (spring). The average values of OS in winter, spring, summer and autumn were 78.04, 81.06, 71.780 and 59.66 % respectively, while the annual average with standard deviation was 72.635±16.277 %. The average value of OS in MB1-MB5 stations were 86.800, 80.075, 57.500, 66.500, and 72.100%, respectively. The OS values in the water of Morava e Binçës river were within the recommended values of the GD161 Regulation and the water of this river was classified as first class (Fig. 3).

Chemical Oxygen Demand (COD)

The COD variation was in the range of 18.20–174.00 mg/L. The lowest value was in MB3 station (autumn), while the highest one in MB4 station (autumn). The average values of COD in winter, spring, summer and autumn were 37.840, 32.640, 55.100 and 77.660 mg/L respectively, while the annual average with standard deviation was 50.810±36.042 mg/L. The average value of COD in MB1-MB5 stations were 8644.30, 33.775, 68.50, 82.75 and 36.75 mg/L respectively. Four values of COD exceeded the recommended norm of the GD161 Regulation: 74.000 mg/L (MB3 in summer), 109.900 mg/L (MB3 in autumn), 79.000 mg/L (MB4 in summer) and 174.000 mg/L (MB4 in autumn). The OCD values in the waters of Morava e Binçës river were slightly over the recommended values of GD161 Regulation and the water of this river was classified as fifth class (Fig. 2).

Biochemical Oxygen demand (BOD_5)

The BOD_5 variation was in the range of 3.300–44.000 mg/L. The lowest value was measured in MB1 station (winter), while the highest one in MB4 station (autumn). The average values of $SHBO_5$ in winter, spring, summer and autumn were 316.480, 23.760, 26.660 and 32.540 mg/L respectively, while the annual average with standard deviation was 24.860±11.508 mg/L. The average value of $SHBO_5$ in MB1-MB5 stations were 20.6, 20.7, 31.825, 34.775 and 26.875 mg/L respectively. Out of the twenty measured values, fifteen values of $SHBO_5$ were above the recommended value of the GD161 Regulation, which means that the water based on this parameter is too polluted; therefore it was classified as fifth class (Fig. 2).

Total organic carbon (TOC)

The TOC variation reached 1.200–33.500 mg/L. The lowest value was measured in MB1 station (winter), while the highest one in MB3 station (autumn). The average values of TOC in winter, spring, summer and autumn were 9.460, 17.020, 16.780 and 21.780 mg/L respectively, while the annual average with standard deviation was 16.260±10.662 mg/L. The average value of $SHBO_5$ in MB1-MB5 stations were 3.80, 13.20, 23.15, 23.575 and 19.075 mg/L respectively. The TOC values in the water of Morava e Binçës river were within the recommended values of the GD161 Regulation and the water of this river was classified as first class (Fig. 2).

Nitrates (NO_3^-)

The NO_3^- variation was in the range of 0.200–11.800 mg/L. The lowest value was measured in MB1 station (autumn), while the highest one in MB4 station (autumn). The average values of NO_3^- in winter, spring, summer and autumn were 2.280, 3.340, 5.500 and 4.620 mg/L respectively, while the annual average with standard deviation was 3.935±3.568 mg/L. The average value of NO_3^- in MB1-MB5 stations were 0.55, 1.525, 4.45, 7.95dhe 5.30 mg/L respectively. Only one value of NO_3^- was above the recommended regulation of GD161: 11.800 mg/L (MB4 in autumn). The NO_3^- values in the water of Morava e Binçës river were within the recommended values of the GD161 Regulation and the water of this river was classified as first class (Fig. 2).

Detergents (DET)

The DET variation reached 0.000–0.500 mg/L. During the whole measurement period, the DET values amouted to 0.000 mg/L, except one value of 0.0500 mg/L in MB2 station, during summer, which shows that the water was not polluted with detergents, therefore it is belonged to the first class (Fig. 4).

Phosphates (PO_4^{3-})

The PO_4^{3-} variation was in the range of 0.100–2.600 mg/L. The lowest value was measured in MB1 and MB2 station (winter), while the highest one in MB3 station (autumn). The average values of PO_4^{3-} in winter, spring, summer and autumn were 0.270, 0.280, 0.500 and 1.340 mg/L, respectively, while the annual average with standard deviation was 0.598±0.632 mg/L. The average value of PO_4^{3-} in MB1-MB5 stations were 0.500, 0.4475, 1.235, 0.600 and 0.405 mg/L, respectively. Only one value of PO_4^{3-} was above the recommended regulation of GD161: 11.800 mg/L (MB4 in autumn). Fourteen values of PO_4^{3-} were above the recommended regulations of GD161, which shows that the water of the river were polluted with this pollutant. The PO_4^{3-} values of Morava e Binçës river were not in proportion with the recommended values of the GD161 Regulation and the water of this river classified lower than the fifth class (Fig. 4).

Ammonium ion (NH_4^+)

The NH_4^+ variation was in the range of 0.240–1.600 mg/L. The lowest value was measured in MB1 (winter), while the highest one in MB6 station (autumn). The average values of NH_4^+ in winter, spring, summer and autumn were 0.742, 0.882, 0.938dhe 0.926mg/L, respectively, while the annual average with standard deviation was 3.935±3.568 mg/L. The average value of NH_4^+ in MB1-MB5 stations were 0.1700, 0.7775, 1.5225, 1.0425 and 0.6675 mg/L respectively. Only one value of PO_4^{3-} was above the recommended regulation of GD161: 11.800 mg/L (MB4 in autumn). Generally, the NH_4^+ values in the water of Morava e Binçës river were within the reccomended values of the GD161 Regulation and the water of this river was classified as the second category (Fig. 4).

Nitrites (NO_2^-)

The NO_2^- variation was in the range of 0.010–1.100 mg/L. The lowest value was measured in MB1 (winter, spring and autumn), while the highest one in MB4 station (autumn). The average values of NO_2^- in winter, spring, summer and autumn were 0.074, 0.248, 0.844 and 0.460 mg/L, respectively, while the annual average with standard deviation was 0.407±0.353 mg/L. The average value of NO_2^- in MB1-MB5 stations were 0.350, 0.2575, 0.530, 0.5725 and 0.4925 mg/L, respectively. Ten values of NO_2^- were above the recommended values of theGD161 Regulation, which indicates that the waters were polluted with this compound, therefore they were ranked lower than the fifth category (Fig. 4).

Sulphates (SO_4^{2-})

The SO_4^{2-} variation was in the range of 4.00–204.00 mg/L. The lowest value was measured in MB1 (winter), while the highest one in MB5 station (autumn). The average values of SO_4^{2-} in winter, spring, summer and autumn were 3.264, 17.322, 34.456 and 70.968 mg/L respectively, while the annual average with standard deviation was 34.003±48.856 mg/L. The average value of SO_4^{2-} in the stations MB1–MB5 were 8.8400, 10.0276, 22.125, 36.3750 and 95.4250 mg/L, respectively. The SO_4^{2-} values in the water of Morava e Binçës river were within the recommended values of the GD161 Regu-

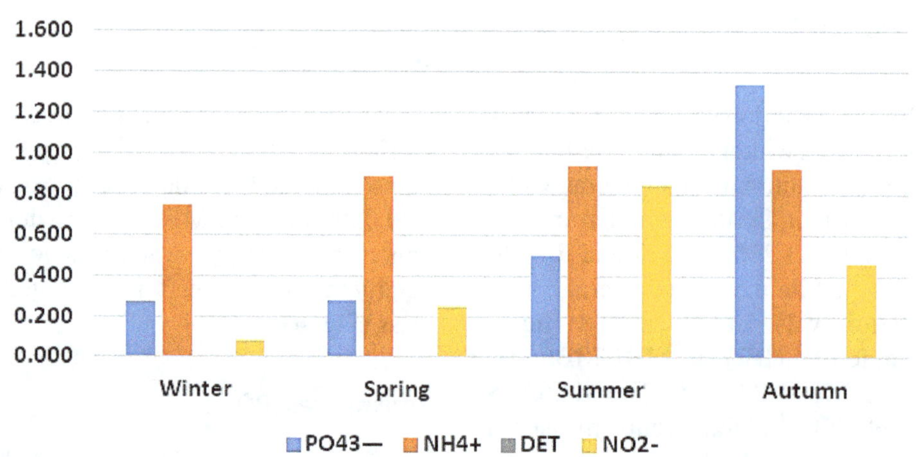

Figure 4. Annual average values of phosphates, ammoniium ion, detergents and nitrites

lation and the water of this river was classified as first class (Fig. 2).

Chlorides (Cl^-)

The Cl^- variation was in the range of 1.100–10.650 mg/L. The lowest value was measured in MB1 (winter), while the highest one in MB3 station (winter). The average values of Cl^- in winter, spring, summer and autumn were 6.468, 5.150, 5.934 and 6.018 mg/L respectively, while the annual average with standard deviation was 34.003±48.856 mg/L. The average value of Cl^- in MB1-MB5 stations were 2.0500, 4.765, 8.7300, 7.4550 and 6.6875 mg/L respectively. The Cl^- values in the water of Morava e Binçës river were within the recommended values of the GD161 Regulation and the water of this river was classified as first class (Fig. 2).

Correlation coefficients of physico-chemical parameters of the Morava e Binçës river

The correlation coefficients between the parameters of the waters in the Morava River in Binçë are shown in Table 1. The results show that some values of correlation coefficient have been more significant. The most significant positive correlations were observed between: EC-WSM ($r = 0.9693$), BOD_5-TOC ($r = 0.9150$), and TUR-BOD_5 ($r = 0.8425$). The most emphasized negative correlation was noted between: OS-TOC ($r = -0.8182$), OS-phosphates ($r = -0.7668$), TUR-OS ($r = -0.7148$), and WSM-OS ($r = -0.7060$).

Water quality assessment of Morava e Binçës river with WQI

The WQI calculation was done by using Water Quality Index Desktop computer program which was developed in 2017 by the authors [Ramadani et *al.*, 2017] according to the literature [CCME, 2001]. The frequency results for the F1, F2, F3 frequencies and WQI for five stations are given in Table 2 and Figure 5.

In MB1 station, the WQI value reached 88, since 3 parameters (BOD_5, phosphates and nitrites) and 5 tests have failed: 1 test of BOD_5; 3 tests of phosphates; and 1 test of nitrites. In MB2 station, WQI value reached, 86 since 3 parameters (BOD_5, phosphates and nitrites) and 7 tests have failed: 3 tests of BOD_5; 3 tests of phosphates; and 1 test of nitrites. In MB3 station, the WQI amounted to 67 with 7 failed parameters (TUR, DO, COD, BOD_5, TOC, phosphates and nitrites) and 16 tests: 2 tests of TUR; 1 test of DO; 2 tests of COD; 3 tests of BOD_5; 1 test of TOC; 4 tests of phosphates; and 3 tests of nitrites. In MB4 station, the WQI value reached 65, since 8 parameters: (TUR, DO, COD, BOD_5, TOC, nitrates, phosphates and nitrites) and 17 tests have failed: 3 TUR tests; 1 test of DO; 2 tests of COD; 4 tests of BOD_5; 1 test of TOC; 1 test of nitrates; 3 tests of phosphates; and 2 tests of nitrites. In MB5 station, the WQI value reached 82, because 4 parameters: (TUR, BOD_5, phosphates and nitrites) and 9 tests have failed: 1 test of TUR; 4 tests of BOD_5; 2 tests of phosphates; and 2 tests of nitrites.

Table 1. Correlation of the parameters

Correlations	WT	TUR	EC	WSM	pH	OT	OS	COD	BOD_5	TOC	NO_3^-	DET	PO_4^{3-}	NH_4^+	NO_2^-	SO_4^{2-}	Cl^-
WT	1.0000																
TUR	0.4390	1.0000															
EC	0.4254	0.6464	1.0000														
WSM	0.3324	0.6098	0.9693	1.0000													
pH	0.1832	0.0779	-0.3403	-0.3251	1.0000												
DO	0.2471	-0.1597	0.3835	0.3022	-0.2296	1.0000											
OS	-0.0887	-0.7148	-0.6809	-0.7060	0.2393	0.2943	1.0000										
COD	0.1181	0.7196	0.5558	0.5426	-0.0997	-0.1768	-0.7311	1.0000									
BOD_5	0.3859	0.8425	0.7365	0.7246	0.0444	-0.1621	-0.8190	0.7163	1.0000								
TOC	0.2806	0.8293	0.7207	0.7034	-0.0273	-0.1647	-0.8182	0.5379	0.9150	1.0000							
NO_3^-	0.3253	0.6407	0.8217	0.7871	-0.2594	0.3475	-0.5067	0.5737	0.6630	0.6195	1.0000						
DET	0.4575	-0.1195	-0.1042	-0.1098	-0.0617	-0.0600	0.2309	-0.0216	-0.0749	-0.2088	-0.2002	1.0000					
PO_4^{3-}	0.0280	0.5097	0.5477	0.5689	-0.0737	-0.2530	-0.7668	0.7245	0.6335	0.5668	0.2183	0.0009	1.0000				
NH_4^+	0.2694	0.4799	0.5680	0.4862	-0.2209	-0.1258	-0.6413	0.4648	0.6728	0.6743	0.4791	-0.0855	0.5584	1.0000			
NO_2^-	0.7020	0.6320	0.8549	0.8056	-0.2564	0.3239	-0.5152	0.4680	0.6118	0.5499	0.7549	0.1091	0.3158	0.4687	1.0000		
SO_4^{2-}	0.0408	0.2597	0.6962	0.8174	-0.1960	0.3512	-0.3731	0.2028	0.3981	0.4157	0.5800	-0.1374	0.3010	0.0501	0.4672	1.0000	
Cl^-	0.0327	0.3623	0.6469	0.6577	-0.4619	0.0743	-0.6242	0.3763	0.6224	0.6611	0.6395	-0.2276	0.3612	0.7380	0.4139	0.4445	1.0000

Table 2. The calculated values of F1, F2, F3 and WQI for five sampling stations

Stations	F1	F2	F3	WQI
MB1	17.64706	7.352942	5.318159	88
MB2	17.64706	10.29412	9.483876	86
MB3	41.17647	23.52941	31.89383	67
MB4	47.05882	26.47059	24.69702	65
MB5	23.52941	13.23529	14.03916	82
Average WQI				77.60

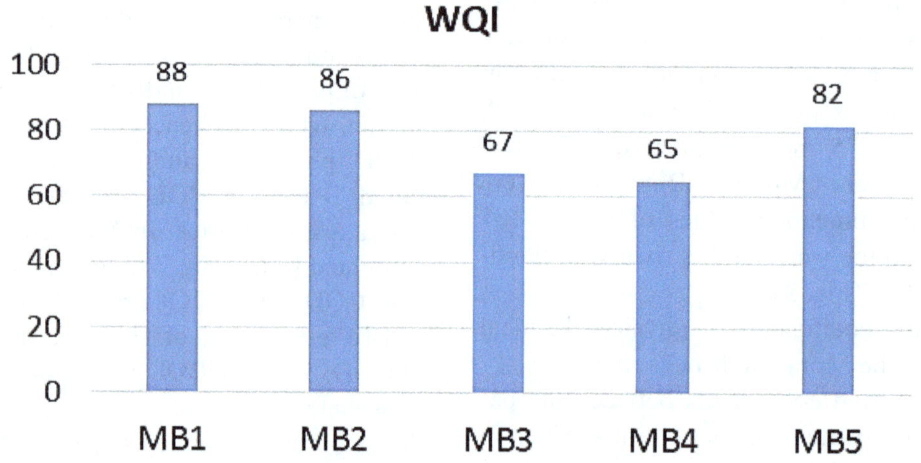

Figure 5. Values of WQI in MB1, MB2, MB3, MB4 and MB5 stations

The results show that the water of MB1 station has better water quality with a WQI 88 value and is ranked in good category, while the lowest water quality was found on MB4 station with a value of WQI 65 as a part of fair category. Finally, the average WQI was calculated for the whole measuring period and it resulted with a value of 77.60 showing that the water of Morava e Binçës river fits in the fair category.

CONCLUSIONS

1. This article analysed the waters of Morava e Binçës by using the physicochemical parameters, in four seasons of 2017. The values of measured parameters were compared with the Romanian Republic standards for the assessment of the ecological status of surface waters (GD161). The following conclusions can be drawn from the research and the discussion of results:
2. The waters of Morava e Binçës were more loaded with pollution at MB3, MB4 and MB5 sampling stations, but the pollutions are not alarming and the river represents a convenient aquasystem for the life of living beings and economic activity.
3. The values of parameters such as: WT, TUR, EC, WSM, pH, DO, OS, TOC, nitrates, DET, ammonium ion, sulfates and chlorides were in consistency with the GD161 recommendations and are categorized as first class;
4. Some of the parameters value such as: COD, BOD_5, phosphates, nitrites exceeded the recommended value of the GD161 regulation and were categorized as of fifth category;
5. The highest positive correlation was found between: EC-WSM (r = 0.9693), BOD_5-TOC (r = 0.9150), whereas the negative one was found between: ng.O_2-TOC (r = – 0.8182) and ng.O_2-phosphates (r = – 0.7668);
6. The Water Quality Index Desktop software is used as a highly efficient tool for calculating WQI according to the guidelines of the Canadian Environmental Ministry;
7. On the basis of the WQI calculations, it was shown that the best quality was found at the MB1 station with the WQI 88 value (good category), the lowest water quality was found at MB4 station with a value of WQI 65 (the category: fair), while the average WQI for the

entire measurement period was 77.60 (the fair category);

8. We propose that the authorities and state institutions should support the river water monitoring as an effective measure to examine their ecological status and protection against pollution.

REFERENCES

1. Anbarasu K., Anbuselvan G. 2017. Physico-chemical parameter analysis of water in Musiri Taluk, Tamil Nadu, India, World News of Natural Sciences (WNOFNS), (6) 36–43.
2. Barakat A., Baghdadi M. E., Rais J., Nadem S. 2012. Assessment of heavy metal in surface sediments of Day river at Beni-Mellal Region, Morocco. Res. J. Environ. Earth Sci., 4; 797–806.
3. Durmishi H. B. 2005. Determination of pollution degree of the river Shkumbin (Pena) with heavy metals and other effluent with spectrometric method and „stripping" voltammetry, Master's Thesis, University of Prishtina, Faculty of Mathematics and Natural Sciences, Department of Chemistry, Prishtina, Kosova, p. 9.
4. CCME. 2001. Canadian water quality guidelines for the protection of aquatic life: CCME Water Quality Index 1.0 User's Manual. Canadian Council of Ministers of the Environment.
5. Durmishi H. B., Ismaili M., Shabani A., Abduli Sh. 2012. Drinking Water Quality Assessment in Tetova Region, American Journal of Environmental Sciences, 8(2), 162–169.
6. Ramadani E., Memeti M., Durmishi H. B. 2017. Water Quality Index: A New Automated Way of Measuring the Quality, International Journal on Information Technologies & Security, 3, 43–52.
7. Krishnan R. R., Dharmaraj K., Kumari B. D. R. 2007. A comparative study on the physicochemical and bacterial analysis of drinking, borewell and sewage water in the three different places of Sivakasi, J. En. Biol. 28, 105–108.
8. Ismaili M., Durmishi B. 2006. Society and Management of Environment, University Book, South East European University, Tetova, R. of Macedonia, 21.
9. Ministry of Environment and Spatial Planning of Kosova (MESP). Environmental Protection Agency 2015. Report on the State of Water. Prishtinë, 42.
10. Ministry of Environment and Spatial Planning. Department of Waters (MMPH) 2010. Hydrography of Kosova's Rivers, Republic of Kosova, 8–11.

16

Heavy Metal Concentrations along the Banks of the Sitnica River and in Four Types of Herbaceous Plants

Albona Shala Abazi[1], Fatbardh Sallaku[3], Pajtim Bytyçi[4], Mimoza Hyseni Spahiu[1*], Fadil Millaku[2]

[1] University "Haxhi Zeka", Faculty of Management in Tourism, Hospitality and the Environment, Str. UCK, Peje 30 000, Republic of Kosovo

[2] University "Haxhi Zeka", Faculty of Agrobusiness, Str. UCK, Peje 30 000, Republic of Kosovo

[3] Agricultural University of Tirana, Albania

[4] South East European University, Faculty of Contemporary Sciences and Technologies, Ilindenska 335, Macedonia

* Corresponding author's e-mail: mimoza.hysenispahiu@unhz.eu

ABSTRACT

This paper presents the results of the heavy metals accumulation (Cd, Cr, Pb, Cu, Zn, Mn, and Fe) in soils and plants (*Typha angustifolia* L., *Urtica dioica* L., *Ranunculus sardous* Crantz and *Rumex Crispus* L.) collected along the banks of the Sitnica River, which flows through various cities of Kosovo and is quite stressed due to the discharges of urban, industrial, and agricultural pollutants. The aim of this research was to verify the accumulation of heavy metals in water, soil and plants, since vegetables that used as food are cultivated in the vicinity of sample collection. The analyses were conducted with applied methodology in the analytic laboratory of Kosovo Agriculture Institute by using MP-AES 4200 for Cd and Cr, ASS-Scientific for Zn, Cu, Fe, Mn and Pb in plant samples, whereas 4100 MP-AES was used for determination of metals in soil. The results show variations in different metal values found in the samples of the analyzed soils and plants. The lowest values accumulation of the metals analyzed were recorded at A1 (Devetak Station) near the river source and did not contain anthropogenic influences, whereas the highest values were recorded at A3 (Kuzmin Station), which belongs to the middle flow of the Sitnica River that is polluted by wastewater (urban) discharges.

Keywords: concentration, metals, plant species, soils, Sitnica River.

INTRODUCTION

The creation of nature is an undeniable phenomenon which has established all methods of preserving natural equilibrium, but the participation of human activity in different natural processes has made it necessary to apply different methods for the preservation and protection of many processes in nature [Dreshaj 2012]. Over-exploitation of nature and uncontrolled use of natural resources, including improper treatment of industrial waste, has caused the contamination of world ecosystems by toxic metals (Pb, Cd, Cu, Zn, Ni, Mn and Fe, Cr) [Bilinski et al. 2010].

Heavy metals are one of the most dangerous groups due to their toxicity, concentration under natural conditions and the ability to be incorporated into the food chain [Zhang et al. 2010]. These chemical elements are usually waste products that come as a result of continuous anthropogenic industrial activities and their emission into the surrounding environment can result in contamination. Under certain environmental conditions, metals may accumulate in toxic proportions and may cause ecological damages [Shala 2015].

The accumulation of heavy metals (HMs) in natural ecosystems poses threat to human health and biodiversity due to their persistence and toxicity [Brut 2014, Singh et al. 2011, Dias and

Nayak 2016, Sedo et al. 2017]. The heavy metal concentrations in plants are directly associated with their concentrations in soils, but their levels significantly differ among plant species, and can be even affected by genotypes within the same species [Öztürk et al. 2011].

Their concentration may vary depending on the type and they can be used as bio-indicators or bio-monitors for quality assessment in aquatic and terrestrial ecosystems [Güleryüz et al. 2008].

Historical pollution of unchecked emissions of heavy metals may persist for decades. Many studies have focused on the heavy metal accumulation capacities of terrestrial plants [González and González-Chávez 2006] and aquatic merged or submerged macrophytes [Güleryüz et al. 2008] in contaminated sites.

The Sitnica River is one of the main rivers in the Republic of Kosovo; its whole length (approximately 90 km) and the catchment area are located in the territory of the Republic of Kosovo. Despite the decrease in the water quality and water levels during these years, this river was used for recreation as well as for irrigation of crops.It is a field stream and with frequent occurrences of changing the river bed from floods, this river flows through numerous settlements during its course, and in this case, considerable amounts of polluted water are discharged into its bed. In order to examine the contamination of flora and vegetation on the banks of the Sitnica River, this study has analyzed the concentration of heavy metals in four herbicidal plants: *Typhaangustifolia* L., *Urticadioica* L., *Ranunculussardous Crantz* and *Rumex Crispus* L., in four different places, which had a wide spread throughout the river flow. Soil research has been done along the course of this river as well.

The aim of this study was to investigate the concentrations of some heavy metals (Cd, Cr, Pb, Cu, Zn, Mn and Fe) on the soil, on the ground and underground parts of the 4 plant species that grow throughout the course of Sitnica River. The data analyzed shows that the soil samples and plants have accumulated a large amount of metal in all sampling locations. The highest concentrations for almost all of the analyzed metals were found in Kuzmin (the most striking locality wickedness from urban and industrial pollution, not to overlook the agricultural one).

This research can be useful for drafting a strategy on improving and managing the surrounding environment around the river Sitnica.

MATERIALS AND METHODS

Study Area

Sitnica River has a surface of 2.861 km^2 and drains on the major part of the hollow of Kosovo. As such, it does not possess a sustainable river bank; during the floods the river covers a major part of the field along its flow, causing huge damages. Sitnica does not have a source, but with such a nomination, it starts at the place where the brook of Shtime covers the brook of Sazlis from the left hand, near the Robovc village. Topolia is considered its sourcing wing, which springs on the western hillsides of Dermanit Peak (1364 m).

From Topilla (1280 m) up to the estuary of Sitnica to Ibër the quote is 497.2 m, the absolute fall is 782.8 m and 7.2% the relative fall [Kastrati 2014]. The river Sitnica is characterized by a quiet hydrologic regime with the average speed of flow 12.9 m^3/sec. [Demaku 2012].

The history shows that the establishment and development of dwellings is closely tied to the river Sitnica, and that the river represents an important factor for the development of those areas. Until lately, the river used to be a source of water for different purposes (fishing, watering and relaxing). Recently, the river Sitnica became a natural recipient of great volume of contaminated water (waters discharged from thermal power plants of Kosovo, waste waters from cities, waters of smaller dwelling areas through which the river goes) which do not have an adequate solution for their cleaning.

Undoubtedly, it can be said that the River Sitnica is endangered from numerous agricultural surfaces, dwellings with no adequate solution of avoiding or treatment of waste waters as well as important industrial departments with no adequate solution for the treatment of waters after their use at certain units [Shala, 2015].

Sample site and sampling

The sites where greater effect of pollution was expected, i.e. near the areas where urban, industrial and agriculture discharges occur, were selected for soil and herbaceous sampling.

Sampling was realized on four types of localities along the river flow (Fig. 1):

1. The first station in Devetak (referred point) – an untouched area of pollution.
2. The second station in Gracka – an influenced locality of agriculture infiltration discharges.

3. The third station in Kuzmin – an influenced locality of urban discharges.
4. The fourth station in Plementin – an influenced locality of black industrial water discharges.

From each type of them, a large number of individuals samples have been taken with the purpose to select one as representative as possible. Originally, they were washed with pure water, and then it was determined how fresh they were. The determination has been done in the field and at the Department of Biology, University of Prishtina, always based on floristic international and domestic literature. Plants have been dried naturally in a cool place (in shadow) and after the drying process the underground part of the plant was separated from the top one, and finally they were ground and weighted.

The samples of herbaceous plants have been gathered in two different time periods, i.e. in summer and autumn, and then the following parameters have been analyzed: Cd, Cr, Pb, Cu, Zn, Mn and Fe. The analyses have been conducted using applied methods in the analytic laboratory of the Agriculture Institute of Kosovo, which are referred to as effective cost analyses of large and small elements as well as stains on food products using MP – AES 4200 for Cd and Cr and ASS-Scientific for Zn, Cu, Fe, Mn, and Pb.

The sample preparation has been conducted in accordance with A.O.A.C 17th edition, 2000 Official Method 999.11.

The soil samples have been gathered only in the autumn, in all sites where herbaceous plants samples were taken, in the depth from 0–30 cm to the point where the major plant root system develops. Analyses have been performed with applied methods in the Analytical Laboratory of Agriculture Institute of Kosovo, used for determination of metals on the ground using 4100 MP-AES (Terrance hettipathirana; Agilent Technologies; Melbourne, Australia; Agriculture). The sample preparation was realized using US EPA Method 3050B (sample preparation procedure relating to the acid digestion of sediments, sludge, and soils).

Statistical data analysis

The statistical data processing was conducted using SPSS 22, the data was mirrored with the arithmetic average and standard deviation (±). Moreover, the correlation of some metals on the ground and plants (Pearson correlation) on the scale of 95% (P 0.05) and 99% of significance (P 0.01), was carried out. Microsoft Excel 2010 charts were used to present metals on plants statistically.

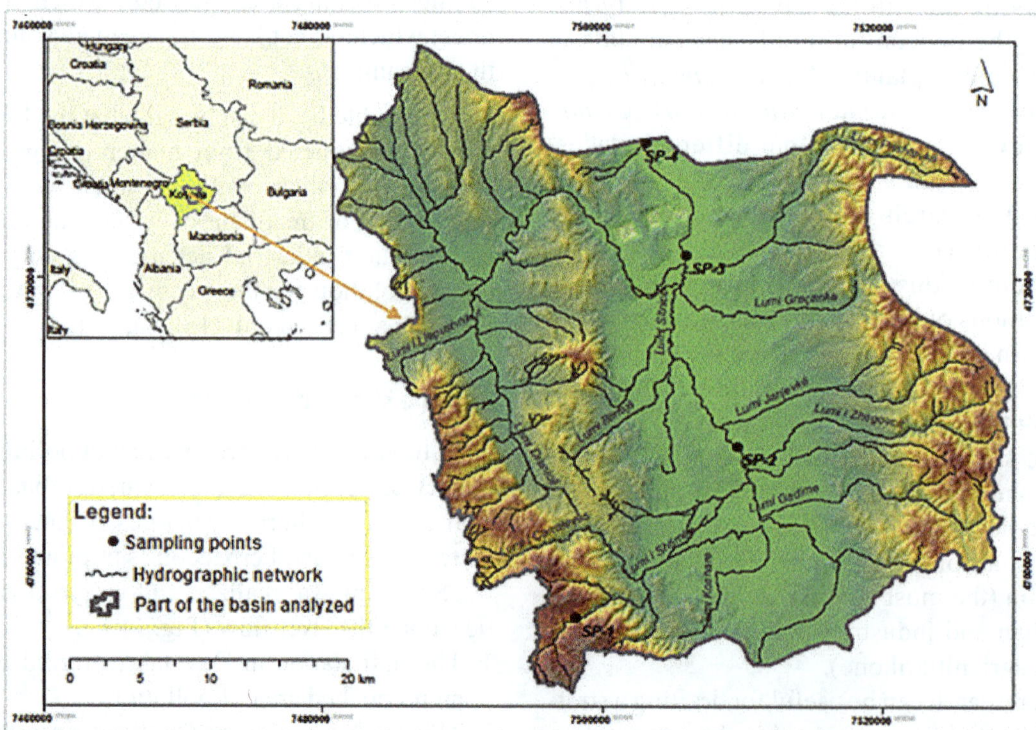

Figure 1. Location map of the study area

RESULTS AND DISCUSSION

In order to realize the goal of the study, as well as conduct the assessment and analysis of the environmental effects along the Sitnica river flow, different representative groups of samples from environmental components were selected and analyzed, including soil (30 cm depth) and herbaceous plants.

The findings of the metals content on the analyzed samples of soil along the Sitnica river flow were presented in the following table (Table 1), where the details regarding the sample sites and their code, monitoring parameters, symbols, measurement units, recorded values for every analyzed parameter and the comparison with maximal allowed values according to the domestic standards, their average values and the standard deviation can be found.

Table 1 also indicates that Cd in the soil samples showed the values from 0.5–1.2 mg/kg. Compared to the domestic standard maximum allowed values, they were within the allowed limits in all localities. The values of Cr have varied from 13.1–106 mg/kg, at Vojnovc, Grcakë, Kuzmin stations, the results of Cr measurements have shown exceeded allowed maximum values. The variations of Pb range from 15.9 to 331 mg/kg, when high contamination of lead is detected, the problems caused in the environment and human beings are unavoidable, particularly if humans are exposed continuously against lead or the environments with lead for long periods of time [Alushllari 2013]. The values exceeding the allowed standards of this element were found in Grackë and Kuzmin. Even the findings of Zn, at the monitoring station of Kuzmin exceeded the allowed values, according to the standards in Kosovo. The values of Cu varied from 11.5–50 mg/kg and have been within the allowed limits. The recorded values for Fe varied from 566.7 to 2033.3 mg/kg, whereas for Mn they ranged from 237.5 to 1340 mg/kg.

Tables 2–5 present the findings obtained from all representative samples of vegetation that were analyzed for the content of seven metals on the ground PM (stem/leaf) and underground PN (root) of four plant species, according to the samples realized during two different seasons, according to the first period (summer season) and the second period (autumn season) of monitoring, respectively.

The data from the table show that Cd during the first monitoring period in all samples of plants has been (<0.3) under the detection level of the used method, while in the second monitoring period the highest value scored 0.73 mg/kg on the ground surface of Rumex crispus. The lowest Pb values equaled 1.8 mg/kg in stem and leaves at Urtica dioica during autumn season, while the highest values reached 25.7 mg/kg on stem and leaves of Rumex crispus plant during summer. The values of Cu varied from 2.6 mg/kg (roots of Rumex crispus) to 14.9 mg/kg (roots of Ranunculus sardous) during the second monitoring period, respectively, in autumn. The highest values of Cr were recorded during the summer season – 14.8 mg/kg in stem/leaves of Rumex crispus, while the lowest value was recorded during the autumn season – 1.2 mg/kg in stem/leaf of Ranunculus sardous. The content of Zn varied from 18.6 to 85 mg/kg at different plant species and at different monitoring periods.

In all plants analyzed during the two monitoring periods, Fe and Mn dominated with regards to concentration, compared to other metals. Their values were different: Fe ranged from 331.6 to 1928.3 mg/kg on roots of Rumex crispus during different seasons, while Mn reached 33–165.3 mg/kg.

Table 1. Findings gained from the analysis of the representative soil samples

Metals, mg/kg	AMV local	Devetak A1	Grackë A2	Kuzmin A3	Plemetin A4	Average and standard deviation (±)
Cd	2	0.5	1.2	1.2	0.6	1.78 ±2.88
Cr	50	13.1	57.7	106.0	98.4	60.08 ±30.92
Pb	50	15.9	58.0	331.0	25.0	210.25 ±393.23
Cu	100	11.5	40.4	50.0	38.5	36.53 ±20.22
Zn	300	23.1	101.9	361.1	97.2	206.58 ±287.99
Mn	-	237.5	975.0	1340.0	812.5	834.0 ±533.36
Fe	-	566.7	1733.3	2033.3	1600.0	1754.16 ±1363.35

AMV – Allowed maximal values.

Table 2. Content of metals in vegetation at the first location (Devetak)

Devetak		*Urtica dioica*		*Ranunculus sardous*		*Rumex crispus*	
Metals	Period	PM	PN	PM	PN	PM	PN
Cd	I	<0.3	<0.3	-	-	<0.3	<0.3
	II	0.71	<0.3	<0.3	<0.3	0.73	<0.3
Pb	I	8.9	8.7	-	-	25.7	6.5
	II	1.8	4.3	4.3	5.6	4.1	2.6
Cu	I	6.8	5.4	-	-	6.5	11
	II	5.7	4.9	9.1	14.9	6.1	2.6
Cr	I	6	13.5	-	-	14.8	5
	II	3.4	10	1.2	5.6	5	3.7
Zn	I	34	19	-	-	18.6	29
	II	19.7	38.1	72.1	85	75.7	40.4
Mn	I	133.7	123.2	-	-	165.3	149.7
	II	80.27	71.09	46.8	75.6	55.1	33
Fe	I	1785	1543.8	-	-	1049.1	1928.3
	II	332.3	755.8	575.6	1436.8	661.7	331.6

Table 3. Content of metals in vegetation at the second location (Grackë)

Gracka		*Typha angustifolia*		*Urtica dioica*		*Ranunculuc sardous*		*Rumex crispus*	
Metals	Period	PM	PN	PM	PN	PM	PN	PM	PN
Cd	I	<0.3	<0.3	<0.3	<0.3	<0.3	<0.3	<0.3	<0.3
	II	<0.3	0.22	0.88	1.03	<0.3	<0.3	<0.3	<0.3
Pb	I	3.8	8.4	5.6	6.3	2.8	3.2	3.9	7.5
	II	2.5	9.8	1.8	6.6	5.4	4.7	2.4	7.8
Cu	I	2.4	3.6	4.7	4.1	6.8	4.2	5.8	5.3
	II	5.2	13.1	5.1	2.7	11.3	14.5	6	4.7
Cr	I	1.3	5.2	4.5	7.9	5.7	12.6	4.2	4.3
	II	0.9	4.8	5.5	9.3	1.6	2.9	2.2	2.9
Zn	I	13.7	12.8	33.9	14.8	27.5	13.4	36.7	29
	II	16.7	23.1	38.4	32.9	94.1	33.2	72	30.9
Mn	I	208.3	116.3	113.7	191.8	59.3	130	60.3	473.4
	II	180.83	103.5	243.31	388.16	53.6	81.9	37.3	67.6
Fe	I	290.7	703.5	452	1535.7	405.8	797.4	660.2	1115.9
	II	104.2	663.1	470.5	884.8	407.7	852	534.8	701.7

The findings pertaining to the metal content of the analyzed samples of four plant species in Grackë station were presented in the Table 3 provided above. In this location, there are no large urban and industrial discharges, but rather agricultural infiltrations; therefore, it is known as an agricultural location. From the data shown in the table, it can be considered that the content of the analyzed metals in the underground and surface organs varies with different values, specifically: Pb (1.8–9.8 mg/kg), Cu (2.4–14.5 mg/kg), Cr (0.9–12.6 mg/kg), Zn (12.8–94.1 mg/kg), Mn (37.3–473.4 mg/kg) and Fe (104.2–1535.7 mg/kg).

The highest content of the majority of metals is as follows: Cd, Pb, Cu, Cr, Mn and Fe were found in underground parts of the analyzed plants, particularly in roots, while regarding Zn, the highest values were recorded in for ground parts (stem/leaf) of Ranunculus sardous and Urtica dioica. Even at this studied location, different values of metal concentration were noticed starting from Cd, which was (<0.3) under the detection level during the first period in all plant samples, while on the second monitoring period the highest value of it scored 1.03 mg/kg on the underground part of plant Urtica dioica.

The table above shows the highest values of Fe (107.9–2967.8 mg/kg) and Mn (39.4–410.5 mg/kg) at all analyzed plants during the two monitoring periods compared to other metals. The Cd value, even at this monitoring station during the first period, was < 0.3 under

Table 4. Content of metals in vegetation at third location (Kuzmin)

Kuzmin		*Typhaangustifolia*		*Urticadioica*		*Ranunculucsardous*		*Rumexcrispus*	
Metals	Period	PM	PN	PM	PN	PM	PN	PM	PN
Cd	I	<0.3	<0.3	<0.3	<0.3	<0.3	<0.3	<0.3	<0.3
	II	0.55	0.29	0.84	0.53	2.56	<0.3	0.44	0.73
Pb	I	5.1	13.8	8.2	15.6	9.1	4.9	7.6	10.5
	II	3.8	60.1	14.3	24.4	32.8	33.6	13.1	8
Cu	I	2.8	5.7	5.1	14.3	11	6.8	1.9	3.8
	II	7.3	14.3	5.6	5.3	20.5	14.4	5.4	4.4
Cr	I	8.4	84.5	5.7	12.4	12.5	13.4	4.8	8.8
	II	5	25.1	16.1	15.1	18	15.5	7.8	10.9
Zn	I	19.3	24	33	83.8	47	29	21.3	34.3
	II	23.6	18.4	54.9	61.9	108.3	65.4	60	28
Mn	I	286.4	410.5	96.8	114.2	111	105.6	46.3	250.3
	II	177.42	75.23	94.3	116.7	309	78.7	60.5	39.4
Fe	I	244.3	1010.7	353.6	2066.6	595.5	1069.2	107.9	595.7
	II	442.4	1008.4	1333.7	1883.5	2967.8	1657.3	735.1	703.4

Table 5. Content of metals in vegetation at fourth location (Plemetin)

Plemetin		*Typha angustifolia*		*Urtica dioica*		*Ranunculuc sardous*		*Rumex crispus*	
Metals	Period	PM	PN	PM	PN	PM	PN	PM	PN
Cd	I	<0.3	<0.3	<0.3	<0.3	<0.3	<0.3	<0.3	0.66
	II	0.28	<0.3	10.4	<0.3	<0.3	0.72	<0.3	<0.3
Pb	I	6.8	31.9	3	12.5	3.6	4.9	10.3	6.7
	II	4	5.4	5.4	4.5	5.4	10.2	4.8	9.2
Cu	I	4.6	8	6.3	0.3	8.6	2.9	5.9	5.1
	II	7.4	2.6	8.6	5.3	10.6	13.2	6.1	6.1
Cr	I	7	39.7	4.1	10.4	5.7	14.9	6.1	10.5
	II	3.7	8.5	5.8	3.8	9.9	17.8	3.3	4.3
Zn	I	19.2	60.1	28.6	46.5	43	14.2	36	32.6
	II	18.3	17.6	50.8	38.6	65.4	66.2	42.2	43.8
Mn	I	137.4	206	41.2	74.9	45.5	29.9	43.3	67.5
	II	260.44	56.16	72.4	41.3	69.3	124.4	46.1	69.3
Fe	I	666	2702.3	477.4	1891.3	790.9	582.7	500.5	524.1
	II	484.7	642.3	688.6	525	813.5	1278.5	558.4	523.5

the detection level of the applied method, while on the second monitoring period the highest value of it was recorded (2.56 mg/kg) in stem and leaves of Ranunculus sardous. The lowest and the highest values for other analyzed metals at Kuzmin station, located in the middle of River Sitnica and polluted by wastewater discharges, the values vary as follows: compared to the station in Grackë, the highest content of the majority of metals such as Cd, Cu, Zn, and Fe in this station was found in the surface parts (stem and leaves), whereas, for Pb, Cr and Mn the highest values were recorded in the underground parts.

In Plementin, the analyzed metals in the underground parts (roots) and surface (stem/leaves) at different plants exhibited different values that vary as follows: Cd (<0.3–10.4 mg/kg) the highest value was recorded in stem and leaves of the Urtica dioica plant during the autumn season, Pb (3–31.9 mg/kg), Cu (0.3–13.2 mg/kg), Cr (3.3–39.7 mg/kg) Zn (14.2–66.2 mg/kg), Mn (29.9–260.4 mg/kg) and Fe (477.4–2702.3 mg/kg).

In this station, the findings indicate that the highest values of the analyzed metals were found in the underground parts, particularly in roots of the vegetation, excluding Cd and Mn which had higher concentration in stem and leaves of the Urtica dioica and Typha angustifolia plants. Compared to the other stations where Cd d was under the detection level during the first period, in Plementin, the concentration of Cd during the first monitoring period in the roots of Rumex crispus is evident.

The following diagrams show the average values of the metal content level in the surface (stem/leaves) and underground parts (roots), of the analyzed plants in all locations.

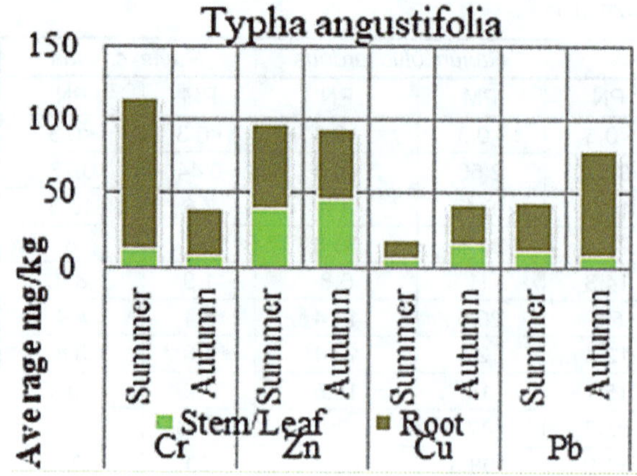

Figure 2. The average values of metals (mg/kg empty weight) accumulated in the samples of Typha angustifolia, during the two monitoring period.

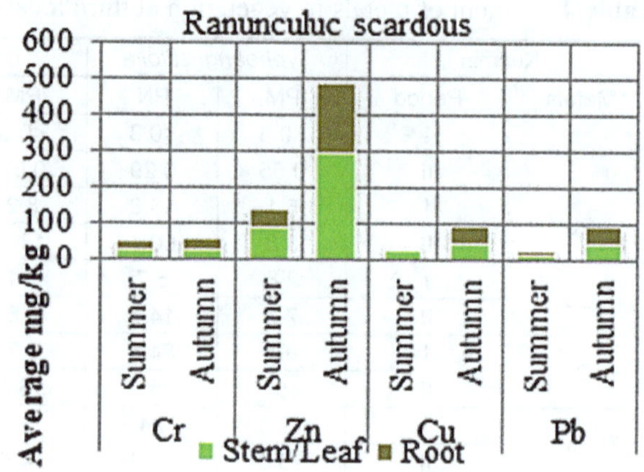

Figure 3. The average values of metals (mg/kg empty weight) accumulated in the samples of Ranunculus sardous, during the two monitoring periods

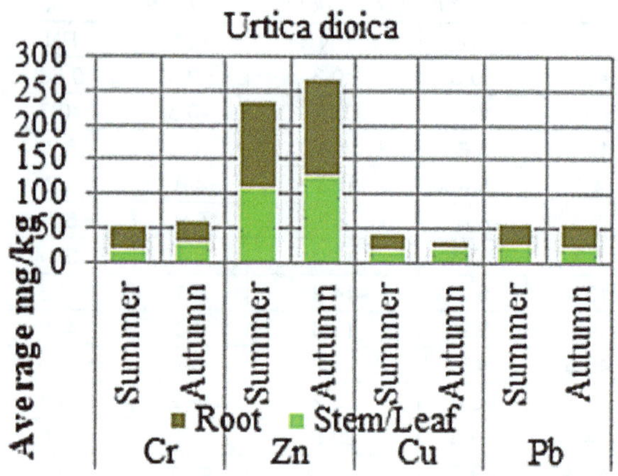

Figure 4. The average values of metals (mg/kg empty weight) accumulated in the samples of Urtica dioica, during the two monitoring periods.

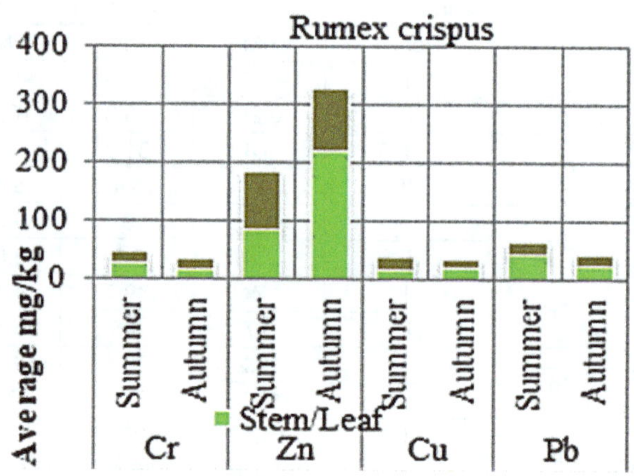

Figure 5. The average values of metals (mg/kg empty weight) accumulated in the samples of Rumex crispus, during the two monitoring periods

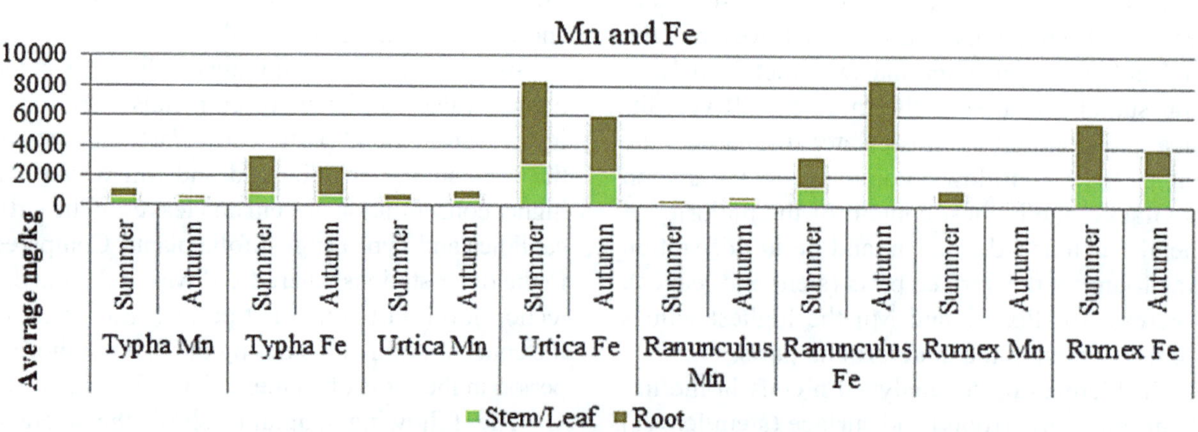

Figure 6. The average values of the Mn and Fe in the analyzed plant samples

Table 6. Correlation of Zn, Mn, and Fe on the herbaceous plant and soil samples in the autumn season

	Zn								
	TAPM	TAPN	UDPM	UDPN	RSPM	RSPN	RCPM	RCPN	Tokë
ZnTAPM	1								
ZnTAPN	-.574	1							
ZnUDPM	.835	-.930	1						
ZnUDPN	,999*	-.544	.571	1					
ZnRSPM	.590	.322	.408	.648	1				
ZnRSPN	.664	-.994	-.320	.269	-.440	1			
ZnRCPM	-.111	.877	-.754	-.202	.285	-.065	1		
ZnRCPN	-.456	-.467	-.323	-.533	-,989*	.523	-.378	1	
Zn Soil	.901	-.163	.529	.918	.896	-.063	.045	-.824	1
	Mn								
MnTAPM	1								
MnTAPN	-.783	1							
MnUDPM	-.569	.957	1						
MnUDPN	-.641	.979	,996**	1					
MnRSPM	-.484	-.166	-.251	-.179	1				
MnRSPN	1,000*	-.766	-.318	-.380	-.262	1			
MnRCPM	-.174	-.477	-.746	-.688	.684	-.317	1		
MnRCPN	.573	.061	.476	.418	-.406	.682	-.858	1	
Mn Soil	-.957	.568	.422	.491	.765	-.548	.194	-.157	1
	Fe								
FeTAPM	1								
FeTAPN	.363	1							
FeUDPM	.619	.957	1						
FeUDPN	.163	.978	.854	1					
FeRSPM	.540	.980	,966*	.925	1				
FeRSPN	.830	.821	.617	.562	.745	1			
FeRCPM	.506	.987	.628	.779	.805	.907	1		
FeRCPN	-.579	.550	.606	.575	.469	-.237	-.059	1	
Fe Soil	-.417	.695	.621	.823	.612	-.008	.312	.869	1

** Significant correlation for P 0.01; * significant correlation for P 0.05; TA – *Typha angustifolia*; UD – *Urtica dioica*; RS – *Ranunculus sardous*; RC – Rumex crispus; PM – foreground parts; PN – underground parts

CONCLUSIONS

1. The most widely spread herbaceous plants along the River Sitnica flow are: *Typha angustifolia* L, *Urtica dioica* L, *Ranunculus Sardous Crantz and Rumex crispus* L. The level of heavy metals concentration in the above-mentioned plants was analyzed. The data analyzed show that plants have attained a considerable quantity of heavy metals in all sampling locations. The highest concentrations for all analyzed metals (Cd, Pb, Cu, Cr, Zn, Mn, and Fe) in plants and soil samples were found in Kuzmin (where industrial and agricultural wastewater is discharged. In contrast, the lowest values for the majority of the analyzed elements were recorded in Devetak (the station close to the river source, with no anthropogenic influences).
2. In general, Fe and Mn accumulated in the analyzed plants along the river Sitnica are found in the highest concentrations;
3. The concentration of metals, in general in *Typha angustifolia* and *Urtica dioica* plants. was higher in the underground parts (roots) during the two monitoring periods (summer and autumn), whereas in *Ranunculus sardous* and *Rumex crispus* plants – the highest content of metals during the two periods was recorded in the surface parts, particularly in stems and leaves.
4. Two out of four selected sampling sites can be considered most polluted because the values of Zn, Pb and Cr analyzed in the samples are over the allowed limit and as such they pose threat.

5. The findings suggest ample evidence regarding the level of pollution with heavy metals, soil and herbaceous plants along the flow of Sitnica river, as a consequence of the anthropogenic activity.

REFERENCES

1. Alushllari M. 2013. The environmental effect and assessment of industrial remains (Pb) in the territory of former Battery factory, Berat. PhD Thesis Faculty of Natural Sciences, Department of Industrial Chemistry, Tirana.
2. Bilinski SF, Gashi F, Bilinsk H. 2010. Establishing of Monitoring Network on Kosovo Rivers (Drini I Bardhë, Morava E Binçës, LepencandSitnica), BALWOIS 2010–Ohrid, Republic of Macedonia.
3. Demaku S. 2012. Heavy metals as soil, water and river pollution near TC KOSOVO. PhD Thesis, Faculty of Mathematical and Natural Sciences, Department of Chemistry, University of Prishtina
4. Dreshaj A. 2012. Chemistry- environmental study of water wells Drini i Bardhë and Ibër in the Republic of Kosovo. PhD Dissertation. Faculty of Natural Sciences, Department of Environmental Industrial Chemistry, University of Tirana, 2012.
5. Öztürk E., Atsan E., Polat T. and Kara K. 2011. Variation in heavy metal concentrations of potato (Solanum tuberosum L.) cultivars. The Journal of Animal & Plant Sciences, 21(2), Page: 235–239.
6. González R.C. and González-Chávez M.C.A. 2006. Metal accumulation in wild plants surrounding mining wastes. Environmental Pollution, 144, 84–92. Medline. DOI10.1016/j.envpol.
7. Dias H.Q. and Nayak G.N. 2016. Geochemistry and bioavailability of mudflats and mangrove sediments and their effect on bioaccumulation in selected organisms within a tropical (Zuari) estuary, Goa, India. Marine Pollution Bulletin, 105(1), 227–236.
8. Burt J.A. 2014. The environmental costs of coastal urbanization in the Arabian Gulf. City, 18(6), 760–770.
9. Kastrati B. 2014. Kosova verilindore: veçoritë fiziko-gjeografike dhe vlerësimi ekonomik i pasurive natyrore. Desertacion i doktoraturës, Tiranë.
10. Sedo K.A., Abido M.S., Salih A.A., Abahussain A. 2017. Assessing Heavy Metals Accumulation in the Leaves and Sediments of Urban Mangroves (Avicennia marina (Forsk.) Vierh.) in Bahrain. International Journal of Ecology, Vol. 2017. Article ID 3978216.
11. Singh R., Gautam N., Mishra A., Gupta R. 2011. Heavy metals and living systems: an overview. Indian Journal of Pharmacology, 43(3), 246–253.
12. Shala A. 2015. The analysis of the industrial and agricultural effects on water quality and several environmental parameters of River Sitnica, Kosovo. PhD Dissertation, UBT, Tirana.
13. Zhang H., Cui B., Xiao R., Zhao H. 2010. Heavy metals in water, soils and plants in riparian wetlands in the Pearl River Estuary, South China. Procedia Environmental Sciences, 2, 1344–1354.

17

Assessment of Diesel Particulate Matter Exposure of Underground Miners in Indonesia

Arif Susanto[1,2], Purwanto Purwanto[1,3], Henna R Sunoko[1,4], Onny Setiani[1,5]

[1] Doctorate Program in Environmental Science, School of Postgraduate Studies, University of Diponegoro, Semarang 50241, Indonesia

[2] Department of Environmental Engineering, Faculty of Civil and Planning Engineering, University of Kebangsaan Republic of Indonesia, Bandung 40263, Indonesia

[3] Department of Chemical Engineering, Faculty of Engineering, University of Diponegoro, Semarang 50275, Indonesia

[4] Department of Medicine, Faculty of Medicine, University of Diponegoro, Semarang 50275, Indonesia

[5] Department of Environmental Health, Faculty of Public Health, University of Diponegoro, Semarang 50275, Indonesia

[*] Corresponding author's e-mail: arifssnt@yahoo.com

ABSTRACT

In Indonesia, there are underground mines for mineral metal such copper (Cu) and gold (Au), built by tunneling towards the mineral location. The purpose of this study was to determine the mapping a concentration of diesel particulate matter (DPM) and assess the impact on health by severity measurement of airflow obstruction of the miners experiencing chronic obstructive pulmonary disease (COPD). The data of DPM were measured with NIOSH method no. 5040 and applied a geostatistical method in mapping concentration at the area of underground mining. A spirometric measurement was conducted to diagnose COPD that is done to the 314 miners. The results showed that the concentrations exceeding the permissible exposure limit (PEL) and spirometric measurement were found for 26 miners (8.3%) who experience COPD (post bronchodilator <0.70). The severity measurement of airflow obstruction of the miners experiencing COPD, severity of airflow limitation for moderate (GOLD 2) was obtained for 14 miners (54%); severe (GOLD 3) for 10 miners (38%) and very severe (GOLD 4) for 2 miners (8%). It can be concluded that the amount of DPM exposure against the severity of airflow limitation with COPD by 0.03, in which the other factors also affect the severity.

Keywords: airflow obstruction; COPD; DPM; geostatistics; PEL; spatial interpolation; severity measurement; underground miners.

INTRODUCTION

Diesel engines (DE) find use in industrial applications, including mining (off-road). Diesel-powered heavy equipment (HE) operating in underground (UG) environments (mines and tunnels) were determined by multiplying the vehicle power by a ventilation rate that was either mandated by regulation(s) or determined empirically from known quantities (Parent, 2007). The process of UG mining requires the application of (HE) such as loader as well as other types of equipment that use diesel engines (DE). Beside loader, haul truck, development jumbo, truck and other mine equipment with DE are found in the underground tunnels.

In Indonesia, Deep Ore Zone (DOZ) is an underground mine (UG) for copper (Cu) and gold (Au). The air pollutant concentration measurement in DOZ is the first step in identifying environmental-health hazards and risk to the miners. (Susanto et al., 2016; Susanto et al., 2017). Diesel fuel (DF) and DE are used increasingly frequently because of the reliability and efficiency of hydrocarbon fuel oxidation. However, there is a concern that the result of diesel engine oxidation is in the form of diesel engine exhaust (DEE) (USEPA, 2002; IPCS, 1996), which is dangerous for miners' health. This concern is related to the small size of respirable diesel particulate matter (DPM) (USEPA, 2002; Kittelson, 1998).

The particles in DEE are DPM consisting of the center of carbon nucleus defined as elemental carbon (EC), and absorbed organic compounds defined as organic carbon (OC), as well as a small number of sulphate, nitrate, metal, and other elements (USEPA, 2002). DEE were classified as carcinogenic to humans (Group 1) (IARC, 2012), the decision was based on National Institute for Occupational Safety and Health (NIOSH) and US National Cancer Institute (NCI) study showed the exposure-response relationships between the respirable EC exposure and lung cancer mortality in UG miners (Attfield et al., 2012; Silverman et al., 2012). The highest levels of EC were reported for enclosed UG mining work sites (Rushton, 2012).

The exposure assessment is the process of measuring or estimating the magnitude, frequency and duration of human exposure to a compound in the environment (Susanto et al., 2016). The human exposure evaluation involved describing the nature and size of the population exposed to air contaminants as well as the magnitude and duration of their exposure. The dose, its duration and timing, the nature and size of the critical measures of exposure were taken into account for risk characterization (Birch et al., 1996; Birch et al., 2004).

DPM are usually less than one micron ($\times 10^{-6}$ m) in size, which causes them to be more easily inhaled and retained in the body. Presently, the United States mining industry is in the process of phasing in stringent regulations relating to diesel particulates in underground mines. The exposure limits for DPM adopted by MSHA for metal or non-metal underground mines (non-coal). The present PEL for DPM as per MSHA (30 CFR 57.5060(b)) is 160 μg TC/m^3, defined as total carbon (TC) content (CDC, 2016).

Chronic obstructive pulmonary disease (COPD) is characterized by persistent respiratory symptoms and airflow limitation, usually caused by a significant exposure to noxious particles or gases. The burden of COPD is predicted to increase due to an additional exposure to environmental risk factors for the development of the disease and greater risk of the disease as the population ages (GOLD, 2017). Indonesia is a country with a high number of smokers, characterized by high COPD prevalence. Unfortunately, the COPD prevalence data are not owned by Indonesia; therefore, it is necessary to study COPD comprehensively so that its prevention can be done well. In 2017, an updated and revised version of the standard-setting Global Initiative for Chronic Obstructive Lung Disease (GOLD) guidelines was released as depicted in table 1 (Classification of COPD based on GOLD 2017). As part of the revision, the definition of COPD was updated to include the importance of environmental disturbances and host factors that contribute to the disease pathophysiology.

The most prevalent environmental risk factor for COPD is cigarette smoking, which is associated with a higher mortality rate, a higher burden of respiratory symptoms and lung function abnormalities, as well as greater decreases in lung function. For the diagnosis of COPD in patients with risk factors, a spirometric assessment is conducted to confirm the presence of persistent airflow limitation. The patients with the severity of airflow limitation post bronchodilator of less than 0.70 forced expiratory volume in one second (FEV_1) / forced vital capacity (FVC) are considered to have a spirometrically confirmed diagnosis (GOLD, 2017).

METHODS

The research was carried out in DOZ, which is a production area of UG located on the coordinate points of 73°32'50" – 73°42'50" Easting; 95°12'50" – 95°22'50" Northing; UTM Zone 53 South. The mining site is situated in Jayawijaya Mountain, East Mimika District, Papua Province, Indonesia. The measurements representing the ambient conditions were taken and a comprehensive chemical analysis was performed. The data of DPM as TC was defined as the sum of EC and OC. Both EC and OC were measured by NIOSH method number 5040.

This study applied spatial analysis by universal kriging for mapping DPM concentration in the area truck haulage drift at DOZ. The data obtained from these calculations were then processed into a frame using a spatial model. It has been most frequently used for providing boundary conditions to the dynamic modeling of the air pollution. In the current study, ArcGIS were used for generating maps. Spirometric measurements were conducted to diagnose COPD and the assessment of airflow limitation based on GOLD, was conducted on the 314 miners that work at the DOZ (Table 1).

Table 1. Classification of COPD based on GOLD

Stage	Spirometric Findings
I. Mild	▪ FEV1/FVC < 0.70 ▪ FEV1 ≥ 80% predicted
II. Moderate	▪ FEV1/FVC < 0.70 ▪ 50% < FEV1 < 80% predicted
III. Severe	▪ FEV1/FVC < 0.70 ▪ 30% ≤ FEV1 ≤ 50%
IV. Very Severe	▪ FEV1/FVC < 0.70 ▪ FEV1 ≤ 30% predicted; or FEV1 ≤ 50% predicted plus chronic respiratory failure

RESULTS AND DISCUSSION

The DOZ mine airflow area providing sufficient airways were required based on the minimum velocity concern in main travelways and dilution of diesel contamination (Stinnette & Souza, 2013). The total airflow demand is 1.498 m^3/s. Airflow to each primary level, undercut, extraction and haulage was based on providing 0.079 $m^3/s/kW$ over diesel equipment and a minimum air velocity of 0.76 m/s in the areas where personnel and non-diesel operate (Susanto et al., 2016). The intake and exhaust airways were required to provide sufficient airflow. It involves multiple parallel drifts on the extraction (production) level to control block-caving mines. Concerning the main trailways, the airflow is required for dilution of DEE contamination. The ventilation system was designed to provide fresh air for all personnel working in truck haulage drifts. Determination of airflow quantity is influenced by mining equipment and wide variations in the characteristics of emissions even amongst vehicles of similar size and power. A number of parameter affects the total airflow required for a given amount of DE.

DEE SURVEY AND MEASUREMENT

DEE is a complex mixture of DF combustion products (table 2 shows the quality analysis for diesel usage in DOZ), and the exact composition of the mixture depends on the nature of engine, operating conditions, lubricating oil, additives, emission control system, and fuel composition (Pronk et al., 2009). DPM generation varies considerably among types, sizes, series manufacturers of DE. DPM behave as an aerosol (Stinnete & Souza, 2013). Being sub-micron in size, they have aerodynamic diameters falling within a range 0.1 to 0.25 µm; thus, their control is similar to other gaseous contaminants. The composition of DPM concentration was taken from the data analysis as depicted in table 3. Its peaks data without could provide information about the spatial changes in the composition of the organic and gas component.

Table 2. Diesel fuel quality analysis results

Test	Method	Units	Limits	Result
Appearance @20°C	Visual	-	Clear & bright	Clear & bright
Ash content	ASTM D482–13	% mass	Max 0.01	<0.001
Water and sediment	ASTM D2709–96(11)e1	% V/V	Max 0.05	<0.005
Micro carbon residue 10% bottom	ASTM D4530–11	% mass	Max 0.16	<0.10
Cetane index	ASTM D4737–10	-	Min 47	47.5
Cloud Point	ASTM D2500–11	°C	Max 0	-3
Cold Filter Plugging Point	IP 309/99	°C	Max -3	-4
ASTM Color	ASTM D1500–12	-	Max 2.5	L1.5
Copper Corrosion (3hrs @100°C)	ASTM D130–12	-	Max 1	1A
Density at 15°C	ASTM D4052–11	Kg/L	0,820–0.870	0.8597
Distillation Range	ASTM D86–12	-	Report	-
50% recovered	ASTM D86–12	°C	Report	280.1
90% recovered	ASTM D86–12	°C	Max 357	332.8
Flash point	ASTM D93–13 (Proc A)	°C	Min 60	87.0
Lubricity – method HFRR	IP 450/00	µm	Max 460	360
Total Acid Number	ASTM D974–12	MgKOH/g	Max 0.25	0.03
Strong Acid Number	ASTM D974–12	mgKOH/g	Nil	Nil
Oxidation Stability	ASTM D2274–10	mg/L	Max 15	5
Sulphur	ASTM D2274–10	% mass	Max 0.05	0.0284
Kinematic Viscosity at 40°C	ASTM D445–12	mm^2/s	1.9 – 5.5	3.270

Table 3. Descriptive statistics of DOZ Truck Haulage for DPM

Area	n	Average	Min-Max	PEL
Access Center Crusher #1–2	2	42	0 – 84	160
Truck Haulage Shop	5	247.2	140 – 527	160
Maintenance Shop	2	164.5	154 – 175	160
Office Area	6	154.5	66 – 428	160
PM Shop	2	161	53 – 269	160
Welding Shop	3	172	90 – 272	160
West Empty Haulage	2	223.5	179 – 568	160
South Full Haulage	2	1151	1055 – 1247	160
Access #1HN to 1JS	2	171	0 – 342	160
Access South Empty	2	591	0 – 1182	160
Access West Full Haulage #10	2	263.5	0 – 527	160
Access West Full Haulage #1–6	2	214	0 – 428	160
Center Crusher #1	2	204	175 – 233	160
Center Crusher #2	2	185.5	160 – 211	160
LP#1 E-1F South	2	595.5	0 – 1191	160
LP#1 G South	2	591.5	0 – 1183	160
LP#1 H South	2	971.5	725 – 1218	160
LP#1 J South	2	198.5	194 – 203	160
LP#1 IE	2	798	759 – 837	160

In order to validate a determination of a miner's exposure based on a personal TC sample, MSHA will analyze a non-citable area sample taken from an area of the mine without organic carbon (OC) interferences. MSHA will multiply the elemental carbon (EC) component of the personal TC sample by the ratio of TC to EC obtained from the area sample to check for non-diesel OC interferences on the personal samples. MSHA uses this procedure when EC on the miner's personal sample is <160 μg/m³ of air times the error factor (EF) for EC, and TC on the miner's personal sample is >160 μg/m³ of air times the EF for TC. MSHA will obtain both personal and area samples using the same type of sampling cassette, previously employed for its compliance sampling. TC from a personal sample must >160 μg/m³ times the EF for TC to result in a miner's overexposure to DPM. The EF for TC is 1.192. Thus, it follows that the miner is not excessively exposed to DPM when TC from the miner's personal sample is ≤160 μg/m³ times the EF for TC.

GEOSTATISTICS – SPATIAL INTERPOLATION

In order to obtain spatial interpolation analysis is using air dispersion model by numerically processing the emission and meteorological data (Zou et al., 2009). Kriging is a commonly used method and represents spatially continuous phenomena. The method has formed the basis for environmental pollution mapping in recent years (Isaaks et al., 2013). This method describes DOZ mine air quality exposure model for predicting DPM concentrations to which the miners are exposed.

Air pollution analysis and points of monitoring and measurement (truck haulage drifts and tunnels) used in this studies of various pollutants from HE operating in UG environments. Air polutant data measurements are used in assessment use of Geographic Information System (GIS) based Kriging method to produce the most accurate and useful information about the spatial interpolation patterns pertaining to DPM concentration in DOZ. Mapping was built in ArcGIS, integrating the relevant data on truck haulage drifts, tunnels and ventilation rate. The results that show DPM can be estimated, mapped and visualized within a GIS framework (Susanto et al., 2017). The spatial distribution of the pollutants is an integral component from risk evaluation aspect concerning health and environment. The technique used for illustrating the spatial distribution provides a meaning for estimating the probability of the attribute value, which exceeds the required threshold in a particular location (Cattle et al., 2002).

The results of the air pollutant concentration mapping was DPM with universal kriging (Figure 2), this method is also useful for identifying metallic (Fonturbel et al., 2011) and par-

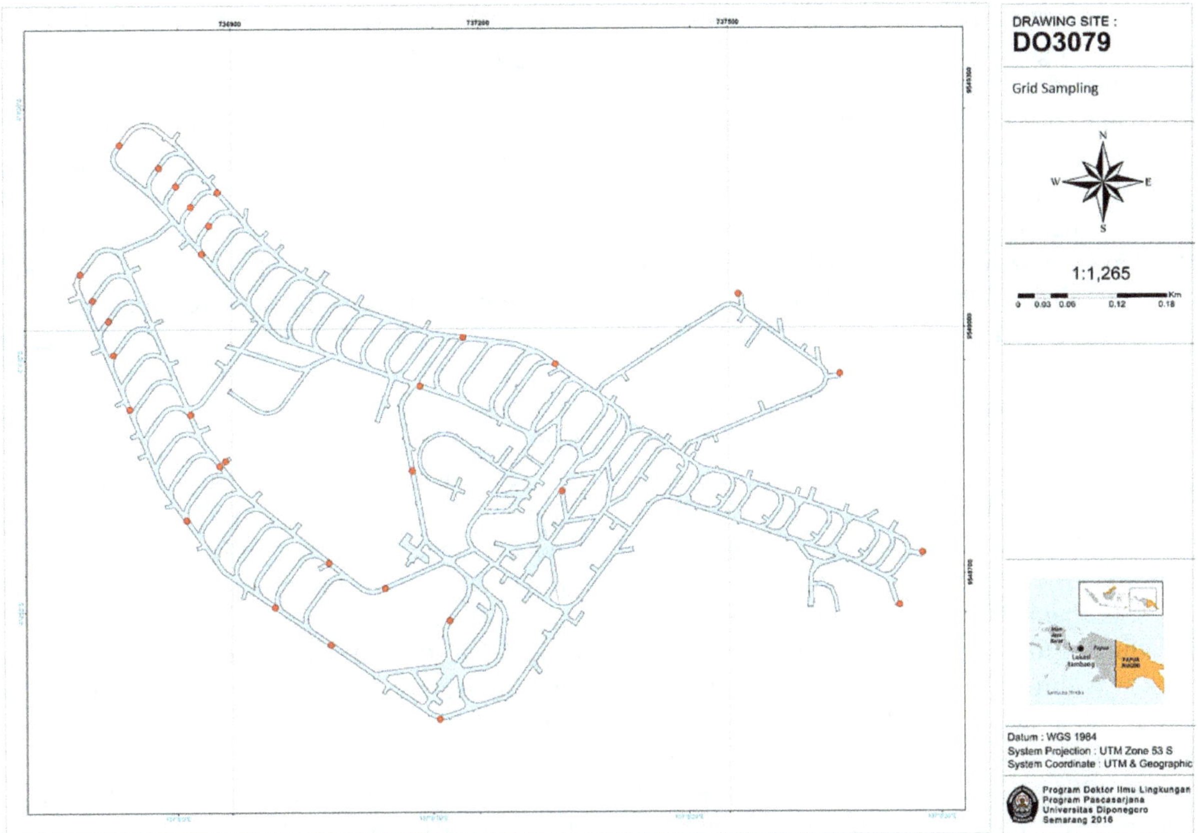

Figure 1. Grid sampling location at truck haulage area for DPM measurement

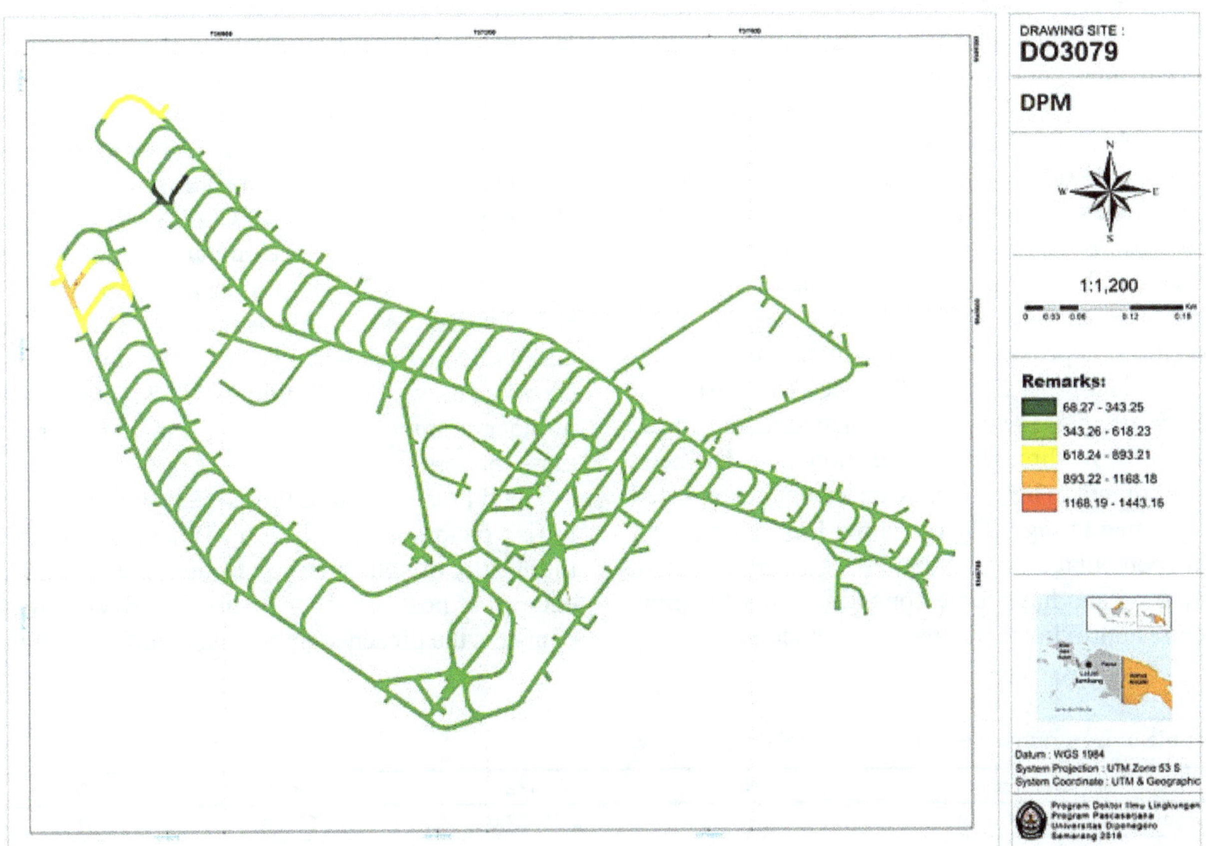

Figure 2. DPM pollutant map at the truck haulage area

ticulate matter (Gulliver et al., 2011; Brauer et al., 2003). In general, it indicated a high concentration; therefore, denoting bad conditions at the truck haulage area, even though no difference in DPM concentration was found. Most areas had the DPM concentration ranging from 343.26 to 618.23 mg/m^3. In some areas, the rates were 618.24 to 1168.18 mg/m^3. Although the air quality inside the DOZ was affected by the balance of the total air intake and exhaust, as the part of ventilation system to supply the needs for air, the quality is still not capable of diluting air pollutants, in particular DPM. The air requirement for all mining workers was 650 individuals with operational factor of 100%, where the total airflow per unit reached 0.03 m^3/s. Besides, the airflow quantity necessary for each unit type of the diesel equipment was characterized by the operational factor depending on the power unit (in horsepower unit), which was different from the unit type affected by operational factor of each unit equipment (Susanto et al., 2016). Figure 2 shows spatially continuous DPM concentrations mapping for points where there are no measurement data which have to be estimated by a spatial interpolation.

The measurement values of both minimum and maximum DPM concentrations were presented in table 3. Minimum airflow value of 0.03 m^3/s/worker and 0.067 m^3/s/kW for DEE dilution as per Indonesian Mining Regulation (Susanto et al., 2016; Susanto et al., 2017). Actually, 0.080 m^3/s/kW is a design value and is higher than the typical MSHA equipment quantities provided for gaseous compliance (Matejicek, 2005). For truck haulage routes, the maximum velocity is 6.1 m/s. The ventilation for diesel shops based on the dilution of exhaust gaseous for two large loaders, requires approximately 40.0 m^3/s. The ventilation of non-diesel shops has been established at 23.5 m^3/s based on the experience at the mine. The airflow through the lube shop areas has been determined to be 28.2 m^3/s, based on the expected equipment usage. The operating factors represent the percentage of time that the equipment will be running and have to be applied in order to determine approximate airflow requirements.

CHRONIC OBSTRUCTIVE PULMONARY DISEASE (COPD)

Environmental exposures to fumes, gases, air pollutants, and occupational dusts contribute to impaired lung function. The most prevalent environmental risk factor for COPD is cigarette smoking, which is associated with a higher mortality rate, a higher burden of respiratory symptoms and lung function abnormalities, and greater decreases in lung function. Spirometric measurement for COPD diagnosis was done to 314 miners that work at the DOZ.

Table 4 shows the frequency distribution of miner age with an average of 37.87 years of age. The age range indicates that the miners are in the productive age. In spirometric measurement other than gender (sex), age is a factor that must be considered (GOLD, 2017). This is in accordance with the 'age range under observation' contained in Figure 3 that is the graph of FEV_1 progression over time.

The height ranges from 1.45 to 1.84 m with the mean of 1.64 m which is the range of average height for Indonesians. The miners usually work in DOZ for 15 years. This is in line with the mining age of DOZ which began operating in 1998. The years of working are in accordance with the amount of time of DEE exposure. The exposure to particles is a factor that affects disease progression, although it is also affected by age and sex, although in this study, gender is not used as an independent variable because all miners in DOZ are male. The prevalence of COPD was higher in smokers and ex-smokers Compared to non-smokers (GOLD, 2017). The number of non-smokers is 142 miners (45.2% of the total number of miners), which is less when compared with the miners who smoke, i.e. 172 miners (54.8%of the total number of miners) – 9.8% higher than miners as non-smokers.

COPD is considered in a history of exposure to risk factors for the disease. Table 5 shows that the results of spirometry measurement, the presence of a post-bronchodilator $FEV_1/FVC < 0.70$ confirms the presence of persistent airflow limita-

Table 4. Descriptive statistics of miners' demography

Description	N	Min-Max	Mean	SD
Age	314	22.0–57.0	37.87	6.611
Height	314	1.45–1.84	1.64	0.056
Working Years	314	2.0–34.0	15.21	6.043

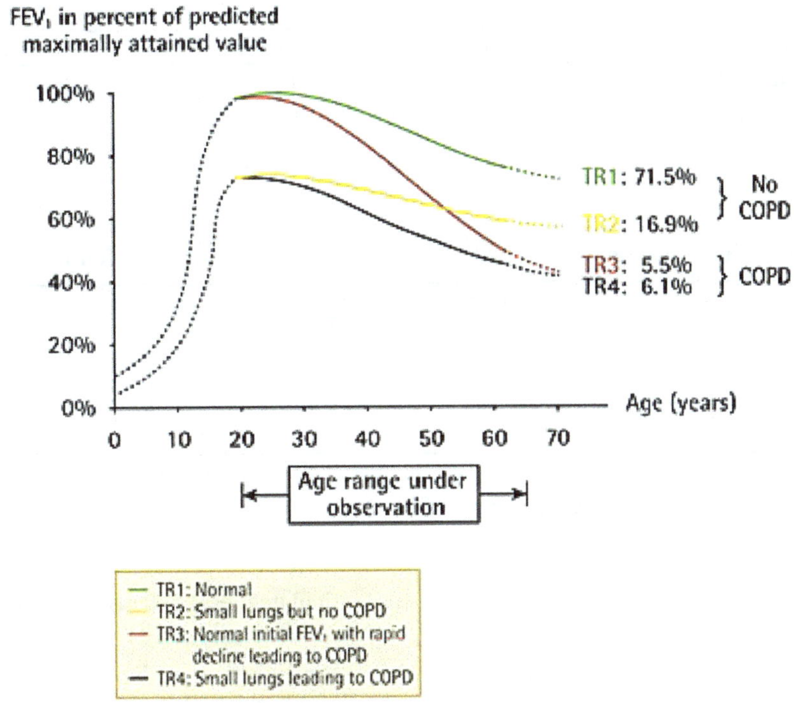

Figure 3. FEV1 progression over time

tion on as many as 26 out of 314 miners (8.3%). Thus, it can be said that exposure to DEE can cause COPD on the miners who worked in DOZ.

ASSESSMENT OF SEVERITY OF AIRFLOW OBSTRUCTION

The 2017 GOLD guidelines introduce an updated ABCD assessment tool for COPD disease severity that requires the formal assessment of disease-specific symptoms and an evaluation of patient history of the exposure to the disease risk factors in order to determine the disease severity independently from the spirometric results (table 6). In order to measure the severity of airflow obstructive toward the miners who experience COPD, a follow-up test was conducted on miners with FEV_1 (% predicted), assessed on the miners with $FEV_1/FVC < 0.70$; the obtained results of the severity of airflow limitation classification for the miners with COPD are: for moderate (GOLD 2) – 14 miners or 54%, severe (GOLD 3) – 10 miners or 38%, and very severe (GOLD 4) – 2 miners or 8%, as described in Table 6.

Following the assessment of airflow limitation mentioned above, ANOVA test was conducted by enter method to determine the amount of working time (exposure) DPM against severity of airflow limitation with COPD and whether other factors including age and smoking habits which can also affect the severity. The obtained results show that the value $R = 0.173$ with value $F = 2.395$ on the value of significance $(\alpha) = 0.05$, so it can be concluded that the GOLD category is influenced by 3% of the length of exposure to DPM, height, smoking habits, and age of miners.

COPD is one of the non-communicable diseases, which is rarely exposed due to lack of information provided. In the United States. the data in 2007 shows that the COPD prevalence 10.1% (SD 4.8) for male is 11.8% (SD 7.9) and for women is 8.5% (SD 5.8) (Buist et al., 2007). Mortality ranks at the fourth position as highest factor that is 18.6 per 100,000 people in 1991 and the death rate increased as much as 32.9% from 1979 up to 1991 (AST, 1995). The COPD prevalence in Southeast Asian countries is estimated as much as 6.3% with the highest occurrence found in Vietnam (6.7%) and China (6.5%) (Yeung et al., 2004). COPD will negatively affect the quality of life, including >40 years that will lead to disability toward the patients. Though they are still in the productive age group, they cannot work optimally due to chronic breathlessness. The co-morbidity of COPD will result in cardiovascular disease, bronchial cancer, lung infections, embolic thrombosis disorder, the presence of asthma, hypertension, osteoporosis, joint pain, depression and axiety (Agusti et al, 2003).

Table 5. Descriptive statistics of spirometric measurement results

Description		Frequency	Percent	Cumulative Percent
Valid	Not COPD	288	91.7	91.7
	COPD	26	8.3	100
	Total	314	100	

Table 6. Results of Severity Measurement of Airflow Limitation Assessment with COPD

Severity of Airflow Limitation	Classification	FEV_1[a]	Frequency	Cumulative Percent
Mild	GOLD 1	≥80 %	0	0
Moderate	GOLD 2	50 – 79 %	14	54
Severe	GOLD 3	30 – 49 %	10	38
Very Severe	GOLD 4	<30 %	2	8

FEV1 indicates forced expiratory volume in 1 second; FVC forced vital capacity
[a] Assessed in the patients with FEV1/FVC less than 0.70

CONCLUSIONS

The survey results that have been conducted in DOZ show that the concentration of DPM ranges from 643.26 to 618.23 μg/m³. The results of spirometric measurements show that 8.3% miners experience COPD (post-bronchodilator <0.70). The severity measurement of airflow obstruction to miners experiencing COPD depends on the amount of working duration (exposure) DPM against severity of airflow limitation with COPD by 3%;the other factors are age and smoking habits, which also affects the severity. From these results, it was concluded that the concentration mapping can be used to evaluate exposure-response relationships. The estimated exposure of the DPM pollutant was recommended to perform as the part of the study on predicting whether the pollutant has any impact on health condition of the mining workers in a long-term exposure.

REFERENCES

1. Agusti, A.G.N., Noguera, A., Sauleda, J., Sala, E., Pons, J. and Busquet, X. 2003. Systemic effect of COPD. Eur Respir J, 21: 347–360.
2. Attfield, M.D., Schleiff, P.L., Lubin, J.H., Blair, A.E., Stewart, P.A., Vermeulen, R., Coble, J.B. and Silverman, D.T. 2012. The diesel exhaust in miner study: a cohort mortality study with emphasis on lung cancer. J Natl Cancer Inst, 104: 869–883.
3. ATS. 1995. Standards for diagnosis and care of patient with COPD. Am J Respir Crit Care Med, 152: S77–120.
4. Birch, M.E. and Cary, R. 1996. Elemental carbon-based method for occupational monitoring of particulate diesel exhaust: methodology and exposure issues. Analyst, 121: 1183–1190.
5. Birch, M.E. and Noll, J.D. 2004. Submicrometer elemental carbon as selective measure of diesel particulate matter in coal mines. J Environ Monit, 6: 799–806.
6. Brauer, M., Hoek, G., Vliet, P.V., Meliefste, K., Fischer, P., Gehring, U., Heinrich, J., Cyrys, J., Bellander, T., Lewne, M. and Brunekreef, B. 2003. Estimating long-term average particulate air pollution concentrations: application of traffic indicators and geographic information systems. Epidemiol, 14: 228–239.
7. Cattle, J.A., McBratney, A.B. and Minasny, B. 2002. Kriging method evaluation for assessing the spatial distribution of urban soil lead contamination. J Environ Qual, 31: 1576–1588.
8. CDC 2003. Diesel particulate matter (as elemental carbon). Centers for Disease Control and Prevention.
9. Chan-Yeung, M., Khaled, N.A, White, N., Ip, M.S. and Tan, W.C. 2004. The burden and impact of COPD in Asia and Africa. Int J Tuberc Lung Dis, 8: 2–14.
10. Fonturbel, F.E., Barbieri, E., Herbas, C., Barbieri, F.L. and Gardon, J. 2011. Indoor metallic pollution related to mining activity in the bolivian altiplano. Environ Pollut, xxx: 1–6.
11. Gulliver, J., Hoogh, K.D., Fecht, D., Vienneau, D. and Briggs, D. 2011. Comparative assessment of GIS-based methods and metrics for estimating long-term exposure to air pollution. Atmospheric Environ, xxx: 1–9.
12. Global Initiative for Chronic Obstructive Lung Disease (GOLD). 2017-a. Global strategy for the diagnosis, management, and prevention of chronic obstructive pulmonary disease. GOLD website. http://goldcopd.org/download/326/. Published Jan-

uary 2017. Accessed April 26, 2017.

13. Global Initiative for Chronic Obstructive Lung Disease (GOLD). 2017-b. Pocket guide to COPD diagnosis, management, and prevention; a guide for health care professionals. GOLD website. http://goldcopd.org/wp-content/uploads/2016/12/wms-GOLD-2017-Pocket-Guide.pdf. Published January 2017. Accessed April 26, 2017.
14. IARC. 2012. Diesel engine exhaust carcinogenic. Press release, 213.
15. IPCS. 1996. Diesel fuel and exhaust emissions. Environmental Health Criteria 171. International Programme on Chemical Safety.
16. Isaaks, E.H., Srivastava, R.M. 1989. An introduction to applied geostatistics. Oxford University Press, Toronto.
17. Matejicek, L. 2005. Spatial modelling of air pollution in urban areas with GIS: a case study on integrated database development. Adv Geosci, 4: 63–68.
18. MSHA. 2013. MSHA/OSHA hazard alert – diesel exhaust / diesel particulate matter. Mine Safety and Health Administration.
19. Parent, M., Rousseau, M., Boffeta, P., Cohen, A. and Siemiatycki J. 2007. Exposure to diesel and gasoline engine emissions and the risk of lung cancer. Am J Epidemiol, 165: 53–62.
20. Rushton, L. 2012. The problem with diesel: editorials. J Natl Cancer Inst, 104: 796–797.
21. Silverman, D.T., Samanic, C,M., Lubin, J.H., Blair, A.E., Stewart, P.A., Vermeulen, R., Coble, J.B., Rothman, N., Schleiff, P.L., Travis, W,D., Ziegler, R.G., Wacholder, S. and Attfield, M.D. 2012. The diesel exhaust in miners study: a nested case-control study of lung cancer and diesel exhaust. J Natl Cancer Inst, 104: 855–868.
22. Stinnette, J.D. and Souza, E.D. 2013. Establishing total airflow requirements for underground metal/non metal mines with tier IV diesel equipment. 23rd World Mining Congress and Expo.
23. Susanto, A., Purwanto, P., Sunoko, H.R. and Setiani, O. 2017. A kriging method for mapping underground mine air pollution. Adv Sci Lett. 23: 2329–2332.
24. Susanto, A., Purwanto, P., Sunoko, H.R. and Setiani, O. 2016. Diesel engine exhaust emissions survey of underground mine in Indonesia. J Ind Pollut Contr. 32: 608–616.
25. USEPA. 2002. Health assessment document for diesel engine exhaust. National Center for Environmental Assessment: US Environmental Protection Agency.
26. Zou, B., Wilson, J.G., Zhan, F.B. and Zeng Y. 2009. Air pollutant exposure assessment methods utilized in epidemiological studies. J Environ Monit, 11: 475–490.

18

Effectiveness of Carbon Monoxide Concentration Reduction on Active Carbon Contact System in Burning Polystyrene Foam

Elanda Fikri[1*], Aprilia Veronica[2]

[1] Department of Environmental Health, Bandung Health Polytechnic, Cimahi Utara, 40514, Doctorate Program of Environmental Studies, Diponegoro University, Semarang, Indonesia

[2] Department of Environmental Health, Bandung Health Polytechnic, Cimahi Utara, Indonesia

[*] Corresponding author's e-mail: elandafikri@yahoo.com

ABSTRACT

One of the wastes generated by PT Beton Elemenindo Perkasa is polystyrene foam waste. Processing of this type of waste is still done with open burning so it has an impact on health and the environmental quality degradation. One of the polystyrene foam waste processing technologies is by constructing a combustion furnace equipped with carbon filter. Activated carbon is one of the air filter media that can absorb harmful gases from the combustion process. The purpose of this research is to identify the effectiveness of carbon monoxide (CO) concentration reduction on active carbon contact system using variation of particle size. This research is a kind of an experimental study involving post test with control design. The number of samples was calculated based on the number of treatments and the number of repetitions in the study. This research used 2 kinds of treatment, including 20 mesh and 30 mesh in 9 repetitions. Independent T-Test statistical analysis showed a significant difference (p-value = 0.001) between the variation of particle size of activated carbon and the CO parameters with the average of the most effective reduction percentage on particle-sized active carbon of 30 mesh was 77.95%.

Keywords: polystyrene foam, particle size, activated carbon, carbon monoxide

INTRODUCTION

National development in the economic field is implemented in order to create a strong economic structure through the development of advanced industries as an economic driving force supported by strength and resilience of resources. According to Law No. 3 in 2014 regarding industry, industry comprises all forms of economic activities that process raw materials and/or utilize industrial resources to produce goods that have added value or higher benefits, including industrial services.

The development of industry in Indonesia is currently increasing, in line with the development of science and for the fulfilment of human needs, but the industry also contributes to the environmental pollution if the waste of industrial activities is not managed properly (Fikri et al., 2015). Industry is always connected with the environmental pollution problem, especially related to the process of industrial activities. Industrial activities initially intended to improve the quality of human life; however, they can be detrimental to the human survival (Danarto et al, 2008).

Broadly speaking, waste comes from the pollution caused by industries and domestic sectors that generate domestic waste. Increased waste production in 2017 cannot yet be balanced with proper processing. Waste becomes a problem especially in the urban areas, where the amount of waste produced can reach about 1,311 tons per day. These types of waste come from household and industrial waste. Waste is divided into two types, namely organic and inorganic waste. The organic waste is a waste that can be broken down by nature. Conversely, the inorganic waste is a waste that is difficult to decompose or even cannot be decomposed naturally. The inorganic waste has become part of the community. One example is styrofoam. Styrofoam is widely used by society as food wrappers. However, it can cause environmental pollution because it is thrown away after use (Wicaksono, 2011).

Polystyrene foam is widely known as Styrofoam, which is often used inappropriately by the public because styrofoam is actually a trade name that has been patented by Dow Chemical company. Styrofoam is intended for use as an insulator on building construction materials, not for food packaging. Polystyrene foam package is chosen as it is lightweight, capable of maintaining the temperature of hot/ cold food and hold it comfortably, keeping the food freshness. Due to these advantages, polystyrene foam packaging is used to package ready-to-eat, fresh foods, or the foods that require further processing. Many fast food restaurants serve their dishes using these packages, as well as other food products such as instant noodles, chicken porridge, meatballs, coffee, and yogurt (Fitrianti et al, 2016).

Solid waste requires serious attention because it can pose a threat during handling until its destruction (Fikri et al., 2016). The lack of attention in waste management resulting from industrial activities is one of the causes of environmental pollution; hence it requires a process of waste treatment to reduce environmental pollution from industrial waste. One way that can be used to overcome these problems is by processing the combustion chamber furnace unit equipped with an air filter. Processing can reduce the volume of solids (Jamilatun et al, 2014).

Beton Elemenindo Perkasa Company is a business engaged in the field of construction. The industry produces organic and inorganic waste that is still not disposed of properly, which is by burning waste in the open space without the any control, so that the burning produces smoke. Domestic waste produced in PT.BEP also includes leftover food, plastic, and styrofoam food container. Waste control, especially pertaining to Styrofoam, can be done physically, chemically, and technically. The physical control involves using styrofoam as a basic material for children's toys and handicraft, while the chemical control can be done by using essential oil from orange peel to break down styrofoam waste; finally, the technical control can be done by carrying out the burning process (Fitrianti et al., 2016). The result of Focus Group Discussion with the company revealed that one of the most effective and feasible ways to do styrofoam waste treatment is by applying the technical solution, i.e. burning. However, it is required to control the exhaust generated from the combustion process.

Burned styrofoam waste can produce harmful gases such as styrene, hydrochloroflourocarbon (HCFC), polycyclic aromatic hydrocarbons (PAHs), carbon black, and carbon monoxide. When styrofoam is being burned, carbon monoxide is released into the air. According to the US Environmental Protection Agency (EPA), carbon monoxide is an undetectable gas that can harm human health. The exposure to carbon monoxide can cause fainting, dizziness, and nausea. Carbon monoxide is one of the most dangerous chemicals as a result of the burning of styrofoam (Kurniati, 2008).

The control of air pollution can be done by using adsorbents such as activated carbon and zeolite (Yuliusman, 2013). Zeolites contain great quantities of impurities that can reduce the surface area; if used directly, they can clarify the smoke. The capability of yeolites to absorb water vapor is greater than absorbing CO so that the adsorption capacity is reduced. Activated carbon is an adsorbent that has the highest CO absorbing ability as it has a better surface area than other adsorbents.

MATERIALS AND METHODS

This type of research is an experimental, with post test design with control group (Postest with control design). In this study, there were two groups of subjects, one was subjected to treatment (experimental group) and the other one was not treated (control group). On the basis of this design, the effect of the treatment on the dependent variable will be tested by comparing the state of dependent variable in the experimental group after treatment with the untreated control group. In this experiment, the subjects were presented with several treatments and then the results were measured (Notoatmodjo, 2012). The strategy in controlling the confounding variables in this study was:

- Temperature – to control the confounding variables, the temperature applied in this research was 150°C.
- Volume of furnace – to control the confounding variables, the furnace volume was adjusted to the amount of polystyrene foam waste.

Sample

The sample size in the study was obtained based on the experimental, completely randomized design. The sample size determined in this study was based on the number of treatments. The treatments used in this study involved variations of the size of active carbon (20 mesh and 30 mesh) with 9 repetitions (Gomez, 2007). The employed polystyrene foam sampling technique was non random purposive sampling, involving polystyrene foam used as food container with the weight about 8 gr/container.

Total polystyrene foam to be taken was:

- (2 treatments × 70 samples of polystyrene foam used as food container) × 9 repetitions + control;
- (140 samples of polystyrene foam used as food container) × 9 repetitions + 9 control × (70 samples of polystyrene foam used as food container);
- (1260 + 630) samples of polystyrene foam used as food container;
- 1890 samples of polystyrene foam used as food container.

Procedures

The first step was designing furnace as a combustion chamber with 25 cm chimney's diameter as well as preparing the tools and materials. The side of the stove is cut to make the door for garbage introduction and place for fuel. The door was welded in order to close the connection. The air filter media was prepared by crushing the activated carbon by using pestle. Crushed pounded carbon was then filtered to the required size of 20 mesh and 30 mesh.

RESULTS AND DISCUSSION

Table 1 shows the lowest temperature is 46.4°C and the highest temperature is 38.1°C; the lowest humidity is 39% and the highest humidity is 30%; and the lowest wind speed is 1.71 m/s and the highest wind speed is 1.07 m/s. The measurement results of carbon monoxide concentration are presented in Table 2.

Table 1. Temperature, humidity, and wind speed

No	Variable	Minimum	Maximum
1	Temperature	38.1 °C	46.40 °C
2	Humidity	30 %	39 %
3	Wind Speed	1.07 m/s	1.71 m/s

Table 2. Carbon monoxide concentration

Repetition	Control (ppm)	20 Mesh (ppm)	30 Mesh (ppm)
1	860	200	161
2	870	295	183
3	879	360	197
4	892	311	192
5	898	319	195
6	918	339	213
7	853	273	172
8	860	341	204
9	871	366	226
Average	878	312	194
Maximum	918	366	226
Minimum	853	200	161
Std Dev	21.16	51.46	19.97

The average concentration of carbon monoxide is 194 ppm, found in the filter media with 30 mesh particle-sized active carbon and the highest average concentration of 878 ppm is in the control. On the basis of the results, the lowest concentration of CO is when air filter medium is passing through size 30 mesh, yielding the concentration of 161 ppm.

The research conducted by Basuki et al (2008) showed that the most effective thickness of air filter media is 15 cm with the percentage of CO reduction of 83.90%. On the basis of the research conducted by Ardiani et al. (2008), the effective use of activated carbon is 15 cm with the percentage of CO reduction of 70.83%. The size of the utilizied activated carbon particles is 20 mesh, 30 mesh, and 40 mesh. On the basis of the previous research, the most effective particle size of the activated carbon is 30 mesh, because 40 mesh particle size produces dust when being passed by activated carbon filter media. Therefore, the variation of particle size in the research includes only 2 kinds, that is 20 mesh and 30 mesh.

Table 3 shows the reduction difference in CO concentration in the air filter media of the active carbon with the particle size of 20 mesh and 30 mesh. The average reduction of CO concentration in air filter media of 20 mesh active carbonwas 566 ppm (64.54%); on the other hand, in the air filter media with 30 mesh active carbon, the reduction amounted to 684 ppm (77.95%). The results of the Independent T-test also showed significant

Table 3. Reduction difference and reduction percentage in carbon monoxide concentration

Repetition	Reduction Difference		Reduction Percentage	
	20 Mesh (ppm)	30 Mesh (ppm)	20 Mesh (ppm)	30 Mesh (ppm)
1	661	699	76.79%	81.29%
2	575	687	66.09%	78.97%
3	519	681	59.03%	77.54%
4	581	700	65.13%	78.51%
5	580	703	64.53%	78.29%
6	578	705	63.02%	76.79%
7	580	681	68.00%	79.84%
8	518	656	60.29%	76.31%
9	505	645	58.02%	74.01%
Average	566	684	64.54%	77.95%

differences in the CO concentration reduction for variations of particle-sized active carbon with p-value of 0.001 (Table 4).

Activated carbon has been widely used as an air filter media in previous studies. On the basis of the research conducted by Maryanto et al. (2009), the average percentage of carbon monoxide reduction was by using activated carbon 45.68%, 70.83% (Ardiani et al., 2016), and 83.90% (Basuki et al., 2008). Chemical adsorbents (in the form of activated carbon / activated charcoal) can theoretically be used to reduce air pollutants. Activated carbon is a carbon compound that has increased its adsorption power by activation process (Jamilatun et al., 2014). In the activation process, the removal of hydrogen, water, and gas from the carbon surface occurs so that there is a physical change on the surface. Activation occurs because the active group formed due to the interaction of free radicals on the carbon surface with atoms such as O_2 and N_2. In the activation process, new pores were also formed due to the erosion of carbon atoms through oxidation/heating.

New pores on activated carbon create a broad and hollow surface with a layered structure. The pores can absorb gas, so that the gas passing through the activated carbon will be bonded and subjected to a pulling force with active carbon pores. On the basis of the previous research, activated carbon can absorb the NOx gas (Jaya, 2014, Verlina, 2014) and CO (Basuki, 2008; Maryanto, 2009; Jaya, 2014; Verlina, 2014; Ardiani, 2016; Nurullita, 2015).

The carbon monoxide gas generated from the furnace combustion process will accumulate on the adsorbent surface. The formation of new pores of activation further expands the surface of the activated carbon creating more areas adsorbing the materials outside. Furthermore, the CO gas

Table 4. Independent test results T-test

Class	Levene's Test for Equality of Variances		t-test for Equality of Means
	F	Sig	Sig (2 tailed)
Equal variances assumed	3.602	0.076	0.001
Equal variances not assumed			0.001

accumulated on the surface of the adsorbent will be absorbed by the adsorbent (Nurullita, 2015).

In the control group, in which no ingredients were given, the CO gas does not absorb but only accumulates in the air. In the treatment group, the filter media were given different size of activated carbon particles, i.e. 20 mesh and 30 mesh. The CO gas has different absorption on each size of activated carbon particles. For that reason, there

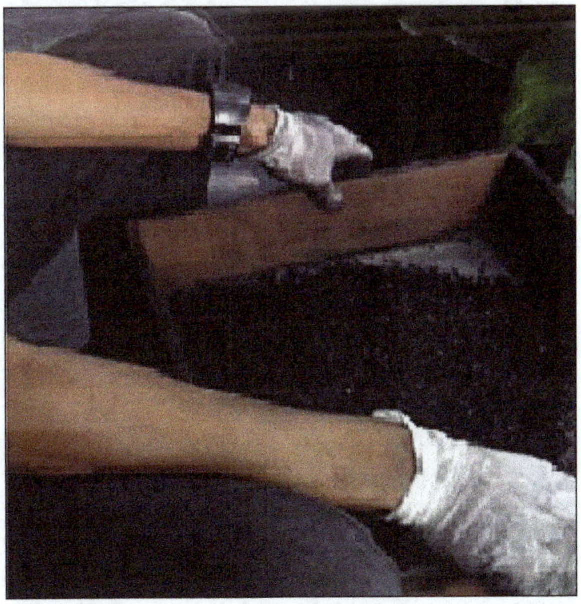

Fig. 1. Active Carbon Filter 20 mesh and 30 mesh

Fig. 2. Furnace

Fig. 3. Size of Activated Carbon Particles 20 Mesh

Fig. 4. Size of Activated Carbon Particles 30 Mesh

is a difference of CO gas concentration on each size of activated carbon particles.

The difference in the CO gas concentration reduction between 20 mesh and 30 mesh particle-sized active carbon is due to different surface area. Therefore, the surface area can affect the absorption of carbon monoxide because the greater the surface area of activated carbon, the more CO will be adsorbed. A filter medium that is capable of absorbing a lot of carbon monoxide (CO) is the one with a particle size of 30 mesh active carbon, hence, the smaller the size of the activated carbon, the higher the absorption or the less CO is produced.

The activated carbon used comes from the coconut shell to absorb the material in the form of steam or gas. The pores present in the type of char are micropores that cause the gas molecules to pass through; however, the molecules of the liquid cannot pass through.

Several factors that may affect the absorption capacity of activated carbon are the physical properties of the activated carbon and the properties corresponding to the number and size of the pores that can be filled by the adsorbate. Each molecule will fill the pores according to its size. The chemistry of activated carbon also affects the absorption of activated carbon because the activating groups on its surface can interact chemically with organic molecules. The presence of Van der Waals forces on the surface of the activated carbon of the adsorbate enables the adsorption process (Khairunisa, 2008).

The chemical properties of the activating groups in the activated carbon are produced at the time of activation. This functional group makes the surface of activated carbon chemically reactive and affects the nature of adsorption (Subekti, 2009). The saturation time of the air filter media used in this study is not known, so the use of air filter media use should be noted because it can affect the effectiveness of carbon monoxide concentration reduction. On the basis of the research by Basuki et al. (2008), the saturation time of activated carbon with the thickness of 15 cm is 45.84 hours.

CONCLUSION

1. There is a difference in the carbon monoxide concentration with variation of particle sizes of activated carbon in polystyrene foam burning process with p-value of 0.001.

2. The average reduction percentage of carbon monoxide concentration in air filter media of 20 mesh particle-sized active carbon was 64.54% (566 ppm) and 30 mesh – 77.95% (684 ppm). This means that the active carbon with the particle size of 30 mesh is quite effective in lowering the concentration of CO in the burning process of polystyrene foam waste.

Acknowledgements

This study was financially supported by Unit Penelitian dan Pengabdian Masyarakat (UPPM) Poltekkes Kemenkes RI Bandung, Ministry of Health, Indonesia scheme.

REFERENCES

1. Ardiani Y., Dwi T., Achmad T. 2016. Reduce the concentration of carbon monoxide from mainstreams cigarette smoke by using coconut sehll activated carbon filter. International Journal of Current Research, 8 (8), 35586–35591.
2. Basuki K. T. 2008. Reduction of CO and NO2 concentrations on exhaust emissions by using local inhibitor TiO2 media on activated carbon. JFN 1(1), 45–64.
3. Danarto Y.C., Samun, T. 2008. The effect of carbon activation of rice husk on the metal adsorption process of Cr (VI). Ekuilibrium, 7(1), 13–16.
4. Fikri, E., Purwanto., Henna, R.S. 2015. Modelling of household hazardous waste (HHW) management in Semarang City (Indonesia) by using life cycle assessment (LCA) approach to reduce greenhouse gas (GHG) emissions. Procedia Environmental Science, 23(2015), 123–129.
5. Fikri, E., Purwanto., Henna, R.S. 2016. Life cycle assessment of household hazardous waste management options for Semarang City, Indonesia. Int. J. Environment and Waste Management, 17(2), 146–157.
6. Fitrianti A.E. 2016. Determination of the essential oil level of the sunkist orange peel (citrus sinensis l. Osbeck) as an alternative to natural styrofoam. Indonesian Journal of Pharmaceutical Science, 3(2), 47–52.
7. Jamilatun, S., Martomo S. 2014. Making of activated charcoal from coconut shell and its application for liquid smoke purification. Spektrum Industri. 12(1), 1–14.
8. Jaya F.T., 2014. Adsorption of CO, NO, and NOx emissions using activated carbon from cocoa fruit leaf waste (Theobroma cacao L.) on four-wheeled vehicles, Makassar. Ph.D. Thesis, Hasanudin University, Makasar.
9. Khairunisa R. 2008. The combination of electrolysis and adsorption techniques uses activated carbon to decrease the concentration of phenol compounds in water. Ph.D. Thesis, Universitas Indonesia, Jakarta.
10. Kurniati. 2008. Utilization of oil palm shells as activated charcoal. Jurnal Penelitian Ilmu Teknik, 8(2), 96–103.
11. Maryanto. 2009. Decrease in carbon monoxide (CO) emissions by addition of activated charcoal to motor vehicles in Yogyakarta. Jurnal Fakultas Kesehatan Masyarakat, 3(3), 198–205.
12. Notoatmodjo S. 2012. Health research methodology. Rineka Cipta, Jakarta.
13. Nurullita U., Mifbakhuddin. 2015. Adsorbtion of indoor carbon monoxide (CO) gas with coconut shell activated carbon and durian skin, semarang. The 2nd University Research Coloquium, 297–306.
14. Subekti P. 2009. Effect of the use of exhaust gas absorbing media on air pollution control devices for diesel-engined vehicles, riau. APTEK, 1(1), 1–11.
15. Verlina W.O.V. 2014. The potential of coconut shell activated charcoal as an adsorbent of CO, NO, and NOx gas emissions in motor vehicles, Makassar. Ph.D. Thesis, Universitas Hasanudin, Makasar.
16. Wicaksono B. 2011. Processing waste styrofoam, orange peel, and sanseveiria fibers into synthetic yarns of economic value. Ph.D. Thesis, Institut Pertanian Bogor, Bogor.
17. Yuliusman W., Wahyu P., Yulianto S.N. 2013. Selection of adsorbents for carbon monoxide adsorption using isothermal langmuir adsorption model. Reaktor, 14(3), 225–233.

19

Epidemiology of Selected Diseases Related to Air Pollution in Krakow

Agnieszka Petryk[1]

[1] Department of Regional Economics, Cracow University of Economics, ul. Rakowicka 27, 31-510 Kraków, Poland, e-mail: agnieszka.petryk@uek.krakow.pl

ABSTRACT

Excessive concentrations of smog toxicants constitute a critical factor influencing the health quality of inhabitants of urban agglomerations. The exposure to elevated concentrations of suspended particulate matter, sulfur dioxide (SO_2), nitrogen dioxide (NO_2) and carbon monoxide (CO) is associated with an increase of adverse health effects. Prolonged exposure to air pollution, especially from low emission, may be linked not only to a greater risk of developing cancer of the respiratory system, or cardiovascular morbidity but also to the increased frequency of admissions to hospitals and emergency units as well as medical consultations in surgeries and outpatient clinics. The issue was investigated using the city of Cracow as an example. Attempts were made to find the connection between air pollution in the city of Cracow and the frequency of hospitalization and medical consultations of patients (registered in the city) due to lung and bronchial malignancy and cardiac dysrhythmia taking into account the data from the years 2010–2016 submitted by the Malopolska Regional Branch of the National Health Fund.

Keywords: air pollution, smog, low emission, morbidity

INTRODUCTION

Air pollutants monitoring, conducted in terms of assessing their harmfulness to human health in accordance with the EU and national regulations, includes the emission of 11 substances [Holnicki et al. 2014]. The study analyzes four selected substances which are especially prevalent in the Cracow agglomeration: suspended particulate matter (PM_{10}, $PM_{2,5}$), sulfur dioxide (SO_2), nitrogen dioxide (NO_2) and carbon monoxide (CO). Particular attention is paid to the threats arising from the emission of suspended particulate matter due to the fact that its concentration levels show the highest values among the four aforementioned toxicants and for this reason they arise public interest. For the conducted analyses, the basic thesis is that air polluted with low emission products poses a potential threat to the health of users and city dwellers. The low-emission sources such as urban transport and other transport pollution, the heating system of detached houses based on the combustion of coal and municipal waste, or local boiler rooms, rather than cogeneration plants, steel plants or other industrial plants, are the main source of air pollution in urban agglomerations [Bagieński and Jaskulska 2016; Report 2010; Petryk 2017a; Petryk 2017b]. This is related to the fact that the emission of pollutants from large industrial, heating, energy and transport sources is subject to numerous legal acts and to strict control, contrary to the uncontrolled practice of burning fuels and waste in household furnaces, which, however, is gradually normalizing [Juda-Rezler 2016]. For this reason, the emission of pollutants from low sources (so-called low emission) has a greater influence on air quality at the ground level of the atmosphere where fuels of low quality are often burnt, rather than from high sources (steel plants, cogeneration plants, etc.) where the quality of fuels used is incomparably higher, and in addition they are obligatorily equipped with high efficiency dust collectors [Juda-Rezler 2016]. In Poland, the most commonly used device for heating detached houses "is a boiler or coal furnace, less often a fireplace or a boiler fuelled with wood or pallets or other types of biomass, whereas gas, oil boilers or electric heating are the rarest" [Degórska 2016]. The sources of low emission show multiple variations, depending on the chemical composition of pollutants and their emission intensity as well as the technical conditions of the emitter (height, diameter, etc.) [Juda-Rezler 2000].

SOURCES AND IMPACT OF AIR POLLUTANTS ON HUMAN HEALTH

The analysis of the related literature does not cast any doubt that the occurrence of some diseases is directly or at least indirectly associated with the pathogenic exposure to air pollution, especially in urban agglomerations. The relationship between the air pollutants and their impact on the health of inhabitants of urban agglomerations, especially metropolitan cities, concerns not only the morbidity of respiratory and immune system diseases or neurological symptoms, but also the prevalence of prematurity, spontaneous abortion, low birth weight and birth defects of newborns that lower their intellectual and immunological potential in subsequent developmental stages [Report on the state of health 2011].

Low emission of suspended particulate matter from the sources of municipal and household sector as well as from the so-called secondary emission related to the negligence of city cleanliness and the growing intensity of car transport is a real problem in Polish cities [Wnuka 2010]. Suspended particulate matter has its toxic impact on significant areas due to its cross-border nature caused by meteorological factors that may stimulate the spread of pollutants by air or restrict air ventilation in rural and urbanized areas [Kozłowska et al. 2011; Wnuka 2010]. Primary particulate matter emission comes from natural and anthropogenic sources, while the particulate matter generated as a result of chemical reactions becomes a secondary toxicant in the atmospheric air [Degórska 2016].

The number of research projects in which confirmation or exclusion of dependence between high concentrations of suspended particulate matter and the higher number diagnoses of specific diseases in people susceptible to short or long-term exposure to these toxicants is increasing. Among the effects of short-term exposure to the absorption of higher concentrations of suspended particulate matter by the body, acute responses of the respiratory system were found in individual age categories. Long-term exposure, even to relatively low concentrations of particulate matter, over many years leads to the development of chronic diseases, including cancer, especially lung cancer, as confirmed by the International Agency for Research on Cancer (IARC) [Skotak 2016].

The prenatal and perinatal exposure to toxic effects of suspended particulate matter may lead to fetal development disorders and cause perinatal complications; it also poses a higher risk of developing allergies, diabetes, asthma, nervous system diseases, etc. in a child's adult life [Skotak 2016]. The risk of recurrent pneumonia is even three times higher in the children whose mothers were exposed to high levels of PM2.5 during pregnancy [Warczyński 2016]. In etiopathogenesis of diseases, both the finest particulate matter fractions and the concentration of its individual constituent substances can be critical. It is believed that it is the content of hydrocarbons or transition metals exert more influence on the development of asthma than the total concentration of particulate matter [Jędrak, Badyda and Konduracka 2016]. Physical activity, which intensifies lung ventilation, exposes the people doing sport under the conditions of higher particulate matter concentration to the inhalation of higher content of the toxicant [Malec and Borowski 2016]. There is no safe level of concentration below which there are no adverse health effects of exposure to this toxicant [Skotak 2016].

The number of medical consultations as well as absenteeism in schools and workplaces rises when the concentration of suspended particulate is relatively low. Higher concentrations may determine the increase in the number of people being hospitalized as well as in the number of deaths [Skotak 2016]. Every year in Poland, more than fifteen times more people die from the diseases related to air pollution than from road accidents. This is estimated at forty-five thousand people a year [Warczyński 2016].

Black smog occurring in industrial areas in the winter is particularly dangerous for human health [Juda-Rezler 2000] due to the high concentrations of sulphur dioxide. Sulphur dioxide (SO_2) of anthropogenic origin is emitted from municipal and household sources, and above all from the energy and industrial sources associated with the extraction, processing and combustion of fossil fuels, in particular, hard coal in Poland [Kordylewski 2008]. Exposure to high concentrations of the colourless gas characterized by a choking odour is fraught with a high risk of airway damage [Kostrz and Satora 2017].

In the case of nitrogen dioxide (NO_2), there is an upward trend in its anthropogenic emission from urban transport sources, with a noticeable reduction in the production of nitrogen oxides in the energy industry [Juda-Rezler 2000]. On the one hand, this is due to the increasing share of cars in urban transport, on the other, it results from the use of modern technologies in the power industry [Juda-Rezler 2000]. Exhaust fumes constitute the main source of carbon monoxide (CO) emission although combustion of fuels in the municipal and household sector, accounting for nearly 50% of its national emissions plays also a significant role [Zakrzewski 1995; Juda-Rezler 2000; Hławiczka 2008].

AIR POLLUTION IN CRACOW

Road transport is a critical source of particulate matter emission. Not only is the dust caused by the combustion of fuels in car engines, but also by the wear of brake pads and the abrasion of tires and road surfaces, including re-suspension, i.e. re-lifting of road dust during road traffic [Degórska 2016]. According to the data of the Department of Municipal Economy and Environmental Protection of the Municipal Office of Cracow, collected on the basis of vehicular traffic measurements, approximately 40,000 vehicles pass through Cracow through transit traffic every day, and the total number of people entering the city daily amounts to 246 thousand. The most traffic-laden are: Kościuszko barrage, the Dębnicki Bridge, as well as ul. Wielicka and Zakopiańska ["Gazeta Krakowska" 2017].

It was estimated that several hundred people die due to air pollution in Cracow per year. "The average number of avoidable deaths per year was estimated for 2001 in relation to the level of pollution reduction (average annual PM2.5). The reduction to the level of 25 μg/m³ would decrease the number of deaths on average by 232, to the level of 20 μg/m³ by 362, to the level of 15 μg/m³ by 492, and to the level of 10 μg/m³ (WHO recommendation) by 612. In addition, it was estimated that a statistical resident of Cracow would live one year longer, on average, if the mean annual concentration of particulate matter PM2.5 could be reduced to 15 μg/m³" [bip.malopolska.pl 2017].

The data from the Krakow Smog Alert indicated that the daily dose of benzo[*a*]pyrene absorbed by residents of Krakow during the year 2011 corresponded to the smoking an average of seven cigarettes [Pankowska, Gorczyca 2015].

The measurements carried out in the air quality control stations at the main communication routes in Cracow showed that the permissible level of PM10 particulate matter concentration had been exceeded in the years 2010–2015 (Table 1). Despite the exceedance of standards, a decrease in the concentration of average annual PM10 in the analyzed five-year period was observed at all measuring points.

In the same period, the values of average annual PM 2.5 concentration also exceeded the permissible standards (Table 2). As in the case of PM 10, a downward trend of PM 2.5 concentrations was noticeable in the years 2011–2015. In 2016, in the Cracow agglomeration, more than 35 exceedances of the permissible standard of PM10 concentrations were observed at all monitoring stations within 24-hour period (Table 3).

Table 1. The concentration of average annual PM 10 at air quality monitoring stations in Cracow in the years 2011–2015

Location of the measuring station	Concentration of average annual PM 10 [μg/m³] Permissible level (since 2005) 40:μg/m³				
	2011	2012	2013	2014	2015
Ul. Bujakowa	54	53	44	46	45
Ul. Bulwarowa	77	66	60	64	68
Al. Krasińskiego	63	51	49	49	52

Source: Prepared on the basis of: Annex No. 1 to Resolution No. XXXII / 451/17 of the Regional Assembly of the Malopolska Region of 23 January 2017.

Table 2. The concentration of average annual PM 2.5 at air quality monitoring stations in Cracow in the years 2011–2015

Location of the measuring station	Concentration of average annual PM 2.5 [μg/m^3] Permissible level (since 2005):25 μg/m^3, (since 2020): 25 μg/m^3				
	2011	2012	2013	2014	2015
Ul. Bujakowa	37	41	33	33	34
Ul. Bulwarowa	42	38	35	32	33
Al. Krasińskiego	55	47	44	45	44

Source: Prepared on the basis of: Annex No. 1 to Resolution No. XXXII / 451/17 of the Regional Assembly of the Malopolska Region of 23 January 2017.

Table 3. List of stations in the area of the Agglomeration of Cracow in 2016 in which there were more than 35 exceedances of levels of permissible concentrations of PM10 – an average of 24 hours (D24) and the number of cases of exceedances at individual stations (health protection)

Zone name	Station code	Number of exceedances D24	The maximum daily concentration value* μg/m^3
The Agglomeration of Cracow	Aleje Krasickiego	165	329
	Ul. Bujaka	78	295
	Ul. Bulwarowa	74	275
	Ul. Dietla	118	314
	Os. Piastów	69	203
	Złoty Róg	85	231

* The number of stations from which the results were used in the assessment for 2016
Source: own study based on the Air quality assessment in zones in Poland for 2016

AIR POLLUTION AND INCIDENCES OF THE RESPIRATORY SYSTEM CANCERS, BRONCHI AND LUNGS CANCERS AS WELL AS CARDIAC DYSRHYTHMIA IN CRACOW

Malignancies caused by environmental pollution, mainly air pollution, constitute nearly 2% of the diagnosed malignancies, although this index is even higher in the populations of the Małopolskie and Silesia region [Kapka et al. 2009]. In the urban population, lung cancer is diagnosed more often than in the rural population [Gładka and Zatoński 2016]. In the population of Cracow, this risk increases by 4.5% in the male group and 10.5% in the female group [Kapka et al. 2009]. Air pollution is a critical factor in the development of lung cancer, which is second only to smoking [Gładka and Zatoński 2016, Gładysz et al. 2010]. It is believed that the abnormal concentration of nitrogen dioxide has the largest pathogenic impact on the occurrence of lung cancer [Gładka and Zatonski 2016]. People exposed to prolonged exposure of suspended particulate matter in the air are also highly susceptible due to the fact that the period of lung cancer development under the conditions of intense particulate matter exposure is by four years shorter than in the case of other carcinogenic substances that pollute the air [Cembrzyńska and in 2012]. Household stoves, in which plastic wastes are burnt, are one of the main emitters of toxicants with a mutagenic and cytotoxic effect [Jędrak et al. 2016].

In the years 2010–2016, men comprised the largest group of the patients treated for malignancy in Cracow; in hospital wards, inpatient emergency departments and admission rooms, the share of men accounted for 62% of the total number of patients treated, while in outpatient clinics it amounted to 60% of the total number of consulted patients (Table 4). Analyzing the diversity of the age structure, it should be stated that men dominated in all age categories. In the years 2011, 2012, 2015 and 2016, no female in the age range of 0–34 applied for treatment in hospital wards, inpatient emergency departments and admission rooms, whereas in outpatient clinics such a situation occurred in 2016. In the male group, the patients in the age range of 0–34 were not treated in hospital wards and in admission rooms in the years 2014 and 2016. There was a noticeable disproportion, both in the female and male group between the number of people in the age range of 0–34 and the total number of people ben-

Table 4. List of patients by sex and age living Cracow treated in the years 2010–2016 for the respiratory system cancers, bronchial and lung malignancy

Name of the disease entity	Year	Total number of medical services	Hospital departments with the inpatient emergency departments and the admission room								Outpatient specialist care clinics contracted by Krakow's hospitals							
			age 0–34		age 35–64		age 65+		age O+		age 0–34		age 35–64		age 65+		age O+	
			M	F	M	F	M	F	M	F	M	F	M	F	M	F	M	F
C-34 (Bronchial and lung malignancy)	2010	4684	2	1	127	67	209	94	338	162	2	2	234	165	352	174	588	341
	2011	4704	1	0	135	69	193	105	329	174	1	2	218	162	353	188	572	352
	2012	5495	2	0	124	67	210	115	336	182	4	2	229	163	366	215	599	380
	2013	5232	1	2	100	80	190	120	291	202	1	2	219	181	371	228	591	411
	2014	5443	0	1	96	74	230	144	326	219	2	3	200	157	402	266	604	426
	2015	5246	2	0	96	70	202	127	300	197	2	1	194	157	406	271	602	429
	2016	5697	0	0	92	65	217	161	309	226	1	0	177	137	430	322	608	459

Source: Own study based on data from the *Malopolska* Provincial Branch of the *National Health Fund* in Krakow obtained applying the public information act (correspondence of 6 July 2017).

efiting from medical services from the other two age ranges of 35–64 and 65+. The morbidity of bronchial and lung malignancies then mainly affected the people over 35 years of age, with a significant predominance of male patients; however, the highest morbidity rate was revealed in people over 65 years of age. In the years 2010–2016, the percentage share of male patients in the age range of 0–34 in the group of all male patients treated for malignant lung cancer and bronchial carcinoma in hospital wards and admission rooms amounted to 0.36%, whereas in outpatient clinics to 0.31%. On the other hand, the percentage share of men in the age range of 65+ treated in the analyzed six-year period in hospital wards and admission rooms accounted for 65%, and in outpatient clinics 64%.

Analyzing the number of medical services provided to patients with bronchial and lung cancer in the reporting period, it should be noted that in comparison to 2010, by 2016 the number of benefits had increased by 1013. In the years 2010–2012, there was an upward trend, which in the next three years was at constant level. Since 2015, there has been a visible, dynamic increase in the number of medical services (Table 4).

The research confirmed that even a short-term exposure to suspended particulate matter may cause electrophysiological changes in the heart that result in cardiac dysrhythmia and intensify susceptibility to this type of heart anomalies [Kowalska and Kocot 2016]. In addition to cardiac dysrhythmia, the consequences of being exposed to air pollution in various time ranges, especially to suspended particulate matter and carbon dioxide, are also indicated by cardiac insufficiency and ischemic heart disease [Kowalska and Krzych 2007; Malec and Borowski 2016]. Obese and older people (over 65 years of age), patients treated for obstructive pulmonary disease or diabetes, and those with chronic respiratory and cardiovascular diseases are listed in the high-risk group [Jędrak et al. 2016 Kowalska and Kocot 2016]. A long-term exposure to low-emission toxicants is also a pathogenic factor of the inflammatory system in young healthy people [*Annex No. 1 to Resolution No. XXXII/451/17 of the Regional Assembly of the Malopolska Region of 23 January 2017*]. Excessive concentrations of smog toxicants may be responsible for a larger number of urgent hospitalizations caused by the onset or exacerbation of cardiovascular disease symptoms [Kowalska and Krzych 2007]. "It was also proven that the increase in the concentration of PM2.5 by only 10ug/m^3 in a short time (<24 hours) r the relative risk of cardiovascular deaths from 0.4% to 1.0%" [Jędrak, Badyda and Konduracka 2016].

In the period from 2010 to 2016 (Table 5), women constituted the largest group of patients treated in Cracow due to dysrhythmias in both hospital wards, inpatient emergency departments and admission rooms (65% of total patients), as well as in outpatient clinics (69%). In the distribution of the age structure, there was a strong dominance of women in all age ranges. In the group of men treated in hospital wards, inpatient emergency departments and in admission rooms, patients over 65 comprised the largest group (37%). In turn, in the group of men treated in outpatient clinics, the majority of patients were in the age range of 0–34 (48%). A small number of men aged 65+ who had medical consultations due to heart rhythm disorders in outpatient clinics (15%) was evident.

Table 5. List of patients by sex and age living Cracow treated in the years 2010–2016 for cardiac dysrhythmia

Name of the disease entity	Year	Total number of medical services	Hospital departments with the inpatient emergency departments and the admission room								Outpatient specialist care clinics contracted by Krakow's hospitals							
			age 0-34		age 35-64		age 65+		age O+		age 0-34		age 35-64		age 65+		age O+	
			M	F	M	F	M	F	M	F	M	F	M	F	M	F	M	F
R00 (cardiac arrhythmia)	2010	414	18	19	14	35	13	32	45	86	25	54	21	60	2	17	48	131
	2011	388	20	32	18	42	16	37	54	111	25	33	10	56	7	14	42	103
	2012	457	17	33	23	42	30	54	70	129	32	39	16	52	5	15	53	106
	2013	587	37	48	34	84	37	58	108	190	27	38	14	61	9	22	50	121
	2014	590	35	50	42	70	45	69	122	189	32	47	26	65	11	15	69	127
	2015	683	35	53	43	116	44	87	122	256	22	43	28	76	9	22	59	141
	2016	611	37	67	30	70	53	76	120	213	24	31	28	71	17	32	69	134

Source: Own study based on data from the *Malopolska* Provincial Branch of the *National Health Fund* in Krakow obtained applying the public information act (correspondence of 6 July 2017).

Among the women subject to treatment in hospital wards and admission rooms, female patients in the age range of 35–64 comprised the largest group (39%); a similar trend was in outpatient clinics (51%). In the analyzed six-year period, a relatively constant upward tendency in the number of medical services provided to the patients with arrhythmias was noticed. In 2016, the number increased by 197 services compared to 2010.

CONCLUSIONS

1. In the years 2010–2016, men constituted the largest group of patients treated for malignancy in Cracow. The morbidity of lung and bronchial malignancy mainly concerned the people over 35 years of age, with a significant majority of male patients. The highest morbidity was found in people over 65 years of age.
2. In the period from 2010 to 2016, women constituted the largest group of patients treated for dysrhythmias. In the distribution of the age structure, strong dominance of women in all age ranges was visible at that time.
3. The latest literature data show that the human exposure to suspended particulate matter (PM10 and PM2.5), sulfur dioxide (SO_2), nitrogen dioxide (NO_2) and carbon monoxide (CO) is associated with an increased risk of cardiac arrhythmia, lung and bronchial malignancy as well as asthma.
4. The analysis of the state of air pollution: PM10 and PM2.5, SO_2, NO_2, CO showed that there were unfavorable weather conditions in the city of Cracow, resulting in long-term excessive concentrations of low-emission toxicants.

REFERENCES

1. Bagieński Z., Jaskulska J. 2016. Limiting the impact of heat supply sources on air quality in areas of intensive low-rise buildings. Atmospheric air. Quality, health, protection. Wroclaw University of Technology. Wroclaw [in Polish].
2. Cembrzyńska J., Krakowiak E., Brewczyński P.Z. 2012. Air pollution with suspended particulate matter PM10 and PM 2.5 in conditions of strong anthropopressure taking of the city of Sosnowiec as an example. Medycyna Środowiskowa, 15(4), 31–38 [in Polish].
3. Cholewiński M., Kamiński M., Pospolita W. 2016. Threats to human health and life from the use of selected fuels in individual heating systems. Kosmos, Problemy Nauk Biologicznych, 65(4), 477–486 [in Polish].
4. Czornik M. 2004. City. Economic aspects of functioning. Publisher of the University of Economics in Katowice [in Polish].
5. Degórska A. 2016. Sources of air pollution with particulate matter. In: K. Juda-Rezler and B.Toczko (Eds). Fine particulate matter in the atmosphere. Compendium of knowledge about air pollution with suspended particulate matter in Poland. Library of Environment Monitoring. Environmental Protection Inspection. Warsaw, pp. 22–26 [in Polish].
6. Gazeta Krakowska, 2017. http://www.gazetakrakowska.pl/motofakty/na-drogach/a/do-krakowa-wjezdza-dziennie-cwierc-miliona-samochodow-raport,12788292/ (retrieved on 18 Februrary 2018).
7. Gładka A., Zatoński T. 2016. The influence of air pollution on respiratory diseases. Kosmos. Problemy Nauk Biologicznych, 65(4), pp. 573–582 [in Polish].
8. Gładysz J., Grzesiak A., Nieradko-Iwanicka B., Borzęcki A. 2010. The impact of air pollution on the health and life expectancy of people. Problemy Hi-

gieny i Epidemiologii, 91(2), 178–180 [in Polish].

9. Hławiczka S. 2008. Heavy metals in the environment. Institute for Ecology of Industrial Areas Ekonomia i Środowisko Publishing Houe, Białystok [in Polish].
10. Holnicki P., Kałuszko A., Stankiweicz K. 2014. Supporting management of air qualityin an urban area. Współczesne Problemy Zarządzania, No 1, 26–27 [in Polish].
11. HWO, 2016. HWO's Global Urban Ambient Air Pollution Database. Update 2016. http://www.who.int/phe/health_topics/en/ (retrieved on 7 August 2017).
12. Jarosiński J. 1996. Clean combustion techniques. Scientific and Technical Publisher, Warsaw [in Polish].
13. Jędrak J., Badyda A.J. Konduracka E. 2016. The impact of air pollution on human health. Sustainable cities. Życie w zdrowej atmosferze, pp. 87–96, [in Polish].
14. Jędrak J., Konduracka E., Badyda A.J., Dąbrowski P. 2016. The impact of air pollution on health. Krakow Smog Alert. Krakow. [in Polish].
15. Juda-Rezler K. [2000]. Environmental Impact of Air Pollution. Warsaw University of Technology Publishers, Warsaw [in Polish].
16. Juda-Rezler K. 2016. Fine particulate matter in the atmosphere. Compendium of knowledge about air pollution with suspended particulate matter in Poland.K. Juda-Rezler and B.Toczko (eds). Library of environment monitoring. Environmental Protection Inspection. Warsaw, pp. 9–10 [in Polish].
17. Kapka L., Zemła B.F., Kozłwoska A., Olewińska E., Pawlas N. 2009. Air quality and the morbidity of lung cancer in selected cities and counties of the Silesia Voivodship. Przegląd Epidemiologiczny, No 63, 439–444 [in Polish].
18. Klich J. 2017. Institutional potential of health care in cities. Part III. Contemporary problems related to urban development. In: M. Kudłacz, J. Hausner (Eds). Funkcjonowanie metropolii w Polsce-Gospodarka, Przestrzeń, Społeczeństwo. CeDeWu Publisher. Warsaw, pp. 237–253 [in Polish].
19. Kordylewski W. 2008. Combustion and Fuel. Wroclaw University of Technology. Wroclaw 2008 [in Polish].
20. Kostrz M., Satora P. 2017. Compounds responsible for air pollution. Inżynieria Ekologiczna, 18, 89–95 [in Polish].
21. Kowalska M., Krzych Ł. 2007. The influence of air polluted with particulate matter and sulfur dioxide on arterial blood pressure – current state of knowledge. Arterial Hypertension, 11(5), 435–442 [in Polish].
22. Kowlaska M., Kocot K. 2016. Short-term exposure to fine suspended particulate matter (PM2.5 and PM10) and the risk of arrhythmias and strokes. Postępy Higieny i Medycyny Doświadczalnej, No 70, 1017–1025 [in Polish].
23. Kozlowska-Szczęsna T. 2004. Influence of the atmospheric environment on health and human well-being. Institute of Geography and Spatial Organization, Warsaw [in Polish].
24. Kozłowska A., Olewińska E., Kolwska-Pawlak A., Pawlas N. 2011. The presence of mutagenic and cytotoxic pollutants in the PM10 and PM2.5 atmospheric aerosol fractions in the town of Sosnowiec. Medycyna Środowiska, 14(4), 21–33.
25. Kozłowska-Szczęsna T., Krawczyk B., Kuchcik M. 2004. Influence of the atmospheric environment on health and human well-being. Stanisław Leszczycki Institute of Geography and Spatial Organization State Academy of Sciences, Warsaw.
26. Lachman P. 2015. Causes of pollutant emission exceedances. Czysta Energia, No 12, 26–29.
27. Malec A., Borowski G. 2016. Particulate matter threats and air monitoring. Inżynieria Ekologiczna, vol. 50, 161–170 [in Polish].
28. Supreme Audit Office (NIK) 2014. Information on the results of the inspection. Air protection against pollution.Warsaw [in Polish].
29. Pałasz J.W. 2016. Low emission from coal combustion and methods of its reduction. Publisher of the Silesian University of Technology. Gliwice [in Polish].
30. Pankowska M., Gorczyca M. 2015. Air protection against pollution – why is it not possible to take a deep breath in Poland? Kontrola Państwowa, 3(362), 60–79 [in Polish].
31. State Environmental Monitoring, 2017. Air quality assessment in zones in Poland for 2016. Environmental Protection Inspection. Attachment A-D. Warsaw [in Polish].
32. Parysek J.J. 2015. The city in terms of system. Ruch prawniczy, ekonomiczny i socjologiczny. Ch. LXXVII/1, 28–40 [in Polish].
33. Petryk A. 2017a. Low-carbon Economy in the Municipality of Sucha Beskidzka. Journal of Ecological Engineering, 18(5), 144–150.
34. Petryk A. 2017b. Development of heating infrastructure as an opportunity to reduce low emission in Krakow. In: T. Kudłacz, P. Brańska (Eds) Studia Komitetu Przestrzennego Zagospodarowania Kraju PAN, vol. 174, 412–421 [in Polish].
35. Air quality plan for the Malopolska Region. Małopolska in a healthy atmosphere. Annex No. 1 to Resolution No. XXXII/451/17 of the Regional Assembly of the Malopolska Region of 23 January 2017. Office of the Marshal of the Małopolskie Voivodship 2017 [in Polish].

36. Report on the state of the environment in Poland, 2010. Chief Inspectorate for Environmental Protection, Library of Environment Monitoring [in Polish].

37. Report on the state of health in the Malopolskie Voivodship, 2011 [in Polish].

38. Regulski J. 1986. Town planning. State Economic Publisher. Warsaw [in Polish].

39. Schwartz J., Laden F., Zanobetti A. 2002. The concentration-response relations between PM25 and daily deaths. Environmental Health Perspectives, 110(10), 1025–1029.

40. Skotak K. 2016. The impact of suspended particulate matter on the environment. Fine particulate matter in the atmosphere. In: K. Juda-Rezler and B.Toczko (Eds). Compendium of knowledge about air pollution with suspended particulate matter in Poland. Library of Environment Monitoring. Environmental Protection Inspection. Warsaw, pp. 75–83 [in Polish].

41. Szymańska D. 2007. Urbanization in the world. PWN, Warsaw [in Polish].

42. Municipality of Krakow (UMK) 2014. Changes in the Study of conditions and directions of spatial development of the city of Krakow. Volume 1, Conditions, Krakow.

43. Justification of the resolution No XVIII/243/16 of Regional Assembly of the Malopolska Region of 15 January 2016 on the introduction, in the area of the Municipality of Krakow, of restrictions on the operation of installations in which combustion takes place on https://bip.malopolska.pl/umwm,a,1159347,uchwala-nr-xviii24316-sejmiku-wojewodztwa-malopolskiego-z-dnia-15-stycznia-2016-r-w-sprawie-wprowadz.html (retrieved on 12 November 2017).

44. Warczyński P. 2016. The impact of air quality on the health of Poles. Sustainable cities. Życie w zdrowej atmosferze, pp. 20–23 [in Polish].

45. Wnuka Z. 2010. Ecology and environmental protection: selected issues. Publishing House of the University of Rzeszów, Rzeszów [in Polish].

46. Zakrzewski S.F. 1995. Outline of environmental toxicology. PWN Warsaw [in Polish].

20

Examination of Dust Concentration in the Interior Air of a Teaching Room

Halina Marczak[1]

[1] Institute of Transport, Internal Combustion Engines and Ecology, Lublin University of Technology, Nadbystrzycka 36, 20-618 Lublin, Poland, e-mail: h.marczak@pollub.pl

ABSTRACT

The issue of indoor air quality is becoming increasingly important. Ensuring optimal air parameters is important for the health of people who are particularly sensitive to impurities contained therein. Finding solutions that improve the quality of indoor air requires a prior detailed recognition of the problem of its pollution. The subject of the research is dust pollution in the air in a teaching laboratory room. The attention was focused on the characteristics of air pollution with the total dust fraction and suspended particulate fractions PM10 and PM2.5. The dust content was measured with a laser meter enabling indirect measurement. It consists in using the phenomenon of damping laser light scattered on the tested air sample at an angle of 90 degrees. As a result of the conducted research, factors influencing the increase of dust concentration in the indoor air were identified. These include the quality of the outdoor air and the source of dust emission in the room. In order to obtain more comprehensive data, it is necessary to continue the research with the possibility of extending the duration of the measurement tests and the test cycle.

Keywords: dust pollution, PM10, respirable fraction, suspended dust fractions, dust concentration in the interior air

INTRODUCTION

Air quality has a direct impact on human health. Poor air quality can cause adverse health effects, especially respiratory and circulatory system diseases. The respiratory system is one of the basic ways of transferring pollutants to the human body.

Dust is one of the pollutants in the air. The harmfulness of dust to the human body depends on its concentration, grain size (the smaller, the more harmful), shape (circular dust is less harmful), chemical composition, crystalline structure, dust solubility in body fluids and time of impact. Dust contamination, depending on its type, can cause irritating, fibrotic effects (hyperplasia of the fibrous tissue in the lungs), sensitizing or toxic reactions. In people with bronchial asthma, exposure to dust pollution may increase the severity of asthma attacks. The respirable fraction of the dust shows the ability to pass through the alveolar walls, then into the pulmonary blood vessels, and then into the entire human blood system. As a consequence, it can affect the development of atherosclerosis and related cardiovascular diseases.

The article presents the results of dust concentration measurements in a teaching room. The main goal of the research is to obtain data and information on air pollution in teaching rooms and determine whether chalk dust is a factor significantly affecting the quality of indoor air.

CHALK DUST AS AN AIR POLLUTANT

The exact explanation of the term dust and the definition of the dust fraction are included in Polish standards regarding air quality [PN-ISO 4225:1999, PN-ISO 7708:2001]. The inhalable fraction and the respirable fraction of dust are also explained in the Regulation of the Minister of Labor and Social Policy on the highest permissible concentrations and intensities of harmful factors in the work environment [Announcement

Dz.U.2017, poz. 1348]. The inhalable fraction is the aerosol fraction penetrating the nose and mouth, which poses a health risk when deposited in the airways. The definition of the inhalable fraction corresponds to the definition of total dust. The respirable fraction is the aerosol fraction entering the respiratory tract, which poses a risk when deposited in the gas exchange area.

Among the harmful dust factors in the work environment which are mentioned in the Regulation, there is a factor: other non-toxic industrial dusts – including those containing free silica below 2%. This category of dusts includes chalk dust. It is included in the group of harmful factors, because it can be irritant or allergic to people. Most often, it has irritating effects on the conjunctiva of the eyes, mucous membranes of the upper respiratory tract and throat. The most harmful to health are fine particles of dust, because they have the ability to penetrate and settle in the area of gas exchange. Dusts with aerodynamic diameters of up to 5 μm (respirable fraction) reach the alveoli [Jakubowski 2018]. They can cause pulmonary disease, including bronchial asthma, when located in the lungs. On the other hand, dust particles with aerodynamic diameters above 5 μm (unresponsible fraction) are retained in the upper respiratory tract or in the tracheobronchial area, from where they are removed with the participation of ciliated and mucus-secreting epithelial cells. Inhaled dust can sometimes interfere with the ability of the respiratory system to remove dust particles, which may be the cause of chronic bronchitis.

Chalk dust may be an air pollutant inside the teaching rooms. The source of emission of this dust is abrasion, especially dry sponge or chalk from the blackboard. Chalk dust belongs to non-toxic dust due to the chemical composition of the chalk. It is a mixture of chalk and gypsum. Writing chalk is a variation of fairly clean limestone. It consists predominantly of calcium carbonate in a form of calcite. The composition of the mass-writing chalk according to Olszewski [1998] is as follows: over 92% $CaCO_3$, from 4 to 5% silica, about 2% clay minerals, about 0.8% other mineral admixtures.

The basic quantitative measure of atmospheric and internal air dustiness (in work rooms and other places) is the dust concentration (S), which is the ratio of the mass or volume of dust particles to the volume or less often the air mass:

$$S = \frac{amount\ of\ dust\ (mass\ or\ volume)}{amount\ of\ air\ (mass\ or\ volume)} \quad (1)$$

On the basis of the dust concentration in the test air, the risk to human health of dust can be assessed.

RESEARCH SUBJECT AND METHODOLOGY

The subject of the research is dust pollution in the air in a laboratory room. The room has an area of about 30 m^2, it is equipped with gravity ventilation and an exhaust fan turned on periodically as needed, most often when working with chemical reagents. The classroom holds classes for groups of students up to 15 people. The duration of one type of class for each group is 2 teaching hours. During the day, classes in the room are carried out for a maximum of 4 groups.

The research included measurements of total (general) dust and suspended dust [Marczak 2017] with a particle diameter of up to 10 μm (PM10) and up to 2.5 μm (PM2.5). The test cycle consisted of 6 measuring days. On the day of measurements, test samples were collected in succession during classes with each group. Two measurements were taken in each group. The first measurement was carried out about 20 minutes after the beginning of the class. Until the first measurement was taken, didactic classes were conducted without using a chalk board. The interval between the first and second measurements was about 20 minutes. In the interval between measurements, chalk was used on a traditional metal plate and the board was often wiped with a sponge, often dry. Square, white, and small-dust chalk made by Mar-Bor company was used.

The measurements were made using a portable laser dust meter called DustTrak DRX Aerosol Monitor model 8533 (TSI Inc.). The dust meter is characterized by a short sampling time of the gas to be tested and allows the determination of mass concentration of particulate matter with particle sizes in the range of 0.1–15 μm. It can be powered by batteries or from the power socket. It measures the real-time dust concentration based on the laser light attenuation scattered on the test sample at an angle of 90 degrees.

The measuring point was located 1.2 m from the floor (breathing zone) and about 3 m from the chalk board (measured perpendicular to it). Before each measurement, the instrument was calibrated. A single measurement lasted 3 minutes. The flow rate of the air tested by the meter was 3 dm^3/min.

The results of the measurements are the average concentrations of the tested dust fractions for 30 seconds averaging time and maximum, minimum and average values recorded during the test. The result data were saved in the internal memory of the meter. For the time of averaging 30 sec measurement results and the duration of the 3-minute test, six averaged values of dust fraction concentrations were stored in the meter's memory. On their basis, the average dust concentration from the measurement test was determined. The applied measurement method provides information on instantaneous concentrations of dust in the examined air.

RESEARCH RESULTS

The obtained results of mass concentration of total dust, PM10 and PM2.5 are presented in Table 1. The set of results includes the minimum, maximum and average values of concentrations of total dust and PM10 fraction as well as average concentrations of PM2.5. The results of the average dust concentration measurements are illustrated in a graph (Figure 1).

Figure 1 shows that on the first and fifth day of measurement, measurements were taken in 2 laboratory groups taking classes one after another. On the second, fourth and sixth day of the measurement, two measurements were taken, because in each of these days only one group of students had educational classes. The third measurement session consisted of 6 measurements – 2 measurements in each of the three laboratory groups.

In the assessment of the uncertainty (scatter) of the results of the measurement of the average concentration of dust, the uncertainty resulting from the use of measuring instruments (uncertainty of calibration) and random uncertainty were taken into account. The standard uncertainty was used as a quantitative measure of the uncertainty of the measurement results.

In calculating the uncertainty of measurement related to measuring instruments, the manufacturer's data with finite accuracy of measuring instruments is used. The accuracy of analog instruments is equal to the elementary space – the distance (in appropriate units) of two consecutive strokes on the scale of the instrument. In digital instruments, an elementary plot is equal to the unit of the smallest decade (unit of the de-

Table 1. Results of measurements of mass concentration of total dust, PM10 and PM2.5

Measurement session (day)	Lab. group	Measurement hour	Concentration min. [mg/m³]		Concentration max. [mg/m³]		Average concentration (from meter) [mg/m³]		
			Total dust (S_{min})	PM10 ($S_{PM10min}$)	Total dust (S_{max})	PM10 ($S_{PM10max}$)	Total dust (S)	PM10 (S_{PM10})	PM2.5 ($S_{PM2.5}$)
1	G1	12:38	0.083	0.037	0.103	0.04	0.094	0.039	0.028
		13:05	0.189	0.113	0.241	0.134	0.209	0.122	0.063
	G2	15:09	0.024	0.018	0.043	0.022	0.032	0.02	0.016
		15:31	0.161	0.085	0.198	0.101	0.178	0.095	0.051
2	G3	10:45	0.081	0.062	0.092	0.068	0.087	0.065	0.057
		11:14	0.178	0.116	0.2	0.129	0.189	0.121	0.085
3	G4	10:42	0.148	0.136	0.168	0.141	0.157	0.139	0.132
		11:11	0.236	0.183	0.256	0.193	0.249	0.19	0.149
	G5	12:25	0.165	0.140	0.171	0.145	0.168	0.142	0.131
		12:49	0.206	0.167	0.225	0.173	0.218	0.171	0.146
	G6	14:25	0.147	0.131	0.17	0.137	0.159	0.134	0.125
		14:50	0.184	0.15	0.241	0.164	0.211	0.157	0.135
4	G7	12:50	0.063	0.045	0.077	0.053	0.067	0.048	0.04
		13:12	0.127	0.086	0.179	0.104	0.152	0.094	0.063
5	G8	10:41	0.047	0.037	0.099	0.045	0.073	0.037	0.025
		11:04	0.145	0.082	0.228	0.118	0.19	0.103	0.054
	G9	12:40	0.034	0.021	0.054	0.029	0.042	0.024	0.017
		13:03	0.170	0.097	0.212	0.108	0.19	0.101	0.05
6	G10	13:40	0.104	0.09	0.116	0.09	0.11	0.092	0.084
		13:59	0.159	0.126	0.22	0.15	0.185	0.136	0.104

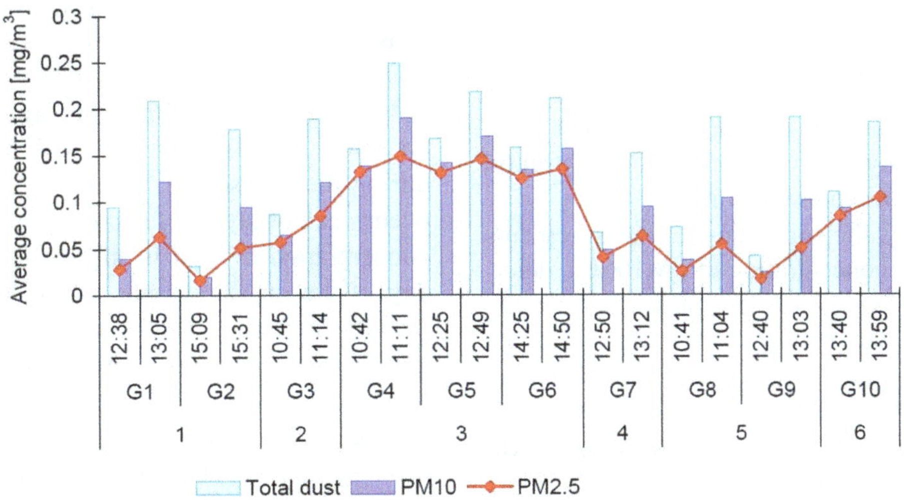

Fig. 1. Medium concentration of dust in the test cycle (12:38 – measurement start time, G1, G2,...,G10 – laboratory groups, 1, 2,...,6 – day of measurements)

cade indicating the smallest measurement value). The dust meter used in the tests has, according to the manufacturer's data, a measuring range of 0.001–150 mg/m³. The elementary plot of the meter is $\Delta_m S = 0.001$ mg/m³. This means that the result of the dust concentration measurement carried out with this meter should not differ from the actual concentration value by more than one elementary scale. In other words, the deviation of the measurement result of the average dust concentration from the actual value can reach up to $\pm\Delta_m S$. This is border uncertainty. In order to convert it to standard uncertainty, it was assumed that the probability of obtaining any value of the total dust concentration in the range ($S-\Delta_m S$; $S+\Delta_m S$) is the same, i.e. it describes the homogeneous distribution (uniform). The standard deviation of the distribution of the homogeneous total dust concentration (D(S)), taking into account the adopted determinations, is determined by the formula:

$$D(S) = \frac{(S + \Delta_m S) - (S - \Delta_m S)}{2\sqrt{3}} = \frac{\Delta_m S}{\sqrt{3}} \tag{2}$$

The standard uncertainty of the measurement of the total dust concentration resulting from the uncertainty of the calibration ($u_m(S)$), equal to the standard deviation of the homogeneous probability distribution, describes the dependence:

$$u_m(S) = \frac{\Delta_m S}{\sqrt{3}} \tag{3}$$

The random uncertainty of the measured value of mean total dust concentration is caused by the instability of external conditions during measurements and the specificity of phenomena occurring in the measuring chamber of the instrument. It is illustrated by the spread of dust concentration results for subsequent averaging times. Random uncertainty ($\Delta_p S$) for each measurement was calculated according to the relationship:

$$\Delta_p S = S_{max} - S_{min} \tag{4}$$

where: S_{min} – the minimum concentration of total dust in the test

S_{max} – maximum concentration of total dust in the test.

The change of random uncertainty $\Delta_p S$ into standard uncertainty was done in the same way as for the uncertainty of calibration. As a result, the standard uncertainty of measuring the average dust concentration resulting from the uncertainty of random measurement ($u_p(S)$) will be equal:

$$u_p(S) = \frac{\Delta_p S}{\sqrt{3}} \tag{5}$$

The standard integer uncertainty was calculated using the following variance property for the independent variables X and Y:

$$D^2(X+Y) = D^2(X) + D^2(Y) \tag{6}$$

Standard deviation will be equal to:

$$D(X+Y) = \sqrt{D^2(X) + D^2(Y)} \tag{7}$$

The uncertainty of the standard total measurement of the mean concentration of total dust is expressed by the following relationship:

$$u(S)=\sqrt{[u_m(S)]^2+[u_p(S)]^2} \qquad (8)$$

where: $u(S)$ – total standard uncertainty.

Standard uncertainty values and total standard uncertainty of average dust concentration measurements are presented in Table 2.

By analogy, the uncertainty of measurements of the average concentration of PM10 was determined, i.e. $u_m(S_{PM10})$ – measurement uncertainty calculated on the basis of the accuracy of the measuring instrument, $u_p(S_{PM10})$ – uncertainty resulting from the statistical distribution of PM10, $u(S_{PM10})$ – uncertainty with the standard total measurement dust concentration of PM10.

The standard uncertainty values and the total standard uncertainty of measurements of the average concentration of PM10 fraction are given in Table 3.

It was observed that the total standard uncertainty shows higher values for the second measurement in each of the G1–G10 groups as compared to the first measurement (Tables 2 and 3). This is due to the larger scatter of instantaneous dust concentration values during the second measurement than during the first measurement.

Before the first measurement, the dustiness in the room was small, although different in individual measuring tests. These differences were caused by a different proportion of secondary pollination in air pollution. Secondary emission consists in raising dust which was previously settled on solid surfaces. Increased activity of students during the course of classes, with simultaneous occurrence of layers of settled dust, increases the concentration of dust in the room. On the third day of the measurements, during the studies in the G4–G6 groups, significantly higher values of total dust concentration were measured in the first measurement compared to the G1, G2 and G7–G10 tests. At the same time, it was found that the results of the first measurement of mean total dust concentration in the G4–G6 trials were similar to each other and were respectively 0.157 mg/m^3, 0.168 mg/m^3 and 0.159 mg/m^3. In the research trials G1, G2, G7-G10, the values of mean total dust concentration ranged from 0.032 mg/m^3 (G2) to 0.11 mg/m^3 (G10). Based on the observations, it can be concluded that the increase in the dust concentration in the first measurement in the G4–G6 trials (3rd day of measurement) was the result of irritating dust settled in the room. This was favored by the increased activity of students in G4–G6 groups in preparation for conducting research as a part of didactic classes.

Table 2. Values of uncertainty of measurements of mean mass concentration of total dust

Measurement session (day)	Lab. group	Measurement hour	Standard uncertainty $u_m(S)$ [mg/m^3]	Random uncertainty $u_p(S)$ [mg/m^3]	Total uncertainty u(S) [mg/m^3]	Measurement result S [mg/m^3]
1	G1	12:38	0.00058	0.01155	0.012	0.094(0.012)
		13:05	0.00058	0.03002	0.030	0.209(0.03)
	G2	15:09	0.00058	0.01097	0.011	0.032(0.011)
		15:31	0.00058	0.02136	0.021	0.178(0.021)
2	G3	10:45	0.00058	0.00635	0.006	0.087(0.006)
		11:14	0.00058	0.01270	0.013	0.189(0.013)
3	G4	10:42	0.00058	0.01155	0.012	0.157(0.002)
		11:11	0.00058	0.01155	0.012	0.249(0.012)
	G5	12:25	0.00058	0.00346	0.004	0.168(0.004)
		12:49	0.00058	0.01097	0.011	0.218(0.011)
	G6	14:25	0.00058	0.01328	0.013	0.159(0.013)
		14:50	0.00058	0.03291	0.033	0.211(0.033)
4	G7	12:50	0.00058	0.00808	0.008	0.067(0.008)
		13:12	0.00058	0.03002	0.030	0.152(0.03)
5	G8	10:41	0.00058	0.03002	0.030	0.073(0.03)
		11:04	0.00058	0.04792	0.048	0.19(0.048)
	G9	12:40	0.00058	0.01155	0.012	0.042(0.012)
		13:03	0.00058	0.02425	0.024	0.19(0.024)
6	G10	13:40	0.00058	0.00693	0.007	0.11(0.007)
		13:59	0.00058	0.03522	0.035	0.185(0.035)

Table 3. Values of uncertainty of measurements of average mass concentration of PM10 dust

Measurement session (day)	Lab. group	Measurement hour	Standard uncertainty $u_m(S)$ [mg/m³]	Random uncertainty $u_p(S)$ [mg/m³]	Total uncertainty u(S) [mg/m³]	Measurement result S [mg/m³]
1	G1	12:38	0.00058	0.00173	0.002	0.039(0.002)
		13:05	0.00058	0.01212	0.012	0.122(0.012)
	G2	15:09	0.00058	0.00231	0.002	0.02(0.002)
		15:31	0.00058	0.00924	0.009	0.095(0.009)
2	G3	10:45	0.00058	0.00346	0.004	0.065(0.004)
		11:14	0.00058	0.00751	0.008	0.121(0.008)
3	G4	10:42	0.00058	0.00289	0.003	0.139(0.003)
		11:11	0.00058	0.00577	0.006	0.19(0.006)
	G5	12:25	0.00058	0.00289	0.003	0.142(0.003)
		12:49	0.00058	0.00346	0.004	0.171(0.004)
	G6	14:25	0.00058	0.00346	0.004	0.134(0.004)
		14:50	0.00058	0.00808	0.008	0.157(0.008)
4	G7	12:50	0.00058	0.00462	0.005	0.048(0.005)
		13:12	0.00058	0.01039	0.01	0.094(0.01)
5	G8	10:41	0.00058	0.00462	0.005	0.037(0.005)
		11:04	0.00058	0.02078	0.021	0.103(0.021)
	G9	12:40	0.00058	0.00462	0.005	0.024(0.005)
		13:03	0.00058	0.00635	0.006	0.101(0.006)
6	G10	13:40	0.00058	0.0	0.001	0.092(0.001)
		13:59	0.00058	0.01386	0.014	0.136(0.014)

A sudden change in environmental conditions in the room – an increase in the content of dust – occurred as a result of writing with chalk and wiping it with a dry sponge of chalk dust from the board. The values of the mass concentration of total dust in the second measurement in the G2 group were higher even by 456.25% from the concentration value in the first measurement (Table 2). The high increase in the total dust concentration in the internal air as a result of chalk dust emission was also recorded during the tests on the fifth day of measurement in the G9 group – about 352% higher than in the first measurement. The smallest impact of chalk dust emission on indoor air quality was found during G5 and G6 measurement tests, in which the concentration of total dust in the second measurement was higher by approx. 30% and approx. 32% respectively than in the first measurement.

The tests showed that the share of the average PM10 fraction in the total dust was from about 41.5% (G1, first measurement) to about 88.5% (G4, first measurement). The PM2.5 fraction constituted from 49.5% (G9, second measurement) to 94.96% (G4, first measurement) of the PM10 dust fraction. A larger share of the average PM2.5 dust concentration in the PM10 dust was observed in the first measurement than in the second measurement in each of the G1–G10 groups. It follows that the dust in the internal air was enriched with PM2.5 fraction before the appearance of the chalk dust source (Figure 2).

On the basis of the presented test results, it can be seen that in each second measurement of test samples G1 to G10 the values of total dust, PM10 and PM2.5 concentrations were higher than in the first measurement. The immediate cause of the increase in the dust content in the indoor air in the didactic room was chalk dust, which arises when writing with a chalk and wiping it with a sponge.

CONCLUSIONS

Before assessing the values of dust concentrations in the indoor air, the measurement results should be compared with the applicable permissible standards or other available data.

The applicable regulations in Poland [Announcement Dz.U. 2017, poz. 1348] determine the values of the highest permissible concentrations of harmful dust agents at work stations. For dust categories: other non-toxic industrial dusts, including those containing free (crystalline) silica below 2%, the maximum permissible concentration of total dust is 10 mg/m³. Taking into account

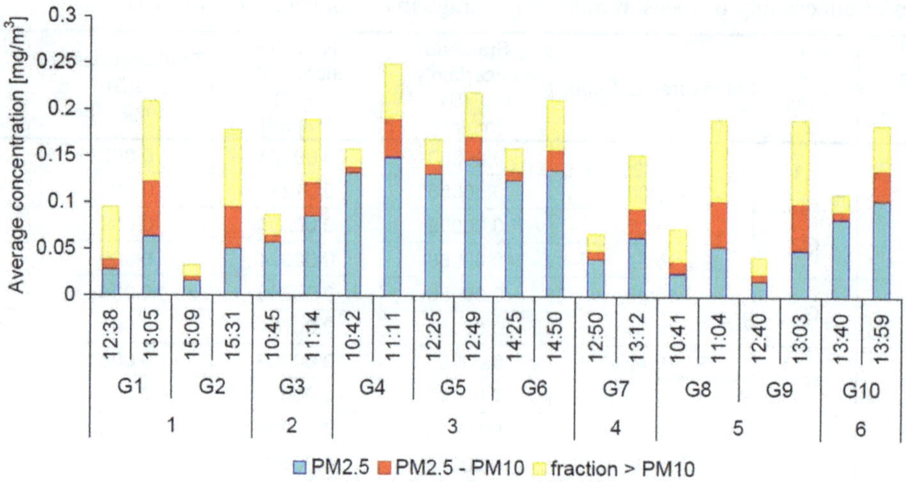

Fig. 2. The average concentration of PM2.5, PM2.5–PM10 and above PM10 dust fraction in the total mean concentration of dust (12:38 – measurement time start, G1, G2,...,G10 – laboratory groups, 1, 2,...,6 – day measurements)

the results of measurements, it can be concluded that the total dust concentration in the tested room does not exceed the given limit. In addition to work positions, there are no legal regulations regarding internal air quality standards.

In EU and Polish law, however, there are acceptable standards for particulate matter PM10 and PM2.5 in the external air [Regulation Dz.U. 2012, poz.1031].

The permissible level for average daily concentrations of PM10 dust is 50 $\mu g/m^3$, and for annual concentrations of 40 $\mu g/m^3$. For PM2.5, the norm is defined only for mean values of annual concentrations – it is 25 $\mu g/m^3$. According to World Health Organization (WHO) recommendations [2006], the daily PM2.5 dust concentration should not exceed 25 $\mu g/m^3$ and annual 10 $\mu g/m^3$. Referring to these standards, it can be said that for 20 measurements of PM2.5, 17 concentrations exceeded the permissible level of 25 $\mu g/m^3$. Analyzing the values of PM10, dust concentrations obtained from measurements, it can be noticed that in 15 measuring samples they were much higher than 50 $\mu g/m^3$.

Research shows that the quality of indoor air in the teaching room is affected by the quality of outdoor air, as well as the source of dust emission in the room. These sources include chalk dust emitted during the use of chalk and the abrasion of chalk dust from the board, as well as dust settled by air movement.

Average concentrations of total dust varied from 0.152–0.294 mg/m^3 with the occurrence of chalk dust emission, while before emission they ranged from 0.042–0.157 mg/m^3. The research also confirmed an increase in the concentration of PM10 and PM2.5 fractions caused by the chalk dust emission source – for PM10 fractions from values in the 0.02–0.142 mg/m^3 range before dust emission to values from 0.094–0.19 mg/m^3 after emission.

For better results, the scope of tests should be extended to include measurements in the conditions of each table wiping with a damp sponge and using the so-called dust-free chalk. In order to assess the inhalation of chalk dust of people writing on the blackboard, it is advisable to measure air samples taken from the breathing zone of these people. Research is also needed to determine the effect of dust concentration, especially the respirable fraction, on the respiratory system of people who stay indoors for a long time with poor air quality. Future tests should extend the duration of the measurement tests and the research cycle to obtain large data sets.

As a result of the research, factors influencing the increase of dust concentration in the indoor air were identified. It allows to indicate actions limiting or eliminating the sources of these factors. These activities include the permanent use of a damp sponge to wipe the traditional board, ensure proper ventilation in the room, or replace traditional chalk boards with dry-wipe magnetic.

REFERENCES

1. Bochentyn B., Kusz B. 2014. Methodology of measurements and uncertainty and measurement errors (in Polish). LAB Laboratoria, Aparatura, Badania, R. 19, 5, 26–35.

2. Gajdek M. 2012. Analysis of measurement results: Methods for estimating measurement uncertainty (in Polish). Politechnika Świętokrzyska, Kielce.
3. Jakubowski J. Health effects of exposure to industrial dust in the work environment (in Polish). http://docplayer.pl/6646363-Zdrowotne-skutki-narazenia-na-pyl-przemyslowy-w-srodowisku-pracy.html (access on 11.02.2018).
4. Marczak H. 2017. Particulate matter in atmospheric air in urban agglomeration. Journal of Ecological Engineering, 18(3), 149–155.
5. Announcement of the Minister of Family, Labor and Social Policy of 7 June 2017 on the publication of a uniform text of the Regulation of the Minister of Labor and Social Policy on the highest permissible concentrations and intensities of factors harmful to health in the work environment (in Polish). Dz.U. 2017, poz. 1348.
6. Olszewski A. 1998. Selected parameters of the microstructure of writing chalk and rocks marl (in Polish). Przegląd Geologiczny, vol. 46, 9, 845–850.
7. PN-ISO 4225:1999. Air quality. General issues. Terminology (in Polish).
8. PN-ISO 7708:2001. Air quality. Definitions of the dust fraction used in the collection of samples for health risk assessment (in Polish).
9. Regulation of the Minister of the Environment of 24 August 2012 on the levels of certain substances in the air (in Polish). Dz.U. 2012, poz.1031.
10. WHO 2006. Air quality guidelines for particulate matter, ozone, nitrogen dioxide and sulfur dioxide. Global Update 2005. Geneva.

21

Effectiveness of Measures taken to Protect Waters by Example of Soda Production Plant in Inowrocław

Adam Gołub[1,*], Janina Piekutin[1]

[1] Bialystok Technical University, Faculty of Civil and Environmental Engineering, ul. Wiejska 45A, 15-350 Białystok, Poland

[*] Corresponding author's e-mail: a.golub@doktoranci.pb.edu.pl

ABSTRACT

The "Soda Mątwy" plant in Inowrocław produces light and heavy soda ash by means of the Solvay method, and the production process is associated with the formation of highly saline waste, which can adversely affect all elements of the environment. In view of the existing threats, attempts are undertaken to limit the quantity and improve the quality of post-production sewage. The aim of the study is to determine the effectiveness of activities to protect the water in the area of the soda production plant in Inowrocław. The assessment of their effectiveness was based on the analysis of the current level of chloride concentration reduction at 4 measurement points around the plant and determination of the trend for further changes. In all piezometers, the concentration of chloride ions has decreased and showed the same tendency in subsequent years, which allows draw conclusions confirming high effectiveness of the activities related to water protection.

Keywords: soda ash; sludge treatment; water quality

INTRODUCTION

There are many industrial areas in the entire country that require reclamation. They constitute a large environmental and ecological threat, because of their storage, i.e. the most common method of waste disposal. Incorrect location and operation of landfills adversely affects all elements of the natural environment (air, plants, groundwater and surface waters, soils). The unsealed bottom of the landfill causes the leachates produced by infiltrating rainwater to penetrate through its subsoil and slopes, becoming a source of long-term pollution of soils and groundwater [Urbańska and Urbański 2012, Wiater 2011]. These leachates affect the receivers in a variety of ways. Some of the components contained in them adversely impact the hydrological conditions, whereas others impair water quality, have a destructive effect on biocenosis and inhibit the self-cleaning processes [Harat and Grmela 2008, Mosiej et al. 2007, Policht-Latawiec et al. 2013]. Therefore, emphasis should be put on limiting the amount of industrial wastewater produced by improving its quality (e.g. using pre-treatment at a given plant or in municipal wastewater treatment plants) and proper water management. The requirement of ensuring proper quality of industrial wastewater was included in the National Program for Municipal Wastewater Treatment [2017].

An example of a strong impact on natural waters includes the plants producing light and heavy soda ash, belonging to the chemical industry. The production of Solvay soda is related to the formation of saline waste [Matthews and Effler 2003, Şener 2008, Steinhauser 2008]. Such plants minimize the negative impact on the natural environment by pre-treating the post-production sewage and preventing their contact with underground and surface water.

This paper presents the diagnosis of groundwater contamination with chloride ions in the "Soda Mątwy" Production Plant in Inowrocław that is a producer of light and heavy soda ash.

The aim of the analysis was to assess the effectiveness of measures limiting the impact of soda industry on the environment.

RESEARCH OBJECT AND METHODS

The Production Plant "Soda Mątwy" in Inowrocław is located within the catchment of the Eastern Noteć river and its tributary (Słony Rów) (Figure 1). The groundwater in the surroundings of the plant is fed from two sources: infiltration of rainwater and inflow from the northern part of Inowrocław. The groundwater flow in the western part of the plant takes place in the south-western direction, and in the south-east (Figure 1).

The post-production residues in the plant contain elevated concentrations various compounds of mainly carbonate, hydroxide and calcium sulfate, magnesium hydroxide and chloride as well as ammonium nitrogen [Matthews and Effler 2003, Şener 2008, Steinhauser 2008]. They are directed to the installation for the production of soda lime, where chlorides are leached out using fresh water. The sludge washed and dehydrated on filter presses, containing less than 3% of chlorides, is transferred for agricultural purposes. Meanwhile, the effluents are discharged to the sedimentary ponds (Figure 2), where their temperature is reduced, and mixing and precipitating the sediments by binding insoluble carbon dioxide from the air occur [Kasikowski et al. 2008, Siwiec et al. 2016]. The collected supernatant water is pumped into the Vistula River [Kasikowski et al. 2007].

In the past, all ponds shown in Figure 2 were used for storing the waste from production and the ashes from a combined heat and power plant. Their total area was over 110 ha. Due to the process of waste utilization from the production launched in 1992, the area of flooded ponds was reduced and the quality of sludge directed to them was improved. Currently, the landfill area in use is around 13 ha.

In order to protect the water, 3 drainage barriers were constructed as well (Figure 2): East (1 and A-I), West (J, K) and Central (L). These barriers, consisting of wells draining the underground water to pressure pipelines to the Vistula river, are aimed at capturing the contaminants that leaked from the area of ponds to groundwater and preventing them from spreading to the adjacent areas [Pietrucin and Czop 2015].

Water for the chloride concentration test, which is an indicator of water pollution around the plants involved in the production of soda ash [Effler and Matthews 2003], was taken from 4 piezometers filtered within the quaternary aquifer (markings 1–4 in Figure 2). The sampling was carried out in accordance with the PN-EN ISO 5667:2008 standard, with the frequency resulting from the plant environmental monitoring program.

For percent values of chloride concentration reduction at each measurement point, a 3-order polynomial trend line was determined along with its equation and the value of coefficient of determination R^2 to determine the change tendencies.

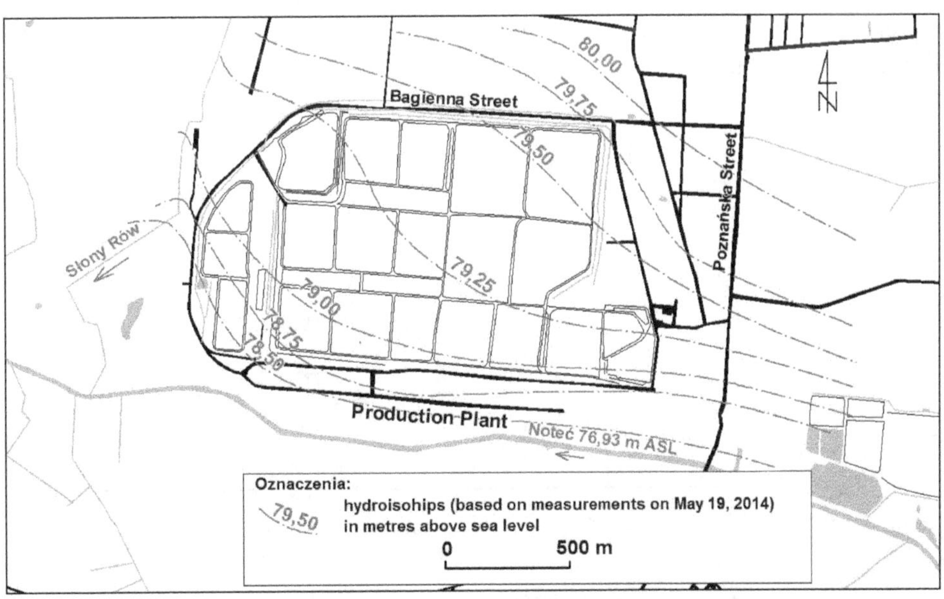

Figure 1. Location of the plant and map of hydroisohips of the Quaternary aquifer

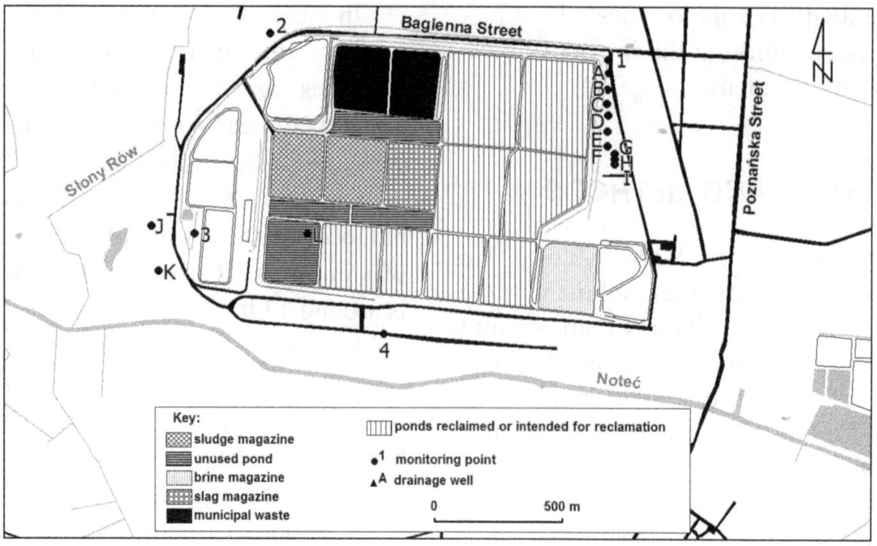

Figure 2. Location of sedimentary ponds, where the 1–4 are measuring points, 1 and A-L are drainage wells

RESULTS AND DISCUSSION

The calculated degrees of chloride ion decrease in individual piezometers over the whole research period, as shown in Figures 3–6. The degree of chloride ions reduction at monitoring point 1 (Figure 3) varies from 1.4 to almost 60%. Over the past 25 years, a regular decline has been noticed and in the last period, about 50% reduction has been achieved. The well has been characterized by irregular work resulting from frequent clogging of the filter ever since it was constructed. An uneven inflow of salinc water and drainage works associated with the reconstruction of the drainage barrier causes large fluctuations in the results between series of measurements. The obtained data suggest that the concentration of chloride ions will drop further in subsequent years. This is due to the trend of line and high R^2 coefficient.

At point 2 (Figure 4), the reduction in the chloride concentration was very rapid, almost reaching 70% in the period from 1995 to 2000. After 2000, however, it was not as intense anymore and amounted to another 20% between 2000 and 2017. During the entire measurement period in this piezometer, the chloride concentration decreased by over 90%. Small variations in the results between the series are characteristic in this case. The coefficient of determination $R^2 = 0.9766$ of the trend line indicates a further drop in the chloride concentration at this point.

At the measurement point 3 (Figure 5), the reduction in chloride concentration is very diverse, and variations between series is substantial. This may result from the varying intensity of pollutants leaching from the soil aeration zone in the area of sedimentary ponds. Therefore, various charges of chlorides reach the piezometer 3, located on the

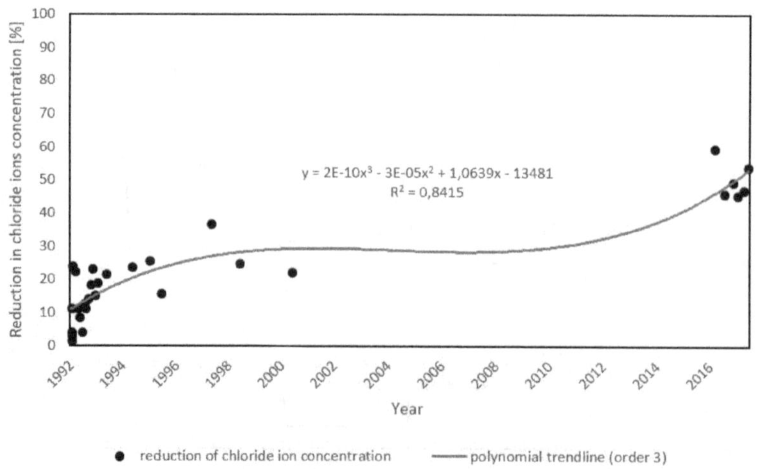

Figure 3. Reduction of chloride ion concentration at point 1

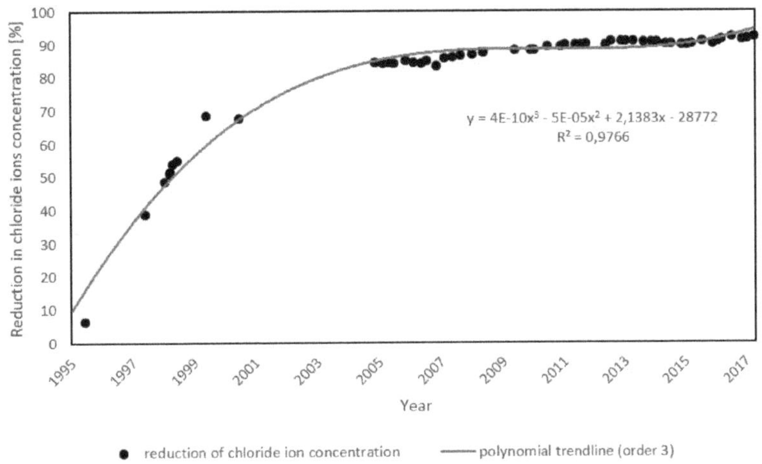

Figure 4. Reduction of chloride ion concentration at point 2

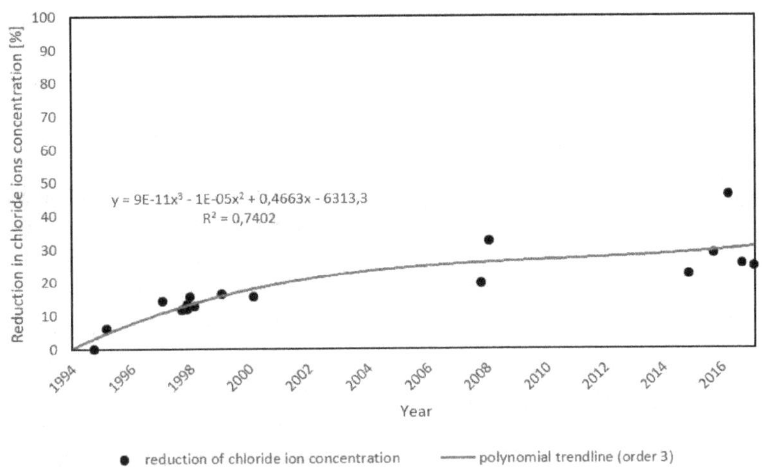

Figure 5. Reduction of chloride ion concentration at point 3

underground water flow from the ponds towards the Noteć river and Słony Rów. The reduction achieved in recent years reaches 30%.

Fluctuations in the reduction of chlorides at point 4 between series are significant; however, there is a clear downward trend (Figure 6). In recent years, the level of chloride ion pollution has decreased by 40%. The coefficient of determination R^2 = 0.9294 indicates a very good model matching [Rabiej 2012], and the trend line suggests further gradual reduction of the pollution level.

The concentration of chloride ions at all described measuring points has decreased since the observation began, which indicates the improvement of the underground water quality in the area of the plant. This is mainly a consequence of the actions that were aimed at reducing the inconvenience of waste to the environment and limiting the release of leachate into waters. These are: launching the process of waste utilization, limiting of its use, sealing and reclamation of sedimentary ponds or construction of drainage barriers.

The pollution accumulated in the soil aeration zone is still an important source of chlorides in water, gradually rinsed with rainwater, infiltrating down the cross profile [Małecki et al. 2016, Wychowaniak and Koda 2014]. It is a process that lasts for many years [Vangelas et al. 2006], but if it happens in such a way as before, the increasingly polluted leaching will have less and less impact on groundwater. A good example of this includes the areas of the former Cracow Soda Works Solvay, which was closed in 1990. According to the studies carried out in 2013, the influence of settlers on the chloride concentration in the Wilga river was significant; however, a reduction in the penetration of ions from settlers into the river along with infiltrating rainwater in comparison to previous years was observed [Gliniak et al. 2014].

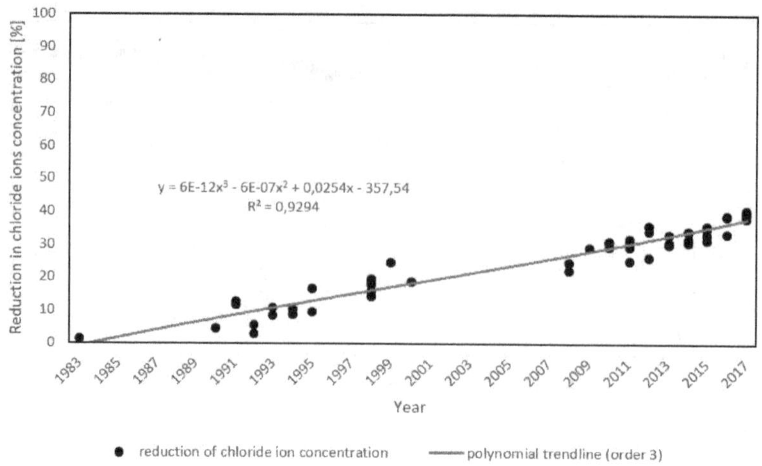

Figure 6. Reduction of chloride ion concentration at point 4.

CONCLUSIONS

1. On the basis of the analysis of the collected material, the following conclusions were drawn:
2. During the entire measurement period (22–34 years, depending on the piezometer), the concentration of chloride ions at measuring points 1–4 was gradually reduced.
3. Chloride ions concentration shows a decreasing trend in the past years.
4. The protective measures undertaken by the plant were effective in terms of limiting the groundwater pollution.

Acknowledgements

The research was carried out as part of the work MB/WBiIS/1/2017 and statutory work S/WBiIŚ/3/2014 and funded by the Ministry of Science and Higher Education.

REFERENCES

1. Effler S.W., Matthews D.A. 2003. Impacts of a Soda Ash Facility on Onondaga Lake and the Seneca River, NY. Lake and Reservoir Management, 19(4), 285–306.
2. Gliniak M., Pawul M., Sobczyk W. 2014. Impact of the transport and postindustrial landfills of Cracow Soda Works "Solvay" on the status and quality of water in Wilga river in Krakow. Logistyka, 4, 4295–4302. (in Polish)
3. Harat A. Grmela A. 2008. Impact of mine water from The Upper Silesian Coal Basin areas on change quality of water in Olza river in years 2000–2007. Natural Environment Monitoring, 9, 57–62.
4. Kasikowski T., Buczkowski R., Cichosz M. 2008. Utilization of synthetic soda-ash industry by-products. International Journal of Production Economics, 112, 971–984. (in Polish)
5. Kasikowski T., Buczkowski R., Cichosz M., Lemanowska E. 2007. Combined distiller waste utilization and combustion gases desulphurisation method. The case study of soda-ash industry. Resources, Conservation and Recycling, 51, 665–690. (in Polish)
6. Małecki Z.J., Wira J., Moshynsky V., Małecka I. 2016. The threat of worsening the quality of surface and underground waters caused by an unrecultivated waste dumping site in Tłokinia Kościelna near Kalisz. Inżynieria Ekologiczna, 46, 77–87 (in Polish).
7. Matthews D.A., Effler S.W. 2003. Decreases in pollutant from residual soda ash production waste. Water, Air, and Soil Pollution, 146, 55–73.
8. Mosiej J., Komorowski H., Kaczmarczyk A., Suska A. 2007. Effect of pollutants discharged from Łódź conurbation on quality of water in Ner and Warta rivers. ACTA Scientiarum Polonorum, 6(2), 19–30 (in Polish).
9. Pietrucin D., Czop M. 2015. Groundwater contamination by organochlorine compounds in the area of industrial waste dump "Zielona" in Bydgoszcz City. Polish Geological Review 63 (10/2), 997–1001 (in Polish).
10. Policht-Latawiec A., Kanownik W., Łukasik D. 2013. Effect of point source pollution on the san river Water quality San. Infrastructure and Ecology of Rural Areas 1/IV/2013, 253–269 (in Polish).
11. Rabiej M. 2012. Statistic with the Statistica program. Helion, Gliwice.
12. Şener S. 2008. Use of solid wastes of the soda ash plant as an adsorbent for the removal of anionic dyes: Equilibrium and kinetic studies. Chemical Engineering Journal, 138, 207–214.

13. Siwiec T., Michel M.M., Reczek L. 2016. The influence of aeration on the change of corrosive aggressiveness of groundwater in relation to concrete and steel. Acta Scientiarum Polonorum Architectura, 15 (1), 95–105 (in Polish).
14. Polish Standard PN-EN ISO 5667:2008 Water quality. Sampling.
15. Steinhauser G. 2008. Cleaner production in the Solvay Process: general strategies and recent developments. Journal of Cleaner Production, 16, 833–841.
16. Updated The National Programme for Municipal Waste Water Treatment (NPMWWT) – AKPOŚK 2017 (in Polish).
17. Urbańska J., Urbański K. 2012. Selected Aspects of Reclamation of Soda Waste Landfill Sites. Geomatics and Environmental Engineering, 6(4), 83–90 (in Polish).
18. Vangelas K.M. Looney B.B., Early T.O., Gilmore T., Chapelle F.H., Adams K.M., Sink C.H. 2006. Monitored natural Attenuation of Clorinated Solvents – Moving beyond reductive dechlorination. Remediation, summer 2006, 5–23.
19. Wiater J. 2011. Influence of municipal landfill site on the ground water quality. Ecological Engineering, 26, 133–146 (in Polish).
20. Wychowaniak D., Koda E. 2014. Assessment of the soil – water system quality in the vicinity of old landfill with a vertical barrier. Remediation, Reclamation and Revitalization. PZITS Oddział Wielkopolski, Poznań (in Polish).

22

Assessment of Ground Ozone Level under the Physiological Strain Conditions

Robert Krzysztof Sobolewski[1], Robert Kalbarczyk[1]

[1] Wrocław University of Environmental and Life Sciences, Institute of Landscape Architecture, Grunwaldzka 55, 50-357 Wrocław, Poland

* Corresponding author's e-mail: robert.sobolewski.lubawka@gmail.com

ABSTRACT

Urban areas are characterised by the impact of negative environmental factors, such as: stress connected with extreme bio-thermal conditions or the presence of high concentrations of air pollutants. This study aims to evaluate the relationship between the hours of O_3 concentrations and the levels of physiological strain (PhS) in Legnica, during the period from December 2013 to November 2014. The hourly concentrations of O_3, NO_2 and meteorological elements used in the study were obtained from the State Environmental Monitoring station in Legnica (Lower Silesia). The evaluation of the bio-thermal conditions was carried out by means of the physiological strain indicator (PhS). The basic statistics were subject to analysis, the frequency of hourly pollution concentrations and the thermal strain were evaluated, as was the Pearson correlation coefficient and multiple regression between O_3 and PhS. A detailed analysis was carried out for the summer months (June-August). The most adverse conditions in terms of pollution with tropospheric ozone and heat strain were noted in July. The strongest relations between O_3 and PhS were observed in June. In winter months (December-February) no significant dependencies were noted between the tested variables. These tests will help to contribute to increasing the current knowledge on evaluating the thermal comfort of urbanised areas and the accompanying aero-sanitary conditions.

Keywords: air pollution, thermal strain, bioclimate, urban climate

INTRODUCTION

Urban areas modify the atmospheric environment, for instance through the changes of the city's thermal conditions as a result of progressive development and loss of vegetation [Nikolopoulou and Lykoudis 2006, Boumans et al. 2014, van Hove 2015], and changes to air chemistry through the presence of particulate and gas pollution, which includes tropospheric ozone [Carnero et al. 2010]. The tests carried out in Seattle showed an approx. 10% increase in A&E visits caused by asthma attacks along with increased levels of O_3 by each 10 ppb [Goodin, Hubbard 2016]. Additionally, among the children aged 6–15, O_3 and $PM_{2.5}$ particulate lead to a decrease in the lung capacity [Chen et al. 2015]. Novack et al. [2016] showed the impact of air pollution on the oxidative stress of a human body. In their tests, they noted an inverse relation between the high concentration of O_3 and a positive relation between suspended particulate PM_{10} and multiplication of blood stem cells. The impact of tropospheric ozone concentration on the respiratory system is confirmed by numerous studies from all over the world: Kinney et al. [1996], Frischer [1999], O'Lenick et al. [2007], Khatri et al. [2001], Tager et al. [2003], Uysal and Schapira [2005]. The research also shows an increase in the death rate by 2% due to hypofunction of the cardiovascular system caused by O_3 pollution on each $10 \mu g \cdot m^{-3}$ [Khaniabadi, et al. 2017]. Tropospheric ozone is a secondary pollutant that arises from a series of photochemical reactions [Clapp and Jenkin 2001, Kleinman 2005, Monks et al. 2015, Tiwari et al. 2015]. Moreover, the solar radiation necessary to start the chain reaction, which includes: NO_X ($NO + NO_2$), CO, VOC's and NMVOCs, is absorbed by the human body. The absorption depends on the factors such as: the intensity and structure of solar radiation, the Sun's altitude (h), ground albedo, body-to-sun orientation and the insulating power and colour of clothing and skin [Błażejczyk 2004]. The highest concentration of O_3 occurs at noon, showing an inverse meridian course from NO_2, NO and NO_x [Escudero et al. 2004, Kalbarczyk et al. 2015, Rozbicka and Rozbicki 2016, Zheng et al. 2017]. In the same hours, the least favourable bio-thermal conditions are observed that impact the human body [Park et al. 2014, Kalbarczyk et al. 2015]. High O_3 concentration is often accompanied by a high air temperature [Vandentorren et al. 2004, Pascal et al. 2012]. Błażejczyk and McGregor [2007], on the basis of 3 bio-thermal indicators: STI, PST, PhS, showed that in London it is possible to explain 20–29% deaths by the certain values of these indicators occuring three days earlier. In Poland, according to Błażejczyka and others [2017], one may expect an increase in mortality on the days characterized by strong and very high heat concentration, with reference to the days when heat stress is observed, by 12% and 47% respectively.

The aim of the study was to evaluate the relationship between the hours of O_3 concentrations and the levels of physiological strain (PhS) in Legnica, during the period from December 2013 to November 2014.

MATERIALS AND METHODS

Legnica is located in the west of Poland, in the Lower Silesia region (51°12′36″N, 16°09′42″E, hs = 113–168 m. a.s.l.). It is the eighth largest (56 km^2) and densely populated (1802 people·km^{-2}) city in the region and the third largest in terms of population (100886 people) after Wrocław and Wałbrzych [Central Statistical Office 2015]. Legnica belongs to the central bioclimatic region of weak stimulusity, characterised by an average air temperature at 12:00 (UTC) of 21.3°C in the summer and 1.9°C in the winter [Błażejczyk and Matzarakis 2007]. This study uses the data of hourly pollution concentrations of O_3, NO_2 and the meteorological data during the period from December 2013 to November 2014. The data was obtained from the State Environmental Monitoring Station in Legnica, marked by the international code PL0190A, indicating a city type station. The data was prepared for universal time (UTC). The initial data verification was carried out on the basis of the correlation between the O_3 hourly concentration and total radiation (Rad, W·m^{-3}), air temperature (Ta, °C), relative air humidity (Rh, %), as well as the wind speed (v, m·s^{-1}) analysis which showed occurrence of two separate data sets. The daily distribution of average values of the meteorological elements showed a sudden drop in the total radiation values between 12:00 and 14:00, depending on the particular month analysed within the researched period, probably caused by the shadowing of the station in the afternoon. The basis of the division into two data sets was the result of carrying out a distribution of average hourly values of meteorological elements for individual decades of each month. The data sets obtained in this way were further used in the correlation analysis.

The evaluation of pollution with the concentrations of O_3 and NO_2 in particular months of the research period was carried out on the basis of descriptive statistics and an analysis of hourly concentration frequencies of O_3 and NO_2 in the adopted classes of pollution levels, consecutively for ozone: 0–40, 41–80, 81–120, >120 µg·m^{-3} and for nitrogen dioxide: 0–20, 21–40, 41–60, 61–80, >80 µg·m^{-3}. The evaluation of the bio-thermal conditions was carried out on the basis of the physiological strain (PhS, dimensionless) (Table 1).

The PhS indicator values were calculated by means of the Bioklima 2.6 software (https://www.igipz.pan.pl/Bioklima-zgik.html). The PhS indicator is expressed as the relation of heat loss from a human body by convection (C) to its loss as a result of perspiration evaporation (E) [Błażejczyk, Matzarakis 2007]:

$$PhS = \frac{C}{E} \quad (1)$$

For each month, there is a frequency related to the occurrence of different concentrations of physiological strain. In turn, the occurrence frequencies of the adopted O_3 concentration in particular classes of physiological strain concentra-

Table 1. Scale of physiological strain (PhS)

PhS	Abbreviation	Scale of physiological strain
< 0.00	EHS	Extreme Hot Strain
0.00 – 0.24	GHS	Great Hot Strain
0.25 – 0.75	MHS	Moderate Hot Strain
0.76 – 1.50	Tn	Thermoneutral
1.51 – 4.00	MCS	Moderate Cold Strain
4.01 – 8.00	GCS	Great Cold Strain
> 8.00	ECS	Extreme Cold Strain

tion were evaluated only for the chosen months characterised by a large and moderate hot strain. For the month of the most frequent extreme conditions in the O_3 concentrations above 120 μg·m^{-3} and hot strain load, an analysis of the daily distribution of O_3, NO_2, values PhS concentration and meteorological elements in subsequent decades was conducted. Due to the asymmetric, right-hand character of the hourly NO_2 concentration schedule, this data was adapted to a normal schedule. In order to do so, the hourly data was stated as a logarithm, which enabled carrying out further statistical analyses. An analysis of the Pearson correlation coefficient between the O_3 and meteorological elements and PhS indicators was carried out for three groups: including the full set of all-day measurements data, morning measurements, that is from sunrise to 13:00 and afternoon measurements data – from 14:00 to sunset. The correlation was evaluated on the level of p=0.05. For the months when the correlation analysis was statistically significant, the equations of multiple regression analyses between tropospheric ozone and conditions of physiological strain in the presence of NO_2 were built.

RESULTS

The descriptive statistics showed that from March to September, the maximum hourly O_3 concentration values in particular months exceeded 127 μg·m^{-3} (Tab. 2) The highest concentration was recorded in July at 159 μg·m^{-3}. In turn, the highest NO_2 concentration in the summer period fell in August and was 100 μg·m^{-3}. In the same month, the median value calculated for the NO_2 emission was equal to 15 μg·m^{-3} In the month when the highest O_3 concentration was documented, the value of the PhS indicator amounted to -0.01, which indicates the occurrence of "extreme hot strain" conditions. In the remaining summer months, i.e. June and August, the maximum values of the PhS indicator showed the conditions of "great hot strain". In the winter months (December-February) the maximum monthly O_3 concentration did not exceed 70 μg·m^{-3}, the highest recorded in January – 69 μg·m^{-3} In the same month, the maximum value of the PhS indicator amounted to 6.06. This confirms the occurrence of "great cold strain" in this month. The maximum concentration of NO_2 was documented in March at 128 μg·m^{-3}; moreover, the maximum concentration above 100 μg·m^{-3} was also recorded in December and February. The value of the third percentile indicates that in July, 25% of all O_3 concentrations were over 98 μg·m^{-3}; in comparison, in January, 25% of all concentrations were within the range of 36–64 μg·m^{-3} The highest median values above 50 μg·m^{-3} occurred in July, May, August, June, and April. Thus, the median value for the spring-summer months was higher than the value of the third percentile for winter months.

In July, around 40% of all cases constituted the hourly concentrations above 80 μg·m^{-3}; out of that concentration >120 μg·m^{-3} amounted to approx. 11.5% (Fig. 1). March was characterised by the least favourable conditions in terms of the frequencies hourly concentrations of NO_2; the concentrations above 80 μg·m^{-3} constituted 2.8% of recorded cases, in turn, in the scope of 60–80

Table 2. Descriptive statistics for individual months of the research period: Dec. 2013 – Nov. 2014

Month	Q_1			Me			Q_3			max			min
	O_3	NO_2	PhS	O_3	NO_2	PhS	O_3	NO_2	PhS	O_3	NO_2	PhS	PhS
	μg·m^{-3}			μg·m^{-3}			μg·m^{-3}			μg·m^{-3}			
Dec	8	12	2.57	19	26	2.80	42	39	3.03	69	108	4.31	2.01
Jan	8	17	2.68	19	24	3.04	36	33	3.78	64	83	6.06	1.8
Feb	13	13	2.47	39	23	2.78	54	40	3.04	79	106	4.3	1.84
Mar	12	12	2.17	41	20	2.53	60	35	2.84	127	128	3.81	0.86
Apr	25	9	1.77	51	14	2.14	72	24	2.49	128	96	3.2	0.87
May	43	6	1.43	61	10	1.83	78	17	2.17	132	78	3.01	0.31
Jun	39	7	1.13	55	12	1.52	75	19	1.84	153	86	2.39	0.09
Jul	40	8	0.58	67	12	1.06	98	20	1.14	159	79	2.32	-0.01
Aug	32	9	1.01	56	15	1.36	78	25	1.70	139	100	2.37	0.15
Sep	18	12	1.38	37	18	1.65	59	28	1.85	154	92	2.46	0.51
Oct	5	13	1.74	22	21	1.94	43	29	2.36	85	82	3.18	1.00
Nov	3	15	2.23	13	19	2.50	27	26	2.71	70	79	3.43	1.39

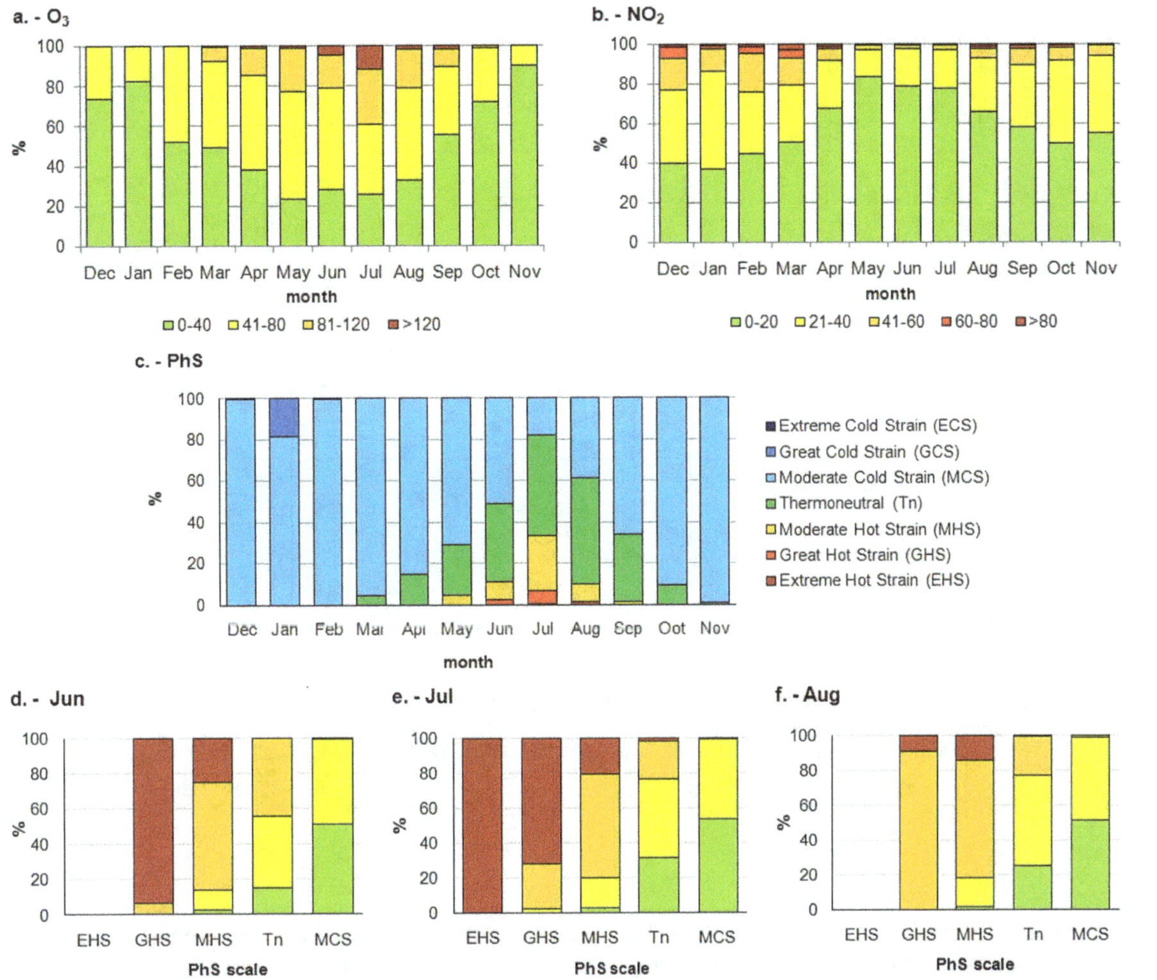

Fig. 1. Occurrence frequencies: the hourly concentrations of gas pollution O_3 and NO_2 (a – b), thermal strain in individual months of the research period (c), and hourly O_3 concentrations in individual classes of physiological strain in June, July, and August (d – f), for the period from December 2013 to November 2014

$\mu g \cdot m^{-3}$ – 4.2%. In July 2014, the conditions connected with hot strain constituted 30% cases. In June and August, the frequencies were comparable, amounting to approx. 11% and 9.7%. In the winter period (December-February), there were almost the same conditions connected with "moderate cold strain", except for January when 18.4% of cases were characterised by "great cold strain". In the remaining months, the strain did not exceed 1%. In the researched period, there were no extreme thermal strain days. Only in July was there a single case of "extreme hot strain". The occurrence of "moderate hot strain" in September was 1.4%, in May approx. 4.5%. The summer period, from June to July, was characterised by the occurrence of thermal conditions from "moderate cold strain" to "extreme hot strain". In July and in June the frequency of O_3 concentrations above 120 $\mu g \cdot m^{-3}$ was connected with the conditions of hot strain. Extreme conditions were recorded only in July, which constituted an isolated case. In June, during the "moderate hot strain" approx. 93% cases involved the concentrations above 120 $\mu g \cdot m^{-3}$. In turn, August was characterised by, in comparison to the remaining summer months, a lower frequency of O_3 concentrations above 120 $\mu g \cdot m^{-3}$.

The average hourly O_3 exceeded the value of 100 $\mu g \cdot m^{-3}$ in the second and third decades of July, from 10:00 to 18:00 and from 11:00 to 17:00 (Fig. 2). NO_2 does not show a high changeability of average daily concentration in particular decades. The highest concentration in the second decade was recorded at night. The daily distribution of average O_3 concentrations shows a reverse distribution in reference to the PhS values. The lower values of PhS indicator show the occurrence of the hot strain in the second decade, where the least favourable PhS conditions are recorded, which overlaps with the high level of O_3

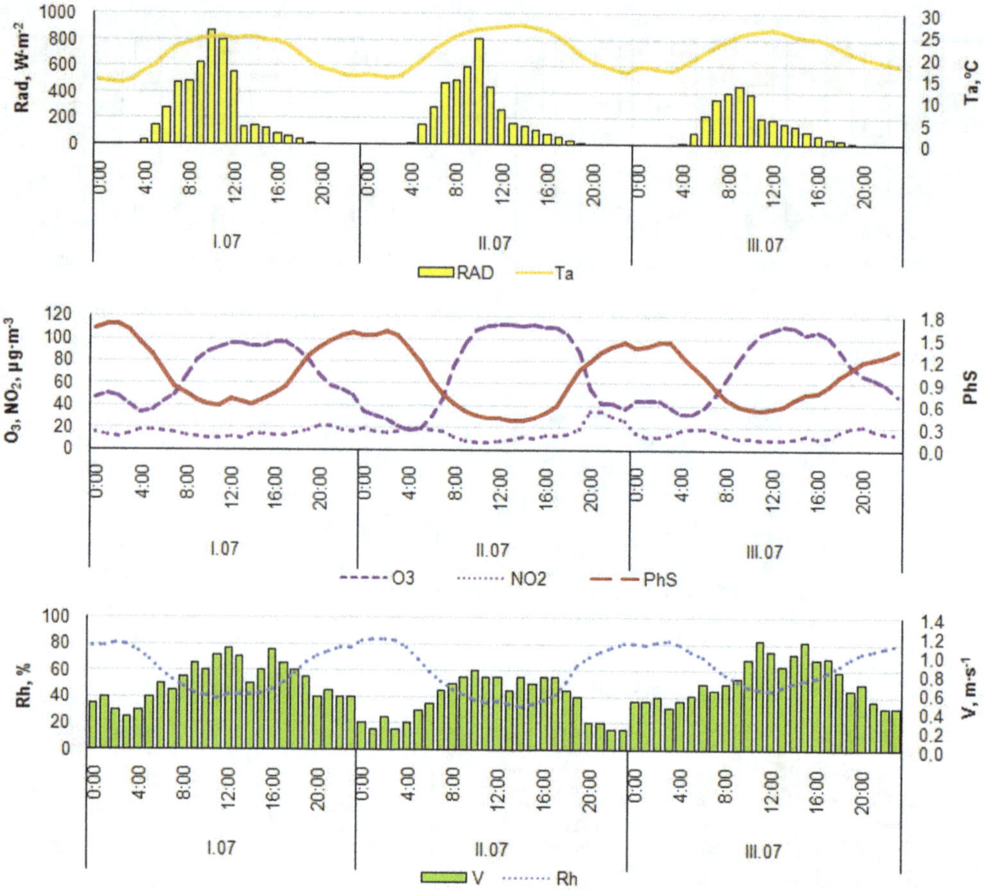

Fig. 2. Daily distribution of average hourly values: O_3, NO_2, PhS and meteorological elements: total radiation Rad, ($W{\cdot}m^{-2}$) air temperature (Ta, °C), relative air humidity (Rh, %), wind speed (V, $m{\cdot}s^{-1}$) in individual decades of July 2014

concentrations. The solar radiation values show a clear, untypical, drop in the afternoon, which is probably caused by the shadowing of the monitoring station. In the third decade, we can notice a slight increase of the air temperature during the night, which may indicate the effect of the urban heat island present in the research period. The increase of the temperature is accompanied by a slight increase of the tropospheric ozone and the drop in the PhS indicator value.

The maximum hourly O_3 concentrations in July above 150 $\mu g{\cdot}m^{-3}$ were recorded between 11:00 and 17:00; they were accompanied by low NO_2 concentrations in the range from 2 to 5 $\mu g{\cdot}m^{-3}$ (Tab. 3). From 9:00 to 17:00, the highest hourly O_3 concentrations occurred during significant moderate and great hot strain. Such conditions were favoured by: low relative air humidity below 34%, air temperature above 29°C, and relatively low wind speed approx. 0.7 $m{\cdot}s^{-1}$. Only between 8:00 and 10:00, the maximum O_3 concentration was accompanied by solar radiation in the range of 520–713 $W{\cdot}m^{-2}$. At 11:00 total radiation accompanying maximum O_3 values amounted to 219 $W{\cdot}m^{-2}$ and gradually decreased in the following hours. The maximum O_3 concentration amounting to 159 $\mu g{\cdot}m^{-3}$ was recorded at 14:00 with the PhS value of 0.09. The least favourable thermal conditions connected with very strong hot strain occurred at 12:00 during the maximum O_3 concentration, amounting to 152 $\mu g{\cdot}m^{-3}$.

In the winter period, there was no significant correlation between the hourly O_3 concentrations and the values of the PhS indicator (Tab. 4) A weak correlation (r=0.16, p=0.01) for the data set from 24 h in January constituted an exception. This proven relation has a positive direction, in contrast to the proven relations in the remaining months, which adopted a negative direction (March-November). The reason for the change in the direction of relations between the O_3 concentrations and the PhS values in January could be a lack of correlation between O_3 and Ta for all possible data sets. The strongest, negative relation (r = -0.87, p=0.01) was recorded in June for the afternoon data set, between 14:00 and 19:00, that

Table 3. Maximum O_3 concentration at particular hours of July 2014 in reference to the accompanying meteorological and bio-thermal conditions

Hour	O_3	NO_2 meteorological and physiological strain conditions during the maximum concentration of O_3						PhS	
		NO_2	Rad	Ta	Rh	V	PhS		
	µg·m^{-3}	µg·m^{-3}	W·m^{-2}	°C	%	m·s^{-1}		max	min
00:00	108	8	0	17.9	60	0.7	1.32	2.08	1.04
01:00	109	7	0	18.5	58	1.4	1.36	2.32	1.08
02:00	106	7	0	17.5	62	1.4	1.49	2.17	1.24
03:00	99	8	1	0.0	60	1.4	1.43	2.09	1.04
04:00	92	9	12	18.5	59	1.4	1.37	1.92	1.07
05:00	92	8	28	21.1	53	1.4	1.07	1.81	0.82
06:00	85	9	328	22.4	47	1.4	0.88	1.69	0.44
07:00	96	7	313	24.3	43	1.4	0.69	1.64	0.28
08:00	107	8	520	25.3	39	1.4	0.58	1.49	0.11
09:00	126	7	629	29.5	34	0.7	0.21	1.53	0.1
10:00	148	4	713	30.0	31	0.7	0.18	1.55	0.06
11:00	153	3	219	30.6	29	0.7	0.14	1.68	0.01
12:00	152	5	185	30.9	28	0.7	0.12	1.72	-0.01
13:00	155	5	148	31.5	27	0.7	0.09	1.54	0.02
14:00	159	2	133	31.5	28	0.7	0.09	1.65	0.01
15:00	157	2	106	31.1	28	0.7	0.11	1.59	0.01
16:00	154	4	79	30.7	29	0.7	0.13	1.56	0.04
17:00	152	2	42	29.4	33	1.4	0.23	1.55	0.14
18:00	147	3	26	26.5	42	0.7	0.46	1.58	0.36
19:00	130	11	7	25.0	55	0.7	0.65	1.7	0.56
20:00	121	13	0	23.0	66	0.7	0.88	1.74	0.81
21:00	125	7	0	22.8	64	0.7	0.89	2.08	0.89
22:00	123	8	0	21.1	76	0.7	1.1	2.14	1.06
23:00	114	8	0	18.8	56	0.7	1.21	2.03	1.09

is sunset. For the data from 24 h, the strongest negative correlations in June and July, were $r = -0.78$ ($p=0.01$) and $r = -0.76$ ($p=0.01$), respectively. In the spring period (March-June) and in September, clear differences in the relations between the morning data sets – from sunrise to 13:00 and in the afternoon – from 14:00 to sunset, were observed. In July and August, the strength of the correlation between O_3 and PhS was comparable both for data from 24 h and from morning and afternoon data sets. In June in the afternoon hours, the strongest correlation between O_3 and Ta ($r = 0.90$, $p=0.01$) was observed. The relation between the hourly O_3 concentrations and total radiation for the data set of afternoon measurements in May and June did not show statistical relevancy on the level of $p=0.05$.

The model of rolling multiple regression was prepared for the months in which there was a significant correlation between O_3 and PhS (Table 5). When examining the model, apart from the PhS indicator, additional hourly NO_2 values in the form of the common logarithm were taken into account. In the adopted models, significant impact of all changeables explaining the size of O_3 immision was confirmed. The greatest fit of the empirical data to the regression function, amounting to 76% and 80%, was obtained for June, for the conditions in the morning and afternoon; in turn, the smallest fit was observed in April (58%, 43%) and in November (50%, 39%). On the basis of the partial correlation factor, stronger connections between O_3 and bio-thermal conditions, other than NO_2 pollutions were proven in the afternoon hours. In the morning hours, a greater impact of NO_2 on O_3 was recorded, in the prepared model.

DISCUSSION

The research carried out between December 2013 and November 2014 showed a significant relationship between the level of tropospheric ozone and unfavourable bio-thermal conditions connected with hot strain. These relations are

Table 4. Relations between the hourly O_3 concentration and individual meteorological variables as well as the values of the PhS indicator

Month	Period of time	RAD W·m^{-2}		Ta °C		Rh %		V m·s^{-1}		PhS	
		N	r	N	r	N	r	N	r	N	r
Dec	23:00 – 00:00	743	n.s.	743	0.43***	743	-0.53***	743	0.60***	743	n.s.
	sun rise – 13:00	184	n.s.	184	0.41***	184	-0.52***	184	0.59***	184	n.s.
	14:00 – sun set	48	n.s.	48	0.46***	48	-0.61***	48	0.57***	48	n.s.
Jan	23:00 – 00:00	742	0.19***	742	n.s.	742	-0.60***	742	0.29***	742	0.16***
	sun rise – 13:00	183	0.38***	183	n.s.	183	-0.58***	183	0.14***	183	n.s.
	14:00- sun set	51	0.55***	51	n.s.	51	-0.58***	51	n.s.	51	n.s.
Feb	23:00 – 00:00	666	0.35***	666	0.56***	666	-0.73***	666	0.57***	666	n.s.
	sun rise – 13:00	178	0.43***	178	0.33***	178	-0.66***	178	0.57***	178	n.s.
	14:00 – sun set	74	0.41***	74	0.26***	74	-0.54***	74	0.46***	74	n.s.
Mar	23:00 – 00:00	690	0.31***	690	0.72***	690	-0.64***	690	0.48***	690	-0.61***
	sun rise – 13:00	212	0.52***	212	0.72***	212	-0.65***	212	0.43***	212	-0.60***
	14:00 – sun set	104	0.20***	104	0.59***	104	-0.66***	104	n.s.	104	-0.54***
Apr	23:00 – 00:00	629	0.33***	629	0.54***	629	-0.70***	627	0.51***	627	-0.43***
	sun rise – 13:00	223	0.53***	223	0.45***	223	-0.66***	223	0.35***	223	-0.40***
	14:00 – sun set	111	0.27***	111	0.59***	111	-0.64***	111	n.s.	111	-0.57***
May	23:00 – 00:00	703	0.24***	703	0.60***	703	-0.67***	703	0.50***	703	-0.59***
	sun rise – 13:00	238	0.46***	238	0.53***	238	-0.69***	238	0.43***	238	-0.54***
	14:00 – sun set	186	n.s.	186	0.69***	186	-0.64***	186	0.36***	186	-0.73***
Jun	23:00 – 00:00	665	0.32***	665	0.81***	665	-0.67***	665	0.56***	665	-0.78***
	Sun rise – 13:00	250	0.72***	250	0.78***	250	-0.67***	250	0.45***	250	-0.77***
	14:00 – sun set	186	n.s.	186	0.90***	186	-0.74***	186	0.43***	186	-0.87***
Jul	23:00 – 00:00	740	0.27***	740	0.76***	740	-0.77***	740	0.46***	740	-0.75***
	sun rise – 13:00	271	0.47***	271	0.79***	271	-0.8***	271	0.36***	271	-0.78***
	14:00 – sun set	196	0.20***	196	0.76***	196	-0.64***	196	0.16***	196	-0.75***
Aug	23:00 – 00:00	743	0.37***	743	0.67***	743	-0.68***	743	0.57***	743	-0.66***
	sun rise – 13:00	246	0.61***	246	0.64***	246	-0.71***	246	0.56***	246	-0.67***
	14:00 – sun set	171	0.29***	171	0.65***	171	-0.52***	171	0.38***	171	-0.65***
Sep	23:00 – 00:00	646	0.36***	646	0.65***	646	-0.73***	646	0.60***	646	-0.58***
	sun rise – 13:00	203	0.68***	203	0.68***	203	-0.76***	203	0.57***	203	-0.64***
	14:00 – sun set	118	0.25***	118	0.73***	118	-0.71***	118	0.47***	118	-0.71***
Oct	23:00 – 00:00	741	0.27***	741	0.54***	741	-0.69***	741	0.63***	741	-0.44***
	sun rise – 13:00	203	0.43***	203	0.58***	203	-0.60***	203	0.62***	203	-0.45***
	14:00 – sun set	93	n.s.	93	0.49***	93	-0.63***	93	0.53***	93	-0.31***
Nov	23:00 – 00:00	720	0.27***	720	0.10**	720	-0.75***	720	0.44***	720	-0.26***
	sun rise – 13:00	156	0.44***	156	0.20**	156	-0.75***	156	0.52***	156	-0.42***
	14:00 – sun set	68	0.42***	68	0.29**	68	-0.77***	68	0.43***	68	-0.39***

confirmed by Paliatsos and Nastos [1999] and Matzarakis et al. [2012], who used the thermal discomfort indicator (DI) in their research. Matzarakis et al. [2012] proved the relationship between the O_3 and DI indicators during heat waves (r=0.51, p=0.01). Our research points to stronger relations between O_3 and PhS than DI in June and July, as is evidenced by the correlation factor calculated for both months, r=0.78 (p=0.01) and r=0.75 (p=0.01), respectively. Papanastasiou et al. [2015] on the basis of the discomfort indicator (DI) and CAQI factor, (taking into consideration pollutants: PM_{10}, NO_2, O_3) showed a deteriorating quality of air during heat waves in Thessaloniki, Athens and Volos in 2001–2010. Kalbarczyk et al. [2016] proved, on the basis of radiation-effective temperature (TRE), a more frequent occurrence of the O_3 concentration levels above 80μg·m^{-3} in July and August, when sensing "cold" and "hot" in comparison to other thermal feelings. In April and May, the high level of O_3 concentrations was noted when feeling "comfortable". The relationship between O_3 and NO_2 was the strongest in the winter months; in turn, the strongest impact

Table 5. Equation of the rolling multiple regression between O_3 and $NO_{2,}$ and PhS for individual months.

Month	Period of time	N	Regression equation			R^2adj.	F	Sy	SD	|SD-Sy|
			intercept	regresion coefficient						
				$\log NO_2$	PhS			$\mu g \cdot m^{-3}$		
Mar	sun rise – 13:00	200	173.35	-55.58 -0.58***	-28.51 -0.52***	0.70	233.32	14.93	26.45	11.52
	14:00 – sun set	95	176.92	-47.37 -0.58***	-25.02 -0.61***	0.62	77.07	14.67	23.76	9.09
Apr	sun rise – 13:00	219	165.64	-59.7 -0.65***	-20.89 -0.37***	0.58	152.79	18.66	28.67	10.01
	14:00 – sun set	111	146.74	-27.46 -34***	-21.33 -0.57***	0.43	43.82	14.28	18.95	4.67
May	sun rise – 13:00	236	146.23	-55.88 -0.55***	-18.61 -0.38***	0.56	152.57	16.65	25.12	8.47
	14:00 – sun set	181	148.02	-36.78 -0.44***	-21.53 0.43***	0.63	153.19	16.61	27.04	10.43
Jun	sun rise – 13:00	248	155.83	-42.74 -0.42***	-44.08 -0.67***	0.76	392.25	14.27	29.4	15.13
	14:00 – sun set	185	156.89	-18.86 -0.21***	-49.50 -0.79***	0.80	365.19	12.63	28.2	15.56
Jul	sun rise – 13:00	265	172.20	-62.53 -0.41***	-47.95 -0.56***	0.73	348.19	20.18	38.82	18.64
	14:00 – sun set	195	161.45	-27.88 -0.27***	-43.38 -0.62***	0.62	160.72	16.85	27.46	10.61
Aug	sun rise – 13:00	246	142.10	-45.93 -0.50***	-30.83 -0.42***	0.64	219.96	17.54	29	11.46
	14:00 – sun set	171	157.04	-42.57 -0.54***	-26.00 -0.44***	0.67	170.10	13.28	23.17	9.89
Sep	sun rise – 13:00	201	176.68	-67.70 -0.56***	-35.04 -0.47***	0.69	227.79	15.00	27.07	12.07
	14:00 – sun set	118	217.73	-63.79 -0.55***	-53.04 -0.53***	0.76	186.53	16.12	32.92	16.80
Oct	sun rise – 13:00	185	141.90	-64.34 -0.67***	-17.94 -0.35***	0.65	167.80	12.74	21.54	8.8
	14:00 – sun set	86	150.98	-58.18 -0.76***	-15.74 -0.3***	0.66	85.51	11.04	20.14	9.10
Nov	sun rise – 13:00	130	142.11	-62.60 -0.57***	-19.43 -0.42***	0.50	64.74	13.02	17.92	4.90
	14:00 – sun set	60	113.11	-51.54 -0.51***	-16.26 -0.38***	0.39	20.04	13.16	16.43	3.27

of tested meteorological elements on the O_3 concentration was recorded in summer. Ramsey et al. [2014] stated that the increase of passable O_3 concentrations during the dry and hot periods can be even three times higher compared to the cold and damp periods. This is confirmed by the occurrence of maximum O_3 concentrations above 150 $\mu g \cdot m^{-3}$ in Legnica during the times of low relative air humidity below 34% and the air temperature above 29°C. Błażejczyk et al. [2017] ascertained an increase of days >32°C UTCI connected with the condition of great hot strain between 1966 and 2012, which may have a direct impact on the interactions between the O_3 pollutions and bio-thermal conditions. The results obtained for Legnica indicate that the direct radiation has the least impact on shaping the relations between ozone pollutants and bio-thermal conditions in 2014, despite the documented impact on shaping the level of tropospheric ozone [Clapp, Jenkin 2001, Kleinman 2005, Monks et. al. 2015, Tiwari et al. 2015]. At the turn of the second and third decade in July, a significant increase of average hourly O_3 concentrations and the loss of average PhS values at night was observed, which could indicate the shaping of the urban heat island effect at that time. Czarnecka and Nidzgorska-Lencewicz [2014] showed a significant connection between the urban heat island in Gdańsk (Poland) and the level of researched pollution concentrations.

CONCLUSIONS

July 2014 was characterised by the least favourable bio-thermal conditions, determined on the basis of the physiological strain indicator (PhS), which were accompanied by an increased risk of occurrence pertaining to the high concentrations of tropospheric ozone above 120 $\mu g \cdot m^{-3}$ in comparison to the remaining months. These conditions mainly occurred during a significant and moderate hot strain in the afternoon hours, especially in the second decade of the month. Low relative air humidity, high air temperature and weak wind speed had an impact on the deterioration of the bio-thermal conditions and the increase of the O_3 concentration in the summer. Despite the least favourable levels of the O_3 concentrations and PhS values in July, the strongest relations were confirmed in June. The tests carried out may contribute to expanding the current knowledge in the area of thermal comfort evaluation of urban areas and the accompanying aero-sanitary conditions.

Acknowledgements

Research was financed from the statutory activity of The Faculty of Environmental Engineering and Geodesy, project no. B030/0082/16.

REFERENCES

1. BioKlima© 2.6., https://www.igipz.pan.pl/Bioklima-zgik.html; 31.09.2016.
2. Błażejczyk A., Błażejczyk, K., Baranowski, J., & Kuchcik, M. 2017. Heat stress mortality and desired adaptation responses of healthcare system in Poland. International Journal of Biometeorology, 1–12.
3. Błażejczyk, K. 2004. Radiation balance in man in various meteorological and geographical conditions. Geographia Polonica, 77(1), 63–76.
4. Błażejczyk K., Matzarakis A. 2007. Assessment of bioclimatic differentiation of Poland, based on the human heat balance. Geographia Polonica, 80(1), 63–82.
5. Boumans R.J., Phillips D.L., Victory W., Fontaine, T.D. 2014. Developing a model for effects of climate change on human health and health–environment interactions: heat stress in Austin, Texas. Urban Climate, 8, 78–99.
6. Carnero J.A., Bolívar J.P., de la Morena B.A. 2010. Surface ozone measurements in the southwest of the Iberian Peninsula (Huelva, Spain). Environmental Science and Pollution Research, 17(2), 355–368.
7. Central Statistical Office 2015, http://stat.gov.pl/; 4.11.2017.
8. Chen, C. H., Chan, C. C., Chen, B. Y., Cheng, T. J., Guo, Y. L. 2015. Effects of particulate air pollution and ozone on lung function in non-asthmatic children. Environmental Research, 137, 40–48.
9. Clapp L. J., Jenkin M. E. 2001. Analysis of the relationship between ambient levels of O_3, NO_2 and NO as a function of NO_x in the UK. Atmospheric Environment, 35(36), 6391–6405.
10. Czarnecka M., Nidzgorska-Lencewicz J. 2014. Intensity of Urban Heat Island and Air Quality in Gdańsk during 2010 Heat Wave. Polish Journal of Environmental Studies, 23(2), 329–340.
11. Escudero M., Lozano A., Hierro J., del Valle J., Mantilla, E. 2014. Urban influence on increasing ozone concentrations in a characteristic Mediterranean agglomeration. Atmospheric Environment, 99, 322–332.
12. Frischer T., Studnicka M., Gartner C., Tauber E., Horak F., Veiter A., Spengler J., Kühr J., Urbanek R. 1999. Lung function growth and ambient ozone: a three-year population study in school children. American Journal of Respiratory and Critical Care Medicine, 160(2), 390–396.
13. Goodin, K., Hubbard M. 2016. Association between asthma hospital visits and ozone concentration in Maricopa County, Arizona (2007–2012). J. Environmental Health, 78(9), 8–13.
14. Kalbarczyk R., Sobolewski R., Kalbarczyk E., 2015: Assessment of human thermal sensations based on bioclimatic indices in a suburban population, Wrocław (SW Poland). Polish Journal of Natural Sciences, 30(2), 185–201.
15. Kalbarczyk R., Sobolewski R., Kalbarczyk E. 2016. Biometeorological determinants of the tropospheric ozone concentration in the suburban conditions of Wroclaw, Poland. Jorunal of Elementology, 21(3), 729–744.
16. Kalbarczyk R., Kalbarczyk E., Niedźwiecka-Filipiak I., Serafin L. 2015. Ozone concentration at ground level depending on the content of NOx and meteorological conditions. Ecological Chemistry

and Engineering S., 22(4), 527–541.

17. Khaniabadi Y.O., Goudarzi G., Daryanoosh S.M., Borgini A., Tittarelli A., De Marco A. 2017. Exposure to PM10, NO2, and O3 and impacts on human health. Environ. Environmental Science and Pollution Research, 24(3), 2781–2789.

18. Khatri S.B., Holguin F.C., Ryan P.B., Mannino D., Erzurum S.C., Teague W.G. 2009. Association of ambient ozone exposure with airway inflammation and allergy in adults with asthma. Journal of Asthma, 46(8), 777–785.

19. Kinney P.L., Thurston G.D., Raizenne, M. 1996. The effects of ambient ozone on lung function in children: a reanalysis of six summer camp studies. Environmental Health Perspectives, 104(2), 170–174.

20. Kleinman L.I. 2005. The dependence of tropospheric ozone production rate on ozone precursors. Atmospheric Environment, 39(3), 575–586.

21. Mavrakis A., Spanou A., Pantavou K., Katavoutas G., Theoharatos G., Christides A., Verouti E. 2012. Biometeorological and air quality assessment in an industrialized area of eastern Mediterranean: the Thriassion Plain, Greece. International Journal of Biometeorology, 56(4), 737–747.

22. Monks P.S., Archibald A.T., Colette A., Cooper O., Coyle M., Derwent R., Fowler D., Granier C., Law K.S., Mills G.E., Stevenson D.S., Tarasova O., Thouret V., von Schneidemesser E., Sommariva R., Wild O., Williams M.L. 2015. Tropospheric ozone and its precursors from the urban to the global scale from air quality to short-lived climate forcer. Atmospheric Chemistry and Physics, 15(15), 8889–8973.

23. Nikolopoulou M., Lykoudis, S. 2006. Thermal comfort in outdoor urban spaces: analysis across different European countries. Building and Environment, 41(11), 1455–1470.

24. Novack L., Yitshak-Sade M., Landau D., Kloog, I., Sarov, B., Karakis I. 2016. Association between ambient air pollution and proliferation of umbilical cord blood cells. Environmental Research, 151, 783–788.

25. O'Lenick C.R., Chang H.H., Kramer M.R., Winquist A., Mulholland J.A. Friberg M.D., Sarnat S. E. 2017. Ozone and childhood respiratory disease in three US cities: evaluation of effect measure modification by neighborhood socioeconomic status using a Bayesian hierarchical approach. Environmental Health, 16(1), 36.

26. Paliatsos A.G., Nastos P.T. 1999. Relation between air pollution episodes and discomfort index in the greater Athens area, Greece. International Journal of Global Nest, 1(2), 91–97.

27. Papanastasiou D.K., Melas D., Kambezidis H.D., 2015. Air quality and thermal comfort levels under extreme hot weather. Atmospheric Research, 152, 4–13.

28. Park S., Tuller S.E., Jo M. 2014. Application of Universal Thermal Climate Index (UTCI) for microclimatic analysis in urban thermal environments. Landscape and Urban Planning 125, 146–155.

29. Pascal M., Wagner V., Chatignoux E., Falq G., Corso M., Blanchard M., Host S., Larrieu S., Laurence P., Declercq, C. 2012. Ozone and short-term mortality in nine French cities: Influence of temperature and season. Atmospheric Environment, 62, 566–572.

30. Ramsey N.R., Klein M.P., Moore III B., 2014. The impact of meterological parameters on urban air quality. Atmos. Environ., 86, 58–67.

31. Rozbicka K., Rozbicki, T. 2016. The "Weekend Effect" on Ozone in the Warsaw Conurbation, Poland. Polish Journal of Environmental Studies, 25(4), 1675–1683.

32. Tager I.B., Balmes J., Lurmann F., Ngo L., Alcorn S., Kunzli N., 2005. Chronic exposure to ambient ozone and lung function in young adults. Epidemiology, 16, 751–759.

33. Tiwari S., Dahiya A., Kumar N. 2015. Investigation into relationships among NO, NO2, NO x, O3, and CO at an urban background site in Delhi, India. Atmospheric Research, 157:,119–126.

34. Uysal N., Schapira, R. M., 2003. Effects of ozone on lung function and lung diseases. Current Opinion in Pulmonary Medicine, 9, 144–150.

35. van Hove L. W.A., Jacobs C.M.J., Heusinkveld B.G., Elbers J.A., van Driel B.L., Holtslag A.A.M. 2015. Temporal and spatial variability of urban heat island and thermal comfort within the Rotterdam agglomeration. Building and Environment, 83, 91–103.

36. Vandentorren S., Suzan F., Medina S., Pascal M., Maulpoix A., Cohen J. C., Ledrans M. 2004. Mortality in 13 French cities during the August 2003 heat wave. American Journal of Public Health, 94(9), 1518–1520.

37. Wolf K., Cyrys J., Harciníková T., Gu J., Kusch T., Hampel R., Schneider A., Peters A. 2017. Land use regression modeling of ultrafine particles, ozone, nitrogen oxides and markers of particulate matter pollution in Augsburg, Germany. Science of the Total Environment, 579, 1531–1540.

38. Zheng S., Singh R.P., Wu Y., Wu C. 2017. A Comparison of Trace Gases and Particulate Matter over Beijing (China) and Delhi (India). Water, Air, & Soil Pollution, 228(5), 181.

23

Bacterial ACC Deaminase Activity in Promoting Plant Growth on Areas Contaminated with Heavy Metals

Anna Grobelak[1*], Paulina Kokot[1], Jakub Świątek[1], Marta Jaskulak[1], Agnieszka Rorat[2]

[1] Czestochowa University of Technology, Faculty of Infrastructure and Environment, Institute of Environmental Engineering, Brzeznicka 60a, 42-200 Czestochowa, Poland

[2] Université de Lille, Sciences et Technologies, Laboratoire de Génie Civil et géo-Environnement, LGCgE EA4515, Bât. SN3, 59655 Villeneuve d´Ascq, France

[*] Corresponding author's e-mail: agrobelak@is.pcz.czest.pl

ABSTRACT

The objective of this study was to explore the possible improvement of plant growth using the activity of the bacterial enzyme ACC (1-aminocyclopropane-1-carboxylate) deaminase (endophytes and rhizobacteria). The beneficial effect of ACC deaminase activity was tested on the plants growing under stress conditions (high concentrations of heavy metals: cadmium, lead, zinc in the soil). The bacteria were isolated from three plants species: *Festuca rubra* L., *Agrostis capillaris* L., *Arabidopsis thaliana* L. Heynh, acquired from the area contaminated with heavy metals. The strains with the highest ACC deaminase activity were used to prepare a bacterial consortium and inoculate the plants. It has been shown that inoculation of plants with ACC producing bacteria has a positive effect on their growth under stress conditions. The bacterial entophytes strains showed a higher activity of ACC deaminase, which resulted in a higher biomass growth of inoculated plants. The PGPB bacteria may limit the toxicity of harmful ions and thus the increase the adaptive properties of plants. Moreover, it was discovered that the bacteria mainly belonging to genus *Bacillus* and *Pseudomonas* had the highest ACC deaminase activity in the environment contaminated with multiple heavy metals. The use of selected microorganisms and plants will provide results in an increasing efficiency of phytoremediation.

Keywords: ACC deaminase, endophytes, heavy metals, plant growth promoting bacteria (PGPB)

INTRODUCTION

Heavy metals are natural components of the Earth's crust. Beside their natural occurrence, most environmental contamination and human exposure result from the anthropogenic activities. Migration of these contaminants into non-contaminated areas as dust or leachates through the soil and spreading of heavy metals containing sewage sludge are a few examples of events contributing towards the contamination of the environment (Kacprzak *et al.* 2014, Kanclerz *et al.* 2016).

Some trace elements, including heavy metals, such as Zn, Cu, Co, Ni and Mn are necessary micronutrients for plant growth, while others have an unknown biological function, such as Cd, Pb, and Hg (Wong-Villareal *et al.* 2016). Excessive level of heavy metal pollution in soil has a harmful effect on biological systems (Salihaj *et al.* 2016). All metals are toxic at higher concentrations, because they cause oxidative stress by formation of free radicals. Another reason why metals may be toxic is that they can replace essential metals in pigments or enzymes, thereby disrupting their function. The hazardous trace elements, especially highly bioavailable ones (Grobelak and Napora, 2015) create unsuitable land for plant growth and destroy the biodiversity (Hadi *et al.* 2014). As a consequence of heavy metal stress, the plant growth is significantly lower than it would be in their absence. Moreover, during its life, a plant is also subjected to a number of non-lethal stresses that limit the plant growth. Certain soil bacteria can help plants to either avoid or partially overcome a variety of environmental factors. These bacteria improve the plant growth, enhance the root development, and thus increase the plant tolerance to metal stress. Some PGPB (Plant Growth Promoting Bacteria) promote the plant growth by lowering the level of ethylene (Ma *et al.* 2009). Ethylene is a plant growth hormone produced endogenously by plants. It is also produced in soil through a variety of biotic and abiotic mechanisms. Under stress conditions, the production of ethylene is accelerated substantially, which adversely affects the root growth and consequently the growth of the plant as a whole organism. Some plant growth promoting bacteria (PGPB) contain an important enzyme, 1-aminocyclopropane-1-carboxylate deaminase (ACCD) that regulates the ethylene production by metabolizing ACC (precursor of ethylene biosynthesis) into alpha-ketobutyrate and ammonia, which can be used up by the plants (Saleem *et al.* 2007). Ethylene biosynthesis takes place in several stages. The first is the conversion of methionine to S-adenosyl-L-methionine, which is transformed by the ACC synthase into 1-aminocyklopropano-1-carboxylic acid. The last step is to convert ACC using ACC oxidase into ethylene. The ACC synthase activity may be increased due to the high concentration of indole-3-acetic acid. As a result, the concentration of 1-aminocyklopropano-1-carboxylic acid, which is a precursor of ethylene, increases in the plant. Inactivation of 1-aminocyklopropano-1-carboxylic acid prevents ethylene biosynthesis, thereby reducing its concentration in the plant. In contrast, ammonia, one of the products of degradation of ACC, is a nitrogen source for the bacteria. With bacterial ACC deaminase, the growth of plant biomass is stimulated and improves their resistance for functioning under environmental stress conditions. (Saleem *et al.* 2007) PGPB rhizosphere bacteria metabolizing ethylene are applied to protect plants against stress factors, such as floods, the presence of organic toxins, heavy metals, as well as high soil salinity (Glick, 2014).

A better understanding of the characteristics of heavy metal resistance and ACC deaminase-producing PGPB bacteria is needed for the development of efficient phytoremediation systems.

The present study was carried out to understand the mechanism of metal stress decrease in plants: *F. rubra* L. and *B. napus* L. by inoculation of ACC deaminase – producing metal resistant bacteria. In this study, the ACCD activity was analyzed for two groups of bacterial isolates: endophytic and rhizosphere. The aim of the study was to assess the effect of inoculation of plants with selected consortia on promoting the plant growth on the land contaminated with heavy metals. The purpose of the study was to find bacteria in soils with over 50-year history of toxic trace elements contamination, which exhibit very high ACC deaminase activity. Moreover, the aim of the study was to assess the bacteria genus characterized by the highest AAC deaminase activity in multiple heavy metal contaminated environment.

MATERIALS AND METHODS

Donor bacteria

The bacteria strains were isolated from: *Festuca rubra* L. (K), *Agrostis capillaris* L. (M). and *Arabidopsis thaliana* L. Heynh (R). The donor plants species were selected because they were characterized by high resistance in a contaminated environment. The three plants species were common in the barren area surrounding a zinc smelter, with toxic trace elements contamination (Pb, Cd, Zn). The endophytes were applied under phytotron conditions to inoculate fescue grass (phytostabilisation potential) and rape (cruciferous plant family similar to *Arabidopsis,* agricultural and phytoextraction potential). The plants were derived from heavy metal contaminated soil from the areas adjacent to the zinc smelter located in the central part of Silesia (Poland). The root samples were collected from plants. The roots were washed, cut and treated with 70% ethanol. In this study, different disinfection times were used to isolate rhizobacteria (disinfection time 0, only sterile distilled water was used) and endophytic bacteria (10 min disinfection followed by rinsing to eliminate ethanol) (Grobelak *et al.* 2014). The disinfected roots were tested for surface sterility via incubation on a solid nutrient agar medium for 3 days at 28°C. The roots were homogenized, serially diluted and placed in differentiating media: Congo-Red agar (CRA) or nitrogen-free base (NFb) media to isolate free-living diazotrophic bacteria, in Luria agar (LA) to isolate the nutritionally demanding bacteria, and Yeast-Extract-Manitol agar (YEMA) to isolate *Rhizobiaceae* bacteria. Next, bacteria were incubated at 30°C for 2 (CRA, NFb and LA) or 7 days (YEMA) to isolate bacteria. A total of 14 rhizosphere isolates and 38 endophytic isolates were collected. Donor bacteria were kept in 40% glycerol at –20°C for maintenance (Sgroy *et al.* 2009). Isolated bacteria were tested for 1-aminocyclopropane-1-carboxylate deaminase (ACCD) activity., Only 6 isolates with the highest activity of ACCD were selected for the further study.

DNA isolation and 16S rRNA gene analysis

The selected endophytic bacteria and rhizobacteria were cultured in an LB medium at 28°C for 24 h. Then, 1.5 ml of the LB medium with bacteria was transferred into the Eppendorf tubes and centrifuged at 45 000 g for 5 minutes at 21°C. The genomic DNA was extracted from the pellet using the GeneMATRIX Bacterial Genomic DNA Kit (EurX, Poland). The 16SrDNA genes were amplified by PCR using the universal sequencing primer 518F as the forward primer (5 -CCA GCA GCC GCG GTA ATACG-3) and the 800R as the reverse primer (5 -TAC CAG GGT ATC TAATCC-3). The amplification was performed using a thermal cycler (Eppendorf). The PCR fragments were frozen and stored at –20°C. The PCR products were separated using 1.5% agarose gel electrophoresis and Sanger sequenced. The 16S rDNA sequences were compared with registered sequences in the Gen Bank database (Fig. 1) using NCBI.

ACCD activity

The bacteria isolates were propagated in an LB liquid medium (to obtain OD600 = 1) and inoculated (1 ml) into 9 ml of M9 medium (5.8 g Na_2HPO_4, 3 g KH_2PO_4, 0.5 g NaCl, 1 g NH_4Cl, 0.25 mM $CaCl_2$, 1 mM $MgSO_4$, 0.15 % glucose, 0.3 µg/ml biotin). The isolates were incubated (3 days, 30°C). Afterwards, the cell suspension was centrifuged, washed in 0.1 M Tris-HCl (pH 7.5) and centrifuged again. For the centrifuged bacterial pellet, 4 ml of the modified M9 medium, enriched with ACC (5 mM) was added. The Eppendorf tubes were incubated in a laboratory shaker (3 days, 30°C). Afterwards, the bacterial suspension was centrifuged (10 min; 10 000 g), then washed twice in 0.1 M Tris-HCl, (pH 7.5) and centrifuged again. The bacterial pellet was diluted in 400 ml of 0.1 M Tris-HCl (pH 8.5). Subsequently, the 20 ml of toluene (5 %) was added and vortexed at the highest speed for 30 seconds. Then, 200 ml of cell extract was added with 20 ml of 0.5 M ACC. The samples were incubated at 30°C for 30 min. After the incubation period, the 1 ml of 0.56 N HCl was added and centrifuged (5 min; 10 000 g). Next, 800 ml HCl (0.56 N) was added to 1 ml of the supernatant, then vortexed and 300 ml 2.4-dinitrophenylhydrazine (0.2%) was added. The samples were incubated (30°C; 30 min) and then 2 ml NaOH (2 N) was added and vortexed. The absorbance of the prepared solution was measured at 540 nm using a scectrophotometer. The calibration curve was made on α-ketobutyrate (KB) in a concentration of 0.1 M to 1 M. The enzyme activity was expressed in mM/h of on α-ketobutyrate (KB) (Sgroy *et al.* 2009). Three replicates were used

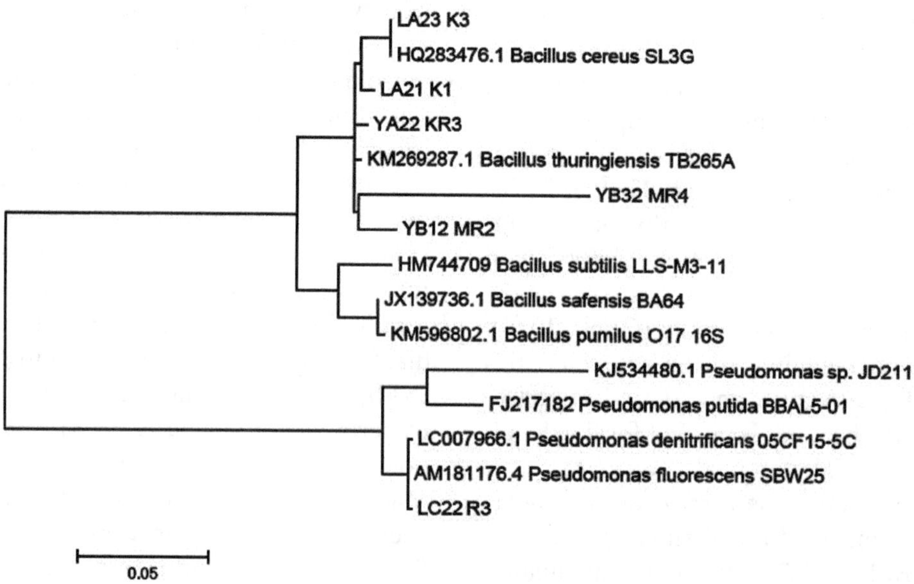

Fig. 1. Phylogenetic tree with two distinct clades corresponding to *Pseudomonas* sp. (endophytic strain R3) and *Bacillus* sp. (endophytic K1 strain and K3 strain identified as *B. cereus* and rhizobacteria strains: MR4, MR2 and KR3 identified as *B. thuringiensis*)

for all study samples. A mathematical analysis of the data received was made in the Excel 2007. An alteration of ACC concentration in samples was obtained by normalization to the average value. Data was also analyzed using the method of one-way analysis of variance (ANOVA).

Preparation of the bacterial consortium

In the experiment, two consortia were used. All bacterial strains were tested in plates on antagonistic activity between the strains. On the basis of the ACC deaminase activity, three isolates with the highest ACCD activity were selected for studying the effects of the strains on plant growth (Fig. 2). The first consortium consisted of three strains of rhizospheric isolates and a second consortium comprised three strains of endophytic isolates. The next phase was to immobilize the selected consortia of sodium alginates solution with a concentration of 2% w/v of LB medium (Luria Bertani) (0.8 g with 40 ml of double distilled H_2O) and calcium chloride solution (2.5% w/v). The sodium alginate was mixed with a suspension of bacteria (bacteria were propagated in LB medium, OD600 nm=1, then centrifuged) and stirred until homogeneous. The suspension of bacteria was transferred into the cylinder and added dropwise to a solution of calcium chloride. It was obtained in the form of a bed of beads with a diameter of 1.5–3 mm. Then, the conditioning time was applied (10 minutes) and beds were filtered on a Buchner funnel (Shin *et al.* 2007). The next phase of the experiment was the application of consortia to the trace elements contamination soil (Grobelak and Napora, 2015). The soil (Pb 9045–1209.7 mg/kg d.w; Cd 8.57–16.8 mg/kg d.w; Zn 773–1100 mg/kg d.w.) was sieved through a sieve with 1 mm mesh size. Due to the low pH of the soil, the lime fertilizer (0.8% weight) was added. Three replicates (pots) of each treatment were prepared. The concentration of bacterial cells calculated for OD600 nm=1.0 was determined for $60\times10^{7.6}$ to $10^{8.2}$ cells per pot (Reed and Glick, 2005). The control soil was also amended with alginate without the bacteria. The seeds were sown directly in the contaminated soil after mixing the soil with alginate beds (with or without bacteria). The experiment was carried out in the growth chamber under controlled conditions, optimal for the growth and development of plants (temperature: day 21°C, night 18°C, humidity approx. 85%). In this study, two plant species were used: *B. napus* L. and grass used in phytoremediation – *F. rubra* L. A plant growth experiment was conducted for 3 months (pot volume 2 l). Then, the obtained biomass gain was determined.

RESULTS AND DISCUSSION

In order to classify the different unknown bacteria, the rhizobia phylogenies based on the 16S rRNA gene were presented (Fig. 1). The 16S rRNA sequence obtained in this way was compared using BLAST tool with nucleotide database. It was found that six sequences with accurate

a) b)

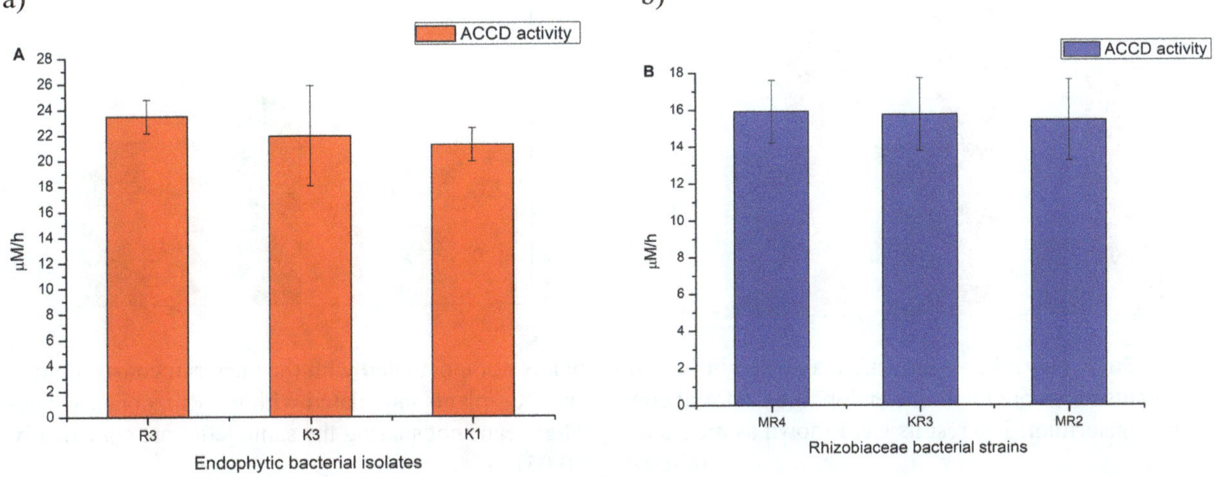

Fig. 2a. ACCD activity of (A) endophytic strains: R3 (23 mM/h); K3 (21.94 mM/h); K1 (21.23 mM/h). The results were shown as mean with SD

Fig. 2b. ACCD activity of rhizospheric strains: MR4 (15.92 mM/h); KR3 (15.76 mM/h); MR2 (15.45 mM/h). The results shown as mean with SD species annotation showing the highest sequence homology proportion and query coverage with lowest the lowest E values were used to develop Neighbor Joining (NJ) tree using MEGA 4.0 for multiple sequence alignment. Phylogenetic tree analysis showed that strains have maximum similarity with: *R3(Pseudomonas sp.), K3 (B. cereus), K1 (Bacillus sp.)* and rhizobacteria strains*: MR4 (Bacillus sp.); KR3 (B. thuringiensis.); MR2 (Bacillus sp.)*. In the biochemical tests, 38 endophytic isolates and 14 isolates of rhizobacteria were tested for the ACCD activity. It was found that 85% of endophytic isolates produced ACC-deaminase. The highest activity of ACC deaminase was observed for R3 isolate (23.49 μM/h); 79% of rhizobacteria isolates produced ACC-deaminase. For these strains, the activity was high and was recorded at the level of (15.92 μM/h) for MR4.

The effect of plant growth promotion was found for the consortia of three isolates. These results are in agreement with the studies where the inoculation effect of the plant growth-promoting bacteria (PGPB) on the growth of plants was determined (Rashid *et al.* 2011). The consortium of bacteria is capable of rapid colonization of the roots during the growing season. It is known that microorganisms isolated from the environment, contaminated with heavy metals, often exhibit tolerance to multiple pollutants because they have adapted to such conditions (Pal *et al.* 2005). According to Glick (2014), ACC may act as a source of nitrogen for some soil bacteria. Recent studies suggest that the mechanism is more complicated and the ACC deaminase activity may be more important in stress minimizing, than in competition among other rhisospheric microorganisms. The investigated soil was also very poor in microorganisms.

The biochemical tests have shown that the endophytic bacteria strains have a greater potential for ACCD production (Fig. 2). The ACC deaminase has been proposed as an agent that plays an important role in the association process of plants and bacteria (Reiter *et al.* 2002). PGPB can significantly increase the growth of plants, especially under difficult conditions or the presence of heavy metals (Khan, 2005). This study has shown that plant inoculation with bacteria consortium (endophytic or rhizobacteria) significantly affected the weight and height of plants (Fig. 3–4).

The endophytic microbes residing inside the host plants, with ACC deaminase activity, can benefit the host plant by reducing stress and increasing the plant growth (Hardoim *et al.* 2008). In the study of Belimov *et al.* (2005), the isolated cadmium-resistant *V. paradoxus* from the rhizosphere of *B. juncea* promoted the plant growth. In general, the metal resistant rhizospheric bacteria have an exceptional ability to protect the host plants from metal toxicity by several possible mechanisms, the best known mechanism being the utilization of ACC by rhizospheric bacteria.

a)

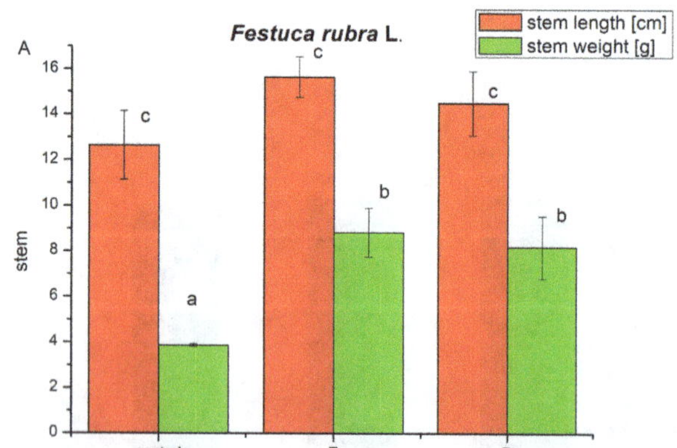

b)

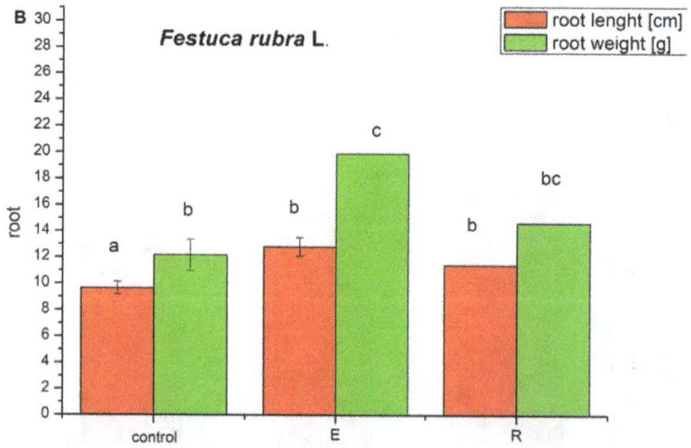

Fig. 3a. *F. rubra* L. – stem length and weight (control – plants not inoculated with the bacterial consortium; E – plants inoculated with the endophitic bacterial consortium; R – plants inoculated with the Rhizobiaceae bacterial consortium. The results were shown as means (SD). The means not sharing the same letter are statistically different ($p<0.05$)

Fig. 3b. *F. rubra* L. – roots length and weight (control – plants not inoculated with the bacterial consortium; E – plants inoculated with the endophitic bacterial consortium; R – plants inoculated with the Rhizobiaceae bacterial consortium. The results were shown as means (SD). The means not sharing the same letter are statistically different ($p<0.05$)

a)

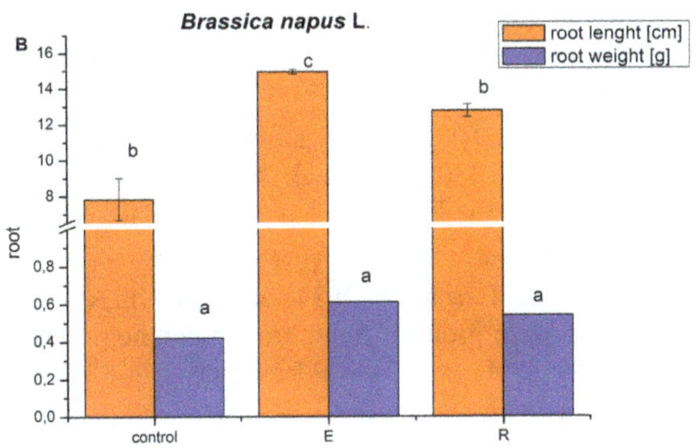

b)

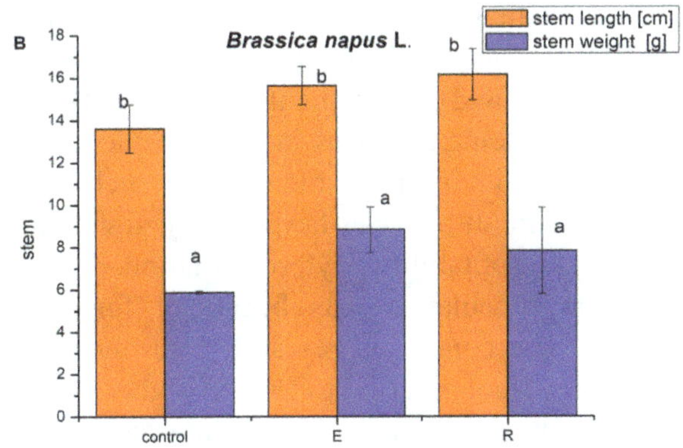

Fig. 4a. *B. napus* L. – stem length and weight (control – plants not inoculated with the bacterial consortium; E – plants inoculated with the endophitic bacterial consortium; R – plants inoculated with the rhizospheric consortium. The results were shown as means (SD). The means not sharing the same letter are statistically different ($p<0.05$)

Fig. 4b. *B. napus* L. – roots length and weight (control – plants not inoculated with the bacterial consortium; E – plants inoculated with the endophitic bacterial consortium; R – plants inoculated with the rhizospheric consortium. The results were shown as means (SD). The means not sharing the same letter are statistically different ($p<0.05$)

Dell'Amico *et al.* (2008) proved that the bacteria having the characteristics of producing ACC deaminase, may enhance the metal accumulation in plant tissue with concurrent stimulation of plant growth. Inoculation with endophytic isolates might have a significant potential for improving the phytoextraction efficiency in metal-contaminated soils (Rajkumar *et al.* 2009). Statistical analysis of the results (*F. rubra* L., *B. napus* L.) showed a significant impact on the plants inoculation (with the endophytic bacterial consortium and rhizobacteria). A one-way ANOVA test showed that the results are statistically significant for both the weight and length of the root and the stem of the tested plants. The highest increase in the weight and length of the stem (Fig. 3) for *F. rubra* L. was obtained after inoculation of the endophytic consortium. These results indicated that the activity of ACCD has an impact on the effective promotion of plant growth, the highest ACCD activity and the highest biomass. Similar pieces of evidence that endophytic bacteria promote plant growth were obtained from crop studies, including potato and clover (Bashan and Bashan, 2005). The highest increase in the weight and length of the root (Fig. 3) of *F. rubra* L. was also obtained after inoculation of the endophytic bacterial consortium. The increase in the above-ground parts (stems) was correlated with better root development (Fig. 5). The role of ACCD in decreasing the ethylene levels by the enzymatic hydrolysis of ACC into KB and ammonia has been presented as one of the major mechanisms of PGPB in promoting the root development and plant growth under metal stress conditions (Madhaiyan *et al.* 2006). The inoculation of rape (*B. napus* L.) (Fig. 4) with a bacteria consortium has an impact on its growth. For an increased stem length, a more pronounced effect was observed after inoculation with a rhizospheric consortium. For the stem weight of rape, higher results were noted after inoculation with endophytic consortium (statistically confirmed impact). As shown in Figure 4, inoculation of plants with a bacterial consortium resulted in the growth of root biomass. The endophytic consortium of bacteria was more effective in promoting root length. Bacterial strains utilizing ACC as a sole source of nitrogen use ACC and enhance the elongation of plant roots (Glick *et al.* 1998). This was also confirmed in this study. The bacteria used in this experiment improved the plant growth and enhanced the dense root system. The obtained results suggest an increase of plant tolerance to metals. In turn, well developed plants are more capable of restoring the plant cover the soils contaminated with trace elements.

CONCLUSIONS

In conclusion, the results indicate that inoculation with ACCD active PGPB may facilitate plant growth and thus increase the phytoremediation efficiency. It was found that the applied bacteria improved the biomass of plants under very difficult soil conditions. PGPB containing ACCD activity could be helpful in sustaining plant growth and development under stress conditions by reducing the stress-induced ethylene production. Achieving plant biomass for such areas is very difficult and can have a very significant impact on profitability phytoremediation.

Plant inoculation was beneficial for plant growth. The conducted research confirmed that among (52) isolates, (6) have much higher ACCD activity. Our research confirmed that such isolates can be used to improve the plant growth in heavy metal contaminated areas. The conducted research showed that high bacterial ACCD activity has an impact on promotion of the growth of roots biomass. It was found that the bacteria screening in soils with over 50 years toxic trace elements contamination have very high ACC deaminase activity. Moreover, it was discovered that the bacteria mainly belonging to genus *Bacillus* and *Pseudomonas* had the highest AAC deaminase activity in the environment contaminated with multiple heavy metals.

Some bacteria can adapt to stress conditions and metal-resistant bacteria can survive in contaminated environment. The use of plant growth promoting bacteria in assisted phytoremediation of metal contaminated sites is necessary.

Acknowledgements

This work was supported by NCN-DEC-2011/03/N/NZ9/02034 grant, TANGO1/266740/NCBR/2015, BS/MN–401–301/17 and BS/PB–401–304/11 grant.

REFERENCES

1. Bashan Y.de-Bashan I.E. 2005. Bacteria/plant growth-promotion. In: Hillel, D. (eds), Encyclopedia of Soils in the Environment, 1. Elsevier, Oxford, UK, 103–115.
2. Belimov A., Hontzeas N., Safronova V.I., Demchinskaya S.V., Piluzza G., Bullitta S., Glick B.R. 2005. Cadmium-tolerant plant growth-promoting bacteria associated with the roots of Indian mustard (Brassica juncea L Czern.), Soil Biol. Biochem., 37, 241–250.
3. Dell'Amico E., Cavalca I., Andreoni V. 2008. Improvement of Brassica napus growth under cadmium stress by cadmium-resistant rhizobacteria. Soil Biol. Biochem., 40, 74–84.
4. Glick B.R., Penrose D.M, Li J. 1998. A model for the lowering of plant ethylene concentrations by plant growth-promoting bacteria. J. Theor. Biol., 190, 63–68.
5. Glick B.R., Cheng Z., Czarny J., Duan J. 2007. Promotion of plant growth by ACC deaminase-producing soil bacteria, Eur J Plant Pathol., 119, 329–339.
6. Glick B.R., Bernard R. 2014. Bacteria with ACC deaminase can promote plant growth and help to feed the world. Microbiological Research. 169, 1, 30–39.
7. Grobelak A., Kacprzak M., Grosser A., Napora A. 2013. Chemofitostabilizacja gleby zanieczyszczonej kadmem, cynkiem i ołowiem. Annual Set The Environment Protection 15(2),, 1982--2002
8. Grobelak A., Napora A., Kacprzak M. 2014. The impact of plant growth promoting bacteria (PGPB), on the development of phytopathogenic fungi. Folia Biologica et Oecologica. 10, 1, 107–112. DOI: 10.2478/fobio-2014–0008.
9. Grobelak A., Napora A. 2015. The Chemophytostabilisation Process of Heavy Metal Polluted Soil. PLoS ONE 10(6), e0129538. doi:10.1371/journal.pone.0129538.
10. Hadi F., Nasir A., Ayaz A. 2014. Enhanced Phytoremediation of Cadmium-Contaminated Soil by Parthenium hysterophorus Plant: Effect of Gibberellic Acid (GA3) and Synthetic Chelator, Alone and in Combinations, Bioremediat J. 18,1, 46–55.
11. Hardoim P.R., van Overbeek L.S, van Elsas J.D. 2008. Properties of bacterial endophytes and their proposed role in plant growth, Trends Microbiol., 16, 463–471.
12. Kacprzak M., Grobelak A., Grosser A., Napora A. 2014. The potential of biosolid application for the phytostabilization of metals. Desalin Water Treat., 52, 3955–3964.
13. Kanclerz J., Borowiak K., Mleczek M., Staniszewski R., Lisiak M. 2016. Phragmites australis and Typha angustifolia as Potential Accumulators of Zinc and Copper in Water Ecosystem at City Area. Annual Set The Environment Protection, 18, 246–257
14. Khan A.G. 2005. Role of soil microbes in the rhizospheres of plants growing on trace metal contaminated soils in phytoremediation. J. Trace Elem. Med. Biol., 18, 355–364.
15. Ma Y., Rajkumar M., Freitas H. 2009. Inoculation of plant growth promoting bacterium Achromobacter xylosoxidans strain Ax 10 for the improvement of copper phytoextraction by Brassica juncea. J. Environ. Manage., 90, 831–837.
16. Madhaiyan M., Poonguzhali S., Ryu J.H., Sa T. M. 2006. Regulation of ethylene levels in canola (Brassica campestris) by 1-aminocyclopropane-1-carboxylate deaminase-containing Methylobacterium fujisawaense, Planta., 224, 268–278.

17. Pal A., Dutta S., Mukherjee P.K., Paul A.K. 2005. Occurrence of heavy metal-resistance in microflora from serpentine soil of Andaman. J. Basic Microbiol., 45, 207–218.

18. Rajkumar M., Ae N., Freitas H. 2009. Endophytic bacteria and their potential to enhance heavy metal phytoextraction. Chemosphere., 77, 153–160.

19. Rashid S., Charles T.C., Glick B.R. 2011. Isolation and characterization of new plant growth-promoting bacterial endophytes. Appl Soil Ecol., 61, 217–224.

20. Reed M.L.E, Glick B.R. 2005. Growth of canola (Brassica napus) in the presence of plant growth-promoting bacteria and either copper or polycyclic aromatic hydrocarbons, Can J of Microbiol., 51, 12, 1061–1069.

21. Reiter B., Pfeifer U., Schwab H., Sessitsch A. 2002. Response of endophytic bacterial communities in potato to infection with Erwinia carotovora subsp., Atrospetica, Appl. Environ. Microbiol., 68, 2261–2268.

22. Saleem M., Arshad M., Hussain S., Bhatti A.S. 2007. Perspective of plant growth promoting rhizobacteria (PGPR) containing ACC deaminase in stress agriculture, J Ind Microbiol Biotechnol., 34, 635–648.

23. Salihaj M., Bani A., Echevarria G. 2016. Heavy metals uptake by hyperaccumulating flora in some serpentine soils of Kosovo, Global NEST Journal, 18, 1, 214–222.

24. Sgroy V., Cassán F., Masciarelli O., Florencia M., Papa D., Lagares A., Luna V. 2009. Isolation and characterization of endophytic plant growth-promoting (PGPB) or stress homeostasis-regulating (PSHB) bacteria associated to the halophyte Prosopis strombulifera. Appl Microbiol Biotechnol., 85, 371–381.

25. Shin S.J., Han J.H., Manning E.J., Collins M.T. 2007. Rapid and reliable method for quantification of Mycobacterium paratuberculosis by use of the BACTEC MGIT 960 System, J Clin Microbiol. 45, 6, 1941–1948.

26. Wong-Villareal A., Reyes-López L., Gonzalez H.C., Gonzalez C.B., Yanez-Ocampo G. 2016. Characterization of bacteria isolation of bacteria from Pinyon Rhizosphere, producing biosurfactants from agro-industrial waste. Pol J Microbiol., 65, 2, 183–189.

24

Application of Steel-Smelting Slags as Material for Reclamation of Degraded Lands

Vera Matveeva[1], Tatiana Lytaeva[1], Aleksandr Danilov[1*]

[1] Geoecology Department, Saint-Petersburg Mining University, 2, 21st Line, Saint-Petersburg, 199106, Russia

* Corresponding author's e-mail: aleksandrsdanilov@gmail.com

ABSTRACT

Presently, the highest level of waste generation is characterized by ferrous metallurgy enterprises, the large-tonnage wastes of which are slags of various industries. Taking into account the negative impact of slags on the environment, the limited opportunities for expanding dump volumes, as well as the depletion of sources of raw mineral raw materials, the aim of research was to develop an alternative method for the utilization of steel-smelting slags of the mining and metallurgical complex. Tthe steel smelting slag of the Novolipetsk metallurgical plant and the sewage sludge of the city water treatment station in Lipetsk were selected as the materials for the study. In the laboratory, the chemical composition of the waste was determined by means of a complex of modern analytical methods, including atomic emission spectrometry with inductively coupled plasma, X-ray fluorescence spectrometry, and atomic absorption spectrometry with electrothermal atomization. The results of the analysis enabled to assess the possibility of sharing waste as a soil additive to improve the quality of the soil. Then, under laboratory conditions, the grass mixture was grown on model soil samples with a soil additive consisting of the investigated waste with a different mass ratio. On the basis of the obtained data on seed germination, the growth of the aerial part of the biomass, the height of the leaves, and also the content of the accumulated metals in the plants, a formulation of an organomineral additive was proposed for poor soils of disturbed lands.

Keywords: mining and metallurgical enterprises, steelmaking slag, sewage sludge, organomineral additive, recultivant, growing of grass

INTRODUCTION

The quality of human habitat is influenced by several factors, including the ones caused by human activities. Waste generation is one of the negative consequences of human life. Today in Russia, about 80 million tons of wastes are stored in dumps and tailing dumps. The enterprises of ferrous metallurgy are characterized by the highest level of waste generation, the annual output of slags at these plants is more than 95 million tons, whereof 79 million tons are blast furnace slag, steel smelting, foundry and ferroalloy productions [Shapovalov et al., 2013]. The area of land occupied by slag heaps in Russia is more than 1 million hectares, with an annual fee for waste disposal reaching tens of millions rubles.

Operation of waste disposal sites leads to environmental degradation and, as a consequence, an increase in the morbidity level. The main reason for the atmospheric air quality decrease is sweeping-away of waste fine fraction from the surface of dumps. This subsequently leads to the contamination of soils and vegetation cover of adjacent territories. Accumulation of pollutants takes place in the upper layer of the soil, which significantly affects the vital activity of plants and biodiversity in the area influenced by the enterprises of the metallurgical industry. Infiltration of atmospheric precipitation and surface runoff through the body of slag dump promotes leaching of pollutants from waste, including heavy metals, which, entering the soil, underground and surface waters, form polluted area and pollution streams. In order to reduce the negative impact of slag dumps on the environment in the iron and steel enterprises recycling system have successfully been implemented. At modern metallurgical plants, blast furnace slags are practically not stored in dumps, but are used as technogenic raw materials in various industries. However, the application of steel smelters in the cement industry and road construction, without special processing, is limited due to the inconstancy of the chemical and mineral compositions, as well as the lack of stability of the emerging structure. As a result, only 20-30% of the total steelmaking slag generated in the Russian Federation is recycled annually [Yaroshenko, 2011].

The development of alternative ways of recycling and processing steelmaking slags has become the goal of research. These methods are aimed at reducing the technogenic load on the environment.

The steel-smelting slag of the Novolipetsk Metallurgical Combine was the subject of research. Annually, the enterprise consumes about 14 million tons of iron ore. One of the main environmental problems of the company is the storage of steelmaking slag in a total area of 16 hectares, while the capacity of slag disposal is about 2 million tons. In this regard, the main tasks of the conducted research were studying the composition and properties of steelmaking slag, as well as reducing its negative impact on the environment by using waste as a product for the reclamation of disturbed lands.

MATERIALS AND METHODS

Insufficient use of metallurgical slag leads not only to the incomplete profit, but also to a rise in the price of the main metallurgical products, due to the high costs of transportation of slags and the maintenance of huge dumps. At present, there are many promising methods for processing steelmaking slags, including obtaining staflux for the sinter production, the production of crushed slag stone for road and industrial construction, the use of waste as fertilizer. [Bobrova et al., 2015, Kogan & Shakhparonova, 2017]. However, all of them still remain unexplored or unclaimed for a number of reasons. Thus, the use of steel-smelting slags in agriculture is limited primarily because of the presence of toxic elements such as As, Pb, Cd, Co, Cr, etc., which can be washed out of the slag and cause the death of plants and soil contamination. In this connection, a series of experimental studies was carried out at the Laboratory of the Mining University devoted to a comprehensive analysis of the chemical composition of the steel smelting slag of the Novolipetsk Metallurgical Combine and the cultivation of grass mixtures on model soil samples with a mineral additive in the form of the studied slag. The primary task was to determine the chemical composition of the material using such modern analytical methods as X-ray fluorescence spectrometry and atomic absorption spectrometry with ETA. The results of the determination of the macrocomponent composition of the sample are shown in Table 1.

As shown in Table 1, the main component of slag is calcium. A phase analysis by X-ray powder diffractometry was carried out to establish the shape of calcium in the waste. The results of the diffractometric analysis showed that the calcium in the sample is mainly portlandite ($CaO{\cdot}H_2O$), calcite ($CaCO_3$), sperrit ($2Ca_2SiO_4{\cdot}CaCO_3$), calcium aluminate ($5CaO{\cdot}3Al_2O_3$), calcium silicate (Ca_2SiO_4). It is known that lime and calcite contribute to the nutrition of plants, reduce the acidity of the soil, loosen it, contributing to the retention of moisture, stimulate the vital activity of useful microorganisms [Stepanova & Shamaraeva, 2012]. Some of the lime is strongly bound in silicate compounds, which hinders its rapid washing, observed with conventional calcareous fertilizers, thereby increasing the absorption capacity of the soil and prolonging the operation of the mineral additive. In addition, calcium silicates have practically the same neutralizing effect as oxide and calcium carbonate. In significant amounts, magnesium in the form of brucite ($Mg(OH)_2$) and periclase (MgO) is detected in the waste, the lack of which is caused by diseases in plants. The elements useful for plants such as iron, silicon and aluminum also entered the macrocomponent composition of the slag. In particular, silica is well absorbed by plants and increases the strength of stems.

Further on the high-tech equipment, the micro-component composition of the steel-smelting slag was determined. The results are shown in Table 2.

Table 1. Chemical composition of steel-smelting slag (Novolipetsk Metallurgical Combine)

Element	Ca	Fe	Mg	Si	Al	Mn	C
Content, %	30.4	18.4	3.4	6.8	1.3	1.2	1.1

Table 2. Microcomponent chemical composition of the steel-smelting slag (Novolipetsk Metallurgical Combine)

Element	Units	Content
Ti	%	0.2
V	%	0.1
S	%	0.1
Cu	mg/kg	71
Zn	mg/kg	640
Pb	mg/kg	120
Cr (III)	mg/kg	38
Cd	mg/kg	5.9
Ni	mg/kg	0.2
Hg	mg/kg	0.1
As	mg/kg	<0.05

The results presented in Table 2 indicate that most of the impurity elements are represented by heavy metals. In terms of significance, heavy metals for plants can be divided into two groups:

- necessary for the life of plants in small concentrations (Co, Cr, Cu, Fe, Mn, Mo, Ni, Zn), become toxic only with a significant increase in their content in soil and plants;
- not participating in the metabolism of plants (Cd, Hg, Pb, V) and toxic even at very low concentrations.

The role of the necessary heavy metals in the vital activity of plants is extremely high. Metals-microelements stimulate growth, synthesis of proteins, fats and carbohydrates, participate in metabolic processes, bind to biologically active substances (hormones, vitamins, proteins), increase the plant immunity and chlorophyll content, as well as stabilize green pigments during aging of chloroplasts [Baisetova & Sartaeva, 2014, Titov & Laidinen, 2007]. That causes the use of a number of heavy metals as microfertilizers. For this reason, it was decided to develop a preparation for the reclamation of disturbed lands under laboratory conditions. As part of the preparation for reclamation, steel smelting slag of the Novolipetsk Metallurgical Combine was used. It was decided to grow the mixture on model soil samples. The production of a highly effective complex additive in the soil involves the use of both mineral and organic components. As a source of organic substances in the experiment, the sewage sludge of the urban aeration station was used. It is assumed that a mixture of sediment and slag will create an organomineral additive that can be used as fertilizers for planting forestry crops along roads, in seed plots of forest and ornamental crops, for reclamation of disturbed lands and slopes of highways and solid waste dumps. As a control sample, the soil of the Lipetsk region was used, which refers to gray forest soils. Before planting the mixture, the annual permissible doses of slag and sludge were determined, which can be introduced into the soil taking into account its properties and actual contamination.

The permissible dose of metals in soil, taking into account its properties and the actual pollution in terms of dry matter, g / ha, was calculated by means of the following formula

$$D_{ad} = (0{,}8 \cdot \mathrm{MPC} - \mathrm{F}) \cdot \mathrm{M} \quad (1)$$

where: 0.8 – the correction factor that reduces regulatory entry of heavy metals into soil by 20%;
MPC – maximum permissible concentration of metals in soil, g/t;
F – the metal content in soil, g/t;
M – the soil arable layer mass in terms of dry matter, t/ha (3000 t/ha).

The annual dose of applying metal to the soil in terms of dry matter for 10 years, t / ha, was determined with the formula:

$$D_{av} = \frac{D_{ad}}{T \cdot C} \quad (2)$$

where: T – the maximum total period of fertilizer application for the same site, years (10 years);
C – the concentration of the metal in the additive, g/t.

Table 3 shows that the maximum permissible concentration of metals in the soil will not be exceeded for all the components under study if the

Table 3. Calculation of the permissible doses of heavy metals, arsenic and the average annual dose of slag and sewage sludge to soil

Element	Admissible doses, kg/ha	Annual dose, t/ha	
		Sewage sludge	Slag
Cd	2.0	152.3	33.6
Cu	158.2	98.9	222.8
As	11.9	438.9	11850.0
Ni	93.0	794.9	46500.0
Hg	4.4	740.0	4440.0
Pb	153.0	327.6	127.5
Zn	262.0	49.2	40.9
Cr (III)	13.4	27.2	35.5

following amounts of sediment and slag are added to the soil in terms of dry matter: 27.2 and 33.6 t/ha, respectively. These values are determined on the basis of the maximum permissible amounts of chromium (III) and cadmium, which may appear in the soil sample after the incorporation of additives.

On the basis of the calculated data, the following types of soil additives were prepared:

- "Type 1" – steelmaking slag;
- "Type 2" – sewage sludge of the urban aeration station;
- "Type 3" – sewage sludge of the urban aeration station + steelmaking slag (1 : 3);
- "Type 4" – sewage sludge of the urban aeration station + steelmaking slag (1 : 1);
- "Type 5" – sewage sludge of the city aeration station + steelmaking slag (3 : 1).

The masses of additives introduced into the soil were determined based on the area where the fertilizer was applied, as well as the moisture values of the steelmaking slag and the sewage sludge. Table 4 presents the estimated doses of sewage sludge from the water treatment plant and steelmaking slag for laboratory research.

In order to prepare the soil additive, the steel-smelting slag was crushed to a size of 0–5 mm. The homogenization of the steelmaking slag and sewage sludge was carried out by mixing. The resulting mixture was applied to the depth of the arable layer (15–20 cm), after which sowing of the lawn grass mixture (60 kg/ha), represented by English bluegrass (perennial tare), was carried out.

English bluegrass is well-adapted to the conditions of a humid temperate climate and refers to the plants of the winter medium-ripening type of development. The main advantage of this herb is its ability to form dense herbage a month after sowing. Particular attention in the process of cultivation was paid to the observance of optimal microclimatic conditions: illumination, temperature and humidity. One of the important factors affecting the growth of plants is the temperature that regulates the activity of enzymes. The optimal temperatures for plant growth are in the range from +18 to +30°C. Illuminance, necessary for photosynthesis in plants, is 600–2000 lux. Sufficient illumination triggers the process of photosynthesis in plants, through which they produce organic compounds from inorganic ones. Humidity also affects the growth and development of plants. Low air humidity increases transpiration and evaporation of water from the substrate. The lower the humidity of the air, the greater the evaporation of water by leaves and soil, and more irrigation is required. On average, the humidity level for normal plant growth should be 60–90% [Zimina & Kukushkin, 2014]. Table 5 presents

Table 4. Estimated doses of sedimentation of sewage sludge and steelmaking slag

Additive type	Fertilized area, cm^2	The annual dose taking into account humidity, g/cm^2	
		sewage sludge	steelmaking slag
1	152	-	57 165
2		71 715	-
3		21 686	41 980
4		40 604	26 201
5		56 950	12 250

Table 5. Microclimate characteristics of laboratory room

Characteristic	Units	Condition
Air temperature	°C	20
Air humidity	%	88
Illumination	lux	770

the microclimatic characteristics of the room in which the grass mixture was grown.

For the normal development of plants, the maintenance of soil processes and the level of soil fertility, the moisture content in the soil and the level of its acidity play an important role. When there is a lack of moisture, the turgor pressure of the cells decreases, their elasticity is lost, the dynamics of biochemical processes decreases, the absorption of carbon dioxide through the stomata decreases, and the inhibitor substances accumulate in the biomass. With an excess of moisture, the oxygen metabolism is disturbed in plants and in the soil-accumulated compounds, which leads to a decrease in the bioproductivity of the soil, as a result of the death of plants [Bayshanova & Kedelbaev, 2016]. In order to maintain the optimum moisture of the model samples, watering was done 1–2 times a week. The volume of added water corresponded to the norm of precipitation in the territory of the Lipetsk region in the warm period of the year. In view of the significant effect of soil acidity on the growth and development of plants, the soil acidity of the model samples was checked during the vegetative period of the grass mixture.

As a result, the main indicators of the experiment were:

- evaluation of seed germination;
- estimation of the increment in the aerial part of the biomass;
- assessment of the content of accumulated metals in grown plants.

RESULTS

Several features may be indicative of ongoing processes in plants growth. These include an increase in mass of plants and parts thereof, the number of cells, linear dimensions (length, height, thickness, and diameter). Separately, each of these signs does not always reliably indicate the presence of growth processes in plants, so a complete representation can be obtained by simultaneously taking into account several indicators. On the fourth day after sowing of the lawn grass mixture, seed germination was observed, while the laboratory seed germination in the control sample was 81%, germination in the remaining samples was 2–6% higher (Table 6), which was due to the introduction of additional substances into the soil.

After the rising of the grass mixture, the height of the leaves was maintained at 10 cm, while the aerial part of the biomass was weighed after each cutting. After 20 weeks, at the end of the growing season, the laboratory experiment was stopped. The value of the aerial part of the biomass is shown in Table 7.

The data shown indicate a decrease in the aerial part of the biomass of the grown grass mixture by 7.5% in the model sample with the addition of "type 1" relative to the control sample (soil) and the likely presence of a factor of plant oppression. A positive effect was observed when the sediment and slag were added together to the soil, based on the observed increase in the aerial part of the bio-

Table 6. Bluegrass seed germination in laboratory conditions

Sample	Seed germination, %
Soil (control sample)	81
Soil + additives «Type 1»	83
Soil + additives «Type 2»	85
Soil + additives «Type 3»	87
Soil + additives «Type 4»	87
Soil + additives «Type 5»	87

Table 7. Value of grass mixture aerial part obtained during the vegetation period

Value of the aerial part of the biomass, g/m²					
Soil (control sample)	Soil + additives «Type 1»	Soil + additives «Type 2»	Soil + additives «Type 3»	Soil + additives «Type 4»	Soil + additives «Type 5»
825.9	763.6	877.0	858.4	987.1	1075.7

mass by 3.9–30.2%, relative to the control sample. During the experiment, the height of the leaves was measured weekly before the cutoff of the aerial biomass. It should be noted that the growth rate of the plants of the grass mixture planted on model samples was practically the same in all cases with minor deviations. The greatest increase was in the plants planted in model samples with the addition of organomineral additives. The introduction of additional substances into the soil only with slag led to the withering of the plants; thus, during the experiment, the smallest proportion of the aerial part of the biomass and the minimal growth of the plant leaves relative to the control sample were recorded in these model samples.

Comparing the obtained data of plant growth, it can be concluded that the deviations from a control reference sample (soil) exceeding 20% were not observed in any of the studied models, indicating that there is no toxic effect of the mineral and toxic additives used on the development of lawn grass mixtures. In addition, it is worth noting that during the growing season, diseases such as chlorosis and necrosis in plants were not observed. Throughout the vegetative period of the plants, the soil acidity of the model samples was monitored. The essence of the method was the extraction of exchangeable cations, nitrates and mobile sulfur from the soil with a solution of potassium chloride with a concentration of 1 mol / dm^3 (1 N) and potentiometric determination of the pH [GOST Soils, 1985]. The results showed that when applying both mineral and organic additives to the soil by the end of the growing season, the pH of the salt extract increased by 1 to 2 units of pH, relative to the reference. The greatest change in soil pH was observed in a model soil sample with the addition of slag, which probably leads to the inhibition of plants. However, in the remaining model samples, there was no apparent difference in plant growth and biomass increment relative to the control sample, which is due to the additional application of organic substances. Since the blue grass used in the studies refers to the plants with a high resistance to changes in soil pH values, in order to study the effect of soil acidity on the germination of plants in more detail, addition was also made to plant meadow fescue, also part of lawn grass mixtures, but more sensitive to increased acidity environment. As a result, the germination of plants in the soil with additives decreased with respect to the control reference sample. Consequently, while choosing the planned grass seed for recultivation, it is necessary to take into account its resistance to changes in the acidity of the medium. At the final stage, studies related to the determination of the content of accumulated metals in grown plants were carried out. This stage has become key since while using industrial waste and municipal services as fertilizers of agricultural soils, it is necessary to carry out systematic monitoring of soil pollution and plant production with heavy metals. At the end of the growing season, the aerial part of the grass cover was cut and dried to an air-dry state in a well-ventilated room. Afterwards, the plant samples were crushed and mineralized by the dry ashing method. In order to determine the accumulated metals in plants, their acid extraction from the mineralized part was carried out. Metals in the ash solutions of plant samples were determined by means of an atomic-emission spectrometer with inductively coupled plasma. The results of the determination of the accumulated metals in plant leaves of the mixture are shown in Table 8.

The analysis of plants, grown using organic and mineral additives, revealed deviations in the content of copper, nickel and zinc, relative to the control soil sample. Thus, the use of steel melting slag as a soil for remediation leads to an increase in the copper content in plants by a factor of 2, nickel by a factor of 1.1, zinc by 1.3 times, with respect to the control sample. As a result of using the sewage sludge of the aeration station for recla-

Table 8. Metal content in the leaves

Sample	Metal content in plant leaves, mg/kg					
	Cd	Cu	As	Ni	Pb	Zn
Soil (control sample)	< 0.5	18.2	<0.05	<3.5	<5.0	26.4
Soil + soil additives Type 1	< 0.5	38.0	<0.05	4.0	<5.0	35.2
Soil + soil additives Type 2	< 0.5	23.3	<0.05	<3.5	<5.0	27.0
Soil + soil additives Type 3	< 0.5	34.5	<0.05	<3.5	<5.0	33.5
Soil + soil additives Type 4	< 0.5	31.0	<0.05	<3.5	<5.0	30.1
Soil + soil additives Type 5	< 0.5	29.5	<0.05	<3.5	<5.0	29.4
Permissible concentration in soil	1.0	66.0	5.0	40.0	65.0	110.0

mation, copper increases by 1.3 times, the content of nickel and zinc remains at the level of the reference sample. While using an organomineral additive with various weight ratios of sediment and slag as a soil for remediation, a minimal increase in the metal content is observed. Due to the fact that the model samples are planned to be used as fertilizers, a comparison of the content of accumulated metals in plants was carried out with the maximum permissible concentrations of metals in the soil. Despite the fact that the metal content in the grown plants on model samples with different additives is higher than the values in the control sample, the maximum permissible concentrations for these elements in the soil were not exceeded. Comparing the obtained data on the germination of plant seeds, the growth of the aerial part of the biomass, the height of the leaves, and also the content of the accumulated metals, it can be concluded that the most favorable conditions for the development of plants are the use as a soil for remediation of an organomineral mixture with a mass ratio of sewage sludge from a city aeration station and steelmaking slag 3: 1. The results of the conducted experiments open the prospect of using steelmaking slag in conjunction with the sewage sludge of urban aeration stations as an organomineral additive to the soil.

CONCLUSION

The carried out laboratory investigations allowed to work out in detail the issue of utilization of wastes represented by steel-smelting slag and sediment from municipal wastewater treatment, as well as to offer an actual solution to the existing problem of utilization of the investigated wastes, consisting in:

- application of steel smelting slag of the Novolipetsk metallurgical plant as a mineral component of the reclaimed land due to the high calcium content (30.4%) represented by portlandite ($CaO \cdot H_2O$), calcite ($CaCO_3$), sperrit ($2Ca_2SiO_4 \cdot CaCO_3$), calcium aluminate ($5CaO \cdot 3Al_2O_3$), calcium silicate (Ca_2SiO_4);
- application of sediment after the treatment of municipal wastewater as an organic constituent of the soil additive;
- development of a recipe for the preparation of a highly effective organomineral additive (reclamation agent) using steelmaking slag and urban sewage sludge in a weight ratio of 1 : 3;
- minimization of heavy metals (Cu, Ni, Zn, Cr(III)) influence on the environment, due to their participation in the metabolism of plants and favorable influence on the cultivated grass mixtures with the application of admissible doses;
- development of a methodology for researching soil additives under laboratory conditions, which includes assessing the germination of seeds, the increase in the aerial part of the biomass, and the content of accumulated metals in the grown plants, and also takes into account the effect of climatic parameters;
- an annual reduction in the volumes of dumps of steel-smelting slags by 3–5% and prevention of storage of newly generated waste.

In general, the introduction of the proposed method for joint utilization of slag from steelmaking and sludge after the treatment of urban wastewater leads to a significant decrease in their impact on the components of the environment, the cost of remediation works in the zone of mining and metallurgical complexes, and also allows taking a colossal step towards zero waste production.

REFERENCES

1. Baiseitova N.M., Sartaeva Ch. M. 2014. Phytotoxic effect of heavy metals in technogenic pollution of the environment. Young Researcher, 2, 382-384.
2. Bayshanova A.E., Kedelbaev B.S. 2016. Problems of soil degradation. Analysis of current state of the fertility of irrigared soils in Kazakhstan. Scientific review. Biological Sciences, 2, 5-13.
3. Bobrova Z.M. et al. 2015. Application of mining and metallurgical waste for rational natural resources management. News of the Ural State Mining University, 4(40), 16-26.
4. GOST Soils. Preparation of salt extract and determination of its pH by CINAO method
5. Kogan V.E., Shakhparonova T.S. 2017. Chemistry as a basis for solving environmental issues. Zapiski Gornogo Instituta, 224, 223-228.
6. Shapovalov N.A. et al. 2013. Rational way to use steel slag. Fundamental Research (Technical Sciences), 1, 439-443.
7. Stepanova L.P., Shamaraeva V.S. 2012. Influence of non-traditional forms for fertilizing nutrients and biological activity of soil. Bulletin of Agrarian Science, 37(4), 7-13.
8. Titov A.F., Laidinen G.F. 2007. Stability of plants to heavy metals. Karelian Research Center of the Russian Academy of Sciences, Petrozavodsk.
9. Yaroshenko U.G. 2011/ Use of secondary resources of ferrous metallurgy: problems and solutions. Metallurgical Heat Technology, 3(18), 164-177.
10. Zimina E.V., Kukushkin I.A. 2014. The use of natural science methods in the study of plant physiology. Bulletin of KemGU, 4(60), 8-13.

Permissions

All chapters in this book were first published in JEE, by Polish Society of Ecological Engineering; hereby published with permission under the Creative Commons Attribution License or equivalent. Every chapter published in this book has been scrutinized by our experts. Their significance has been extensively debated. The topics covered herein carry significant findings which will fuel the growth of the discipline. They may even be implemented as practical applications or may be referred to as a beginning point for another development.

The contributors of this book come from diverse backgrounds, making this book a truly international effort. This book will bring forth new frontiers with its revolutionizing research information and detailed analysis of the nascent developments around the world.

We would like to thank all the contributing authors for lending their expertise to make the book truly unique. They have played a crucial role in the development of this book. Without their invaluable contributions this book wouldn't have been possible. They have made vital efforts to compile up to date information on the varied aspects of this subject to make this book a valuable addition to the collection of many professionals and students.

This book was conceptualized with the vision of imparting up-to-date information and advanced data in this field. To ensure the same, a matchless editorial board was set up. Every individual on the board went through rigorous rounds of assessment to prove their worth. After which they invested a large part of their time researching and compiling the most relevant data for our readers.

The editorial board has been involved in producing this book since its inception. They have spent rigorous hours researching and exploring the diverse topics which have resulted in the successful publishing of this book. They have passed on their knowledge of decades through this book. To expedite this challenging task, the publisher supported the team at every step. A small team of assistant editors was also appointed to further simplify the editing procedure and attain best results for the readers.

Apart from the editorial board, the designing team has also invested a significant amount of their time in understanding the subject and creating the most relevant covers. They scrutinized every image to scout for the most suitable representation of the subject and create an appropriate cover for the book.

The publishing team has been an ardent support to the editorial, designing and production team. Their endless efforts to recruit the best for this project, has resulted in the accomplishment of this book. They are a veteran in the field of academics and their pool of knowledge is as vast as their experience in printing. Their expertise and guidance has proved useful at every step. Their uncompromising quality standards have made this book an exceptional effort. Their encouragement from time to time has been an inspiration for everyone.

The publisher and the editorial board hope that this book will prove to be a valuable piece of knowledge for researchers, students, practitioners and scholars across the globe.

List of Contributors

Magdalena Daria Vaverková, Dana Adamcová, Jan Zloch and Alžbeta Maxianová
Department of Applied and Landscape Ecology, Faculty of AgriSciences, Mendel University in Brno, Zemědělská 1, 613 00 Brno, Czech Republic

Maja Radziemska
Department of Environmental Improvement, Faculty of Civil and Environmental Engineering, Warsaw University of Life Sciences, Nowoursynowska 159, 02-776 Warsaw, Poland

Agnieszka Boas Berg
Shada BV, Kanaal Noord 350, NL-7323 Am Apeldoorn, Holland

Stanislava Voběrková
Department of Chemistry and Biochemistry, Faculty of Agronomy, Mendel University in Brno, Zemědělská 1, 613 00, Czech Republic

Osama Hasib Ghazal
Mechanical Engineering Department, Applied Science Private University, Al-Arab Street 21, Amman 11931 Shafa Badran, Jordan

Petr Karásek, Jana Konečná, Michal Pochop and Josef Kučera
Research Institute for Soil and Water Conservation, Department of Land Consolidations and Land Use Planning, Lidická 25/27, 602 00 Brno, Czech Republic

Jana Podhrázská
Research Institute for Soil and Water Conservation, Department of Land Consolidations and Land Use Planning, Lidická 25/27, 602 00 Brno, Czech Republic
Faculty of Agronomy, Mendel University in Brno, Zemědělská 1, 613 00 Brno, Czech Republic

Andrey Vladimirovich Ivanov, Yuri Dmitrievich Smirnov and Georgiy Igorevich Petrov
Saint Petersburg Mining University, 21st Line of Vasilyevsky Island, 2, Saint-Petersburg, 199106, Russia

Joanna Kazimierowicz
Department of Environmental Engineering Technology and Systems, Faculty of Civil and Environmental Engineering, Bialystok University of Technology, ul. Wiejska 45A, 15-351 Białystok, Poland

Andrzej Mazur
University of Life Sciences in Lublin, Department of Environmental Engineering and Geodesy, ul. Leszczyńskiego 7, 20-069 Lublin

Agnieszka Malec and Gabriel Borowski
Department of Environmental Engineering, Lublin University of Technology, ul. Nadbystrzycka 40B, 20-618 Lublin, Poland

Anna Smurzyńska, Wojciech Czekała, Stanisław Marks, Jakub Mazurkiewicz, Michał Brzoski, Dawid Chełkowski and Kamil Kozłowski
Poznań University of Live Sciences, Faculty of Biosystem Engineering, ul. Wojska Polskiego 50, 60-637 Poznań, Poland

Piotr Hektus
Adam Mickiewicz University in Poznań, Faculty of Socio-economic Geography and Spatial Economy, Krygowskiego 10, 61-680 Poznań, Poland

Monika Gałwa-Widera, Anna Kwarciak-Kozłowska and January Bień
Czestochowa University of Technology, Faculty of Infrastructure and Environment, Institute of Environmental Engineering, ul. Brzeźnicka 60a, 42-200 Czestochowa, Poland

dan Yulinah Trihadiningrum
Department of Environmental Engineering, Institut Teknologi Sepuluh Nopember, Surabaya 60111, Indonesia

R. Y. Perry Burhan
Departement of Chemistry, Institut Teknologi Sepuluh Nopember, Surabaya 60111, Indonesia

Masrullita
Department of Environmental Engineering, Institut Teknologi Sepuluh Nopember, Surabaya 60111, Indonesia
Universitas Malikussaleh ACEH, Aceh 24351, Indonesia

Ferdije Zhushi Etemi
Department of Biology, Faculty of Mathematics and Natural Sciences, University of Prishtina "HasanPrishtina", George Bush, Prishtina 10000, Kosova

Murtezan Ismaili
Southeast European University (SEEU), Ilindenska no 335 Tetova, FYR Macedonia

Pajtim Bytyçi
Department of Biology, Faculty of Mathematics and Natural Sciences, University of Prishtina "HasanPrishtina", George Bush, Prishtina 10000, Kosova
Southeast European University (SEEU), Ilindenska no 335 Tetova, FYR Macedonia

Bujar H. Durmishi
University of Tetova, Faculty of Mathematics and Natural Sciences, Department of Chemistry, Str. Ilindeni nn, 1200 Tetova, FYR Macedonia

Osman Fetoshi
University of Applied Sciences, Faculty of Tourism and Environment, Ferizaj, 70000, Kosova

Hazir Çadraku
University for Business and Technology, Faculty of Engineering Energy, Kalabira 10000, Pristina, Kosova

Albona Shala Abazi
University "Haxhi Zeka", Faculty of Management in Tourism, Hotels and the Environment, Pejë, Kosova

Mirosław Skorbiłowicz, Elżbieta Skorbiłowicz and Małgorzata Górska
Bialystok University of Technology, Faculty of Civil and Environmental Engineering, Department of Technology in Engineering and Environmental Protection, ul. Wiejska 45A, 15-354 Bialystok, Poland

Urszula Wydro, Elżbieta Wołejko, Agata Jabłońska-Trypuć, Andrzej Butarewicz and Tadeusz Łoboda
Bialystok University of Technology, Department of Chemistry, Biology and Biotechnology, Wiejska 45E, 15-351 Białystok. Poland

Agus Tjahjono
Engineering Department, Semarang Merchant Marine Polytechnic (Politeknik Ilmu Pelayaran Semarang), Singosari 2 a Semarang, 50242, Central Java, Indonesia

Djoko Suwarno
Civil Engineering Department, Unika Soegijapranata Chatolic University, Pawiyatan Luhur IV/1 Bendan Dhuwur, Semarang, 50234, Central Java, Indonesia

Milaim Musliu and Murtezan Ismaili
South East European University, Faculty of Contemporary Sciences and Technologies, Tetovë, Republic of Macedonia

Astrit Bilalli
University of Peja "Haxhi Zeka", Faculty of Agribusiness, Pejë, Kosovo

Bujar Durmishi
Faculty of Natural Sciences and Mathematics, State University of Tetova, Ilinden str n.n., Tetovo, Republic of Macedonia

Halil Ibrahimi
University of Prishtina "Hasan Prishtina", Faculty of Mathematics and Natural Sciences, Department of Biology, Prishtina, Kosovo

Mimoza Hyseni Spahiu
University "Haxhi Zeka", Faculty of Management in Tourism, Hospitality and the Environment, Str. UCK, Peje 30 000, Republic of Kosovo

Fadil Millaku
University "Haxhi Zeka", Faculty of Agrobusiness, Str. UCK, Peje 30 000, Republic of Kosovo

Fatbardh Sallaku
Agricultural University of Tirana, Albania

Pajtim Bytyçi
South East European University, Faculty of Contemporary Sciences and Technologies, Ilindenska 335, Macedonia

Arif Susanto
Doctorate Program in Environmental Science, School of Postgraduate Studies, University of Diponegoro, Semarang 50241, Indonesia
Department of Environmental Engineering, Faculty of Civil and Planning Engineering, University of Kebangsaan Republic of Indonesia, Bandung 40263, Indonesia

Purwanto Purwanto
Doctorate Program in Environmental Science, School of Postgraduate Studies, University of Diponegoro, Semarang 50241, Indonesia
Department of Chemical Engineering, Faculty of Engineering, University of Diponegoro, Semarang 50275, Indonesia

Henna R Sunoko
Doctorate Program in Environmental Science, School of Postgraduate Studies, University of Diponegoro, Semarang 50241, Indonesia
Department of Medicine, Faculty of Medicine, University of Diponegoro, Semarang 50275, Indonesia

Onny Setiani
Doctorate Program in Environmental Science, School of Postgraduate Studies, University of Diponegoro, Semarang 50241, Indonesia
Department of Environmental Health, Faculty of Public Health, University of Diponegoro, Semarang 50275, Indonesia

Elanda Fikri
Department of Environmental Health, Bandung Health Polytechnic, Cimahi Utara, 40514, Doctorate Program of Environmental Studies, Diponegoro University, Semarang, Indonesia

Aprilia Veronica
Department of Environmental Health, Bandung Health Polytechnic, Cimahi Utara, Indonesia

Agnieszka Petryk
Department of Regional Economics, Cracow University of Economics, ul. Rakowicka 27, 31-510 Kraków, Poland

Halina Marczak
Institute of Transport, Internal Combustion Engines and Ecology, Lublin University of Technology, Nadbystrzycka 36, 20-618 Lublin, Poland

Adam Gołub and Janina Piekutin
Bialystok Technical University, Faculty of Civil and Environmental Engineering, ul. Wiejska 45A, 15-350 Białystok, Poland

Robert Krzysztof Sobolewski and Robert Kalbarczyk
Wrocław University of Environmental and Life Sciences, Institute of Landscape Architecture, Grunwaldzka 55, 50-357 Wrocław, Poland

Anna Grobelak, Paulina Kokot, Jakub Świątek and Marta Jaskulak
Czestochowa University of Technology, Faculty of Infrastructure and Environment, Institute of Environmental Engineering, Brzeznicka 60a, 42-200 Czestochowa, Poland

Agnieszka Rorat
Université de Lille, Sciences et Technologies, Laboratoire de Génie Civil et géo-Environnement, LGCgE EA4515, Bât. SN3, 59655 Villeneuve d´Ascq, France

Vera Matveeva, Tatiana Lytaeva and Aleksandr Danilov
Geoecology Department, Saint-Petersburg Mining University, 2, 21st Line, Saint-Petersburg, 199106, Russia

Index

A

Aero-sanitary Conditions, 196, 204
Agrostis Capillaris L, 206-207
Amylases Activity, 120-122, 125-127
Amylolytic Bacteria, 120-122, 125-127
Anadara Granosa, 129, 132, 138-139
Arabidopsis Thaliana L., 206-207
Atomic Absorption Spectrophotometry, 96
Automatic Potentiometric Titrator, 39

B

Biofiltration Process, 78-79, 82, 86
Biogas Plants, 46, 49-50, 52-53, 71
Biomass Fermentation, 46, 49
Bottom Sediments, 110-118

C

Carbon Monoxide, 21, 35, 48, 71, 168-174, 176, 179
Chemical Oxygen Demand, 8, 145
Chlorella Algae, 35-36, 42, 44
Chloride Ions, 190, 192-194
Cluster Analysis, 111, 113, 116-117
Coliform Abundance, 129, 131, 134-135, 138
Colorimetric Method, 87, 90, 122

D

Diesel Particulate Matter, 159, 166-167
Dust Concentration Measurement, 185
Dust Fraction, 182, 184, 187-189

E

Endophytes, 206-207, 212-213
Erosive Precipitation, 54, 56, 59

F

Festuca Rubra, 121-122, 206-207
Festuca Rubra L., 206-207
Flame Absorption Spectrometry Method, 110, 112
Forested Loess Slope, 54, 58

G

Geochemical Index, 113
Geotextile Membranes, 3, 13
Granbial, 120-126
Granulometric Composition, 35-36, 41, 44, 55, 120, 122-123

H

Hydrochloroflourocarbon, 169

L

Land Consolidations, 22-28, 31-34
Landfill Leachate, 1, 4, 7-8, 11-14

M

Metallurgical Complex, 214
Methane Emissions, 15, 77
Mineralisation, 46, 49, 82
Municipal Solid Waste, 1-3, 13-14

N

Nitrogen Dioxide, 174, 176-177, 179, 189, 197

O

Odour Pollution, 78, 81
Ordinary Portland Cement, 87
Organomineral Additive, 214, 216, 220
Oxygen Saturation, 143, 145
Ozone Depletion, 70

P

Paint Filter Test, 87, 90, 93
Pedological-ecological Units, 26
Permissible Exposure Limit, 159
Peroxyacetyl Nitrate, 71
Photochemical Smog, 71, 76-77
Physicochemical Parameters, 141-143, 148
Physiological Strain, 196-199, 201, 204
Phytotoxicity, 1-2, 4-5
Plant Growth Promoting Bacteria, 206-207, 212
Poa Pratensis, 121-122
Polychlorinated Dibenzofurans, 13, 48
Polycyclic Aromatic Hydrocarbons, 88, 119, 169, 213
Polystyrene Foam Waste, 168-169, 173
Post-fermentation Sludge, 51

R

Ranunculus Sardous Crantz, 150, 157
Rhizosphere, 120, 125, 207, 210, 213
Rice Husk Ash, 88, 95
Road Acoustic Screens, 67
Rumex Crispus L, 150-151, 157

S

Sawdust, 80, 82, 84-85
Sewage Sludge, 14, 45, 51-53, 78, 94, 102, 120-128, 206, 214, 216-217, 219-220
Soil-vegetative Layer, 37
Stabilization/solidification Process, 87-88, 93
Steelmaking Slag, 214-215, 217, 220
Subsurface Outflow, 54, 56-58
Surface Sediment, 129, 133-135, 137

T

Total Organic Carbon, 120, 123, 126, 145
Toxicity Characteristic Leaching Procedure, 87, 89, 95
Trass Soil, 87, 89-93
Tropospheric Ozone, 196, 198, 200-201, 203-205
Turbidity, 130-131, 134, 136-137, 143
Typha Angustifolia, 150, 154-157, 212

U

Urtica Dioica L, 150, 157

W

Water Pollution Index, 129-131
Water Quality Index, 58, 96-98, 107-109, 131-132, 141-142, 147-149
Wind Erosion, 22-26, 28-34

X

X-ray Fluorescence Spectrometry, 36, 214-215

www.ingramcontent.com/pod-product-compliance
Lightning Source LLC
LaVergne TN
LVHW060556210926
848485LV00005B/1322

* 9 7 8 1 6 4 1 1 6 6 0 3 4 *